Pitman Research Notes in Mathematics Series

Submission of proposals for consideration

Suggestions for publication, in the form of outlines and representative samples, are invited by the Editorial Board for assessment. Intending authors should approach one of the main editors or another member of the Editorial Board, citing the relevant AMS subject classifications. Alternatively, outlines may be sent directly to the publisher's offices. Refereeing is by members of the board and other mathematical authorities in the topic concerned, throughout the world.

Preparation of accepted manuscripts

On acceptance of a proposal, the publisher will supply full instructions for the preparation of manuscripts in a form suitable for direct photo-lithographic reproduction. Specially printed grid sheets are provided and a contribution is offered by the publisher towards the cost of typing. Word processor output, subject to the publisher's approval, is also acceptable.

Illustrations should be prepared by the authors, ready for direct reproduction without further improvement. The use of hand-drawn symbols should be avoided wherever possible, in order to maintain maximum clarity of the text.

The publisher will be pleased to give any guidance necessary during the preparation of a typescript, and will be happy to answer any queries.

Important note

In order to avoid later retyping, intending authors are strongly urged not to begin final preparation of a typescript before receiving the publisher's guidelines and special paper. In this way it is hoped to preserve the uniform appearance of the series.

Longman Scientific & Technical
Longman House
Burnt Mill
Harlow, Essex, UK
(tel (0279) 26721)

W9-AFD-389

Titles in this series

Surveys of some recent results in operator theory VOLUME II

John B Conway &
Bernard B Morrel (Editors)

Indiana University/Indiana University – Purdue University
at Indianapolis

Surveys of some recent results in operator theory VOLUME II

 Longman
Scientific &
Technical

Copublished in the United States with
John Wiley & Sons, Inc., New York

Longman Scientific & Technical
Longman Group UK Limited
Longman House, Burnt Mill, Harlow
Essex CM20 2JE, England
and Associated Companies throughout the world.

Copublished in the United States of America with
John Wiley & Sons, Inc., 605 Third Avenue, New York, NY 10158

First published 1988

AMS Subject Classification: 47A99

ISSN 0269-3674

British Library Cataloguing in Publication Data
Surveys of some recent results in operator
theory. — (Pitman research notes in
mathematics series; ISSN 0269–3674; 192).
Vol. 2
1. Mathematics. Operators
I. Conway, John B   II. Morrel, B.B.
(Bernard Baldwin) *1940–*
515.7′24

ISBN 0-582-00518-3

Printed and bound in Great Britain by
Biddles Ltd, Guildford and King's Lynn

Contents

Preface

This is the second of two volumes entitled *Surveys of Some Recent Results in Operator Theory*. The papers herein are based on series of lectures given at Indiana University as part of the Special Year in Operator Theory held there during the 1985-86 academic year. While both volumes fall under the general rubric of Operator Theory, the first volume was primarily concerned with the study of single linear operators, while this volume is directed toward the study of algebras of bounded linear operators on Hilbert space.

As originally envisioned, each of the lecture series in the Special Year would develop a topic of current interest in Operator Theory, starting from a point accessible to advanced graduate students in the area, and proceeding rapidly to current and future researchers. We feel that the results have far exceeded our expectations.

In any enterprise of this sort, there are many who deserve thanks and recognition. First and foremost we thank the authors: Jim Agler, Raúl E. Curto, Ronald G. Douglas, David R. Larson, Stephen Power and Derek W. Robinson. Thanks also to Hari Bercovici, Ciprian Foiaş, Mort Lowengrub, Joe Stampfli, and Bill Ziemer of Indiana University their support and encouragement. Deep thanks to Jan Want of Texas A&M University for her careful preparation of the manuscript in TEX: thanks to her, its elegant style matches its substance.

Finally, thanks to Indiana University, the National Science Foundation, and the Argonne Universities Trust Fund, without whose financial support the Special Year would never have taken place.

John B. Conway
Bernard B. Morrel

An Abstract Approach to Model Theory

by

Jim Agler

0. Introduction

In these notes we shall summarize our recent results on the uses of hereditary polynomials in operator theory. In order to map out the theory in an economical manner we shall purposely omit proofs of all but the most trivial facts and adopt an expository and at times informal style.

The type of structure that hereditary polynomials are adapted to analyze is that of a *lifting theorem*. If $\mathcal{B}$ and $\mathcal{F}$ are collections of operators with $\mathcal{B} \subseteq \mathcal{F}$ and $\mathcal{B}$ closed with respect to direct sums and unital representations we say that a theorem is a lifting theorem if it has the form

0.1   If $T \in \mathcal{F}$ then there exists $B \in \mathcal{B}$ and $\mathcal{N}$ such that $\mathcal{N}$ is an invariant subspace for B and $T = B|\mathcal{N}$.

One thinks of $\mathcal{B}$ as being a highly distinguished subcollection of $\mathcal{F}$ whose elements possess a highly developed model. The long term goal in the case when 0.1 holds is to use the model for elements of $\mathcal{B}$ to study the elements of F. Two pre-eminent examples are the theory of subnormal operators ($\mathcal{B}$ = normals, $\mathcal{F}$ = subnormals) and the Nagy-Foias, deBranges-Rovnyak theories of contractions ($\mathcal{B}$ = coisometries, $\mathcal{F}$ = contractions).

1. Hereditary Polynomials and Functional Calculus

By a *hereditary polynomial* we shall mean a polynomial in two noncommuting variables x and y of the form

1.1 $$p = \sum c_{ij} y^j x^i, c_{ij} \in C.$$

If A is an algebra with unit and an involution, $a \in A$, and p is as in 1.1 let us agree to define $p(a)$ by the formula

1.2 $$p(a) = \sum c_{ij} a^{*j} a^i.$$

1

Before continuing we make the simple observation that for *any* polynomial $q(x, y)$ there exists a unique hereditary polynomial $p(x, y)$ such that

$$p(z, w) = q(z, w)$$

for all complex numbers z and w. In this situation we *define* $q(a)$ by setting $q(a) = p(a)$. Thus, for example we have

$$(y - x)^m(a) = \sum_{k=0}^{m} (-1)^k \binom{m}{k} a^{*m-k} a^k.$$

Let us agree to denote the set of hereditary polynomials by $\mathcal{P}$. If $\mathcal{M}_n$ is the C^*-algebra of $n \times n$ matrices with complex entries then we can construct

$$\mathcal{P}_n = \mathcal{M}_n \otimes \mathcal{P}.$$

In an obvious way we can extend our hereditary functional calculus to $\mathcal{P}_n$. For $p = \sum M_k \otimes p_k, M_k \in \mathcal{M}_n, p_k \in \mathcal{P}$ and $a \in A$, define $p(a) \in \mathcal{M}_n \otimes A$ by

1.3
$$p(a) = \sum M_k \otimes p_k(a).$$

Needless to say there are a number of generalizations of the hereditary functional calculus defined by 1.2 and 1.3 above. We briefly summarize three.

Let K be a compact subset of the plane, let A be a C^* algebra with unit and assume $a \in A$ with $\sigma(a) \subseteq K$. Let $H(K \times K)$ denote the set of functions that are holomorphic on a neighborhood of $K \times \overline{K}$. For $f \in H(K \times \overline{K})$ define $f(a) \in A$ by

1.4
$$f(a) = \frac{1}{(2\pi i)^2} \int_{\overline{\gamma}} \int_{\gamma} f(\lambda, z)(\lambda - a^*)^{-1}(z - a)^{-1} dz d\lambda.$$

It is easy to check that if $f_k, 1 \leq k \leq m$, are rational functions with poles off K and $g_k, 1 \leq k \leq m$ are rational functions with poles off $\overline{K}$, and $f \in H(K \times \overline{K})$ is defined by $f(z, \lambda) = \sum_{k=1}^{m} f_k(z) g_k(\lambda)$, then provided one chooses the contours γ and $\overline{\gamma}$ appropriately,

$$f(a) = \sum_{k=1}^{m} g_k(T^*) f_k(T).$$

Thus, 1.4 defines an obvious extension of the calculus defined by 1.2 to $H(K \times \overline{K})$.

2

In the next example we let C^∞ denote the complex valued infinitely differentiable functions defined on **R**. We say that an operator $T \in \mathcal{L}(\mathcal{H})$ has a C^∞ *functional calculus* if there exists a constant C, an integer N, and a compact interval $I \subseteq R$ such that

1.5
$$\|q(T)\| \leq C \max_{\substack{1 \leq k \leq N \\ t \in I}} |q^{(k)}(t)|$$

for all polynomials q of one variable. It is easy to see that 1.5 together with the fact that polynomials (of one variable) are dense in C^∞ implies that the map $q \mapsto q(T)$ extends to a continuous map on all of C^∞,

1.6
$$\phi \to \phi(T) \in \mathcal{L}(\mathcal{H}).$$

The map in 1.6 is referred to as the C^∞ functional calculus for T. Observe that if T has a C^∞ functional calculus then the bilinear map defined on $C^\infty \times C^\infty$ by

1.7
$$(\phi, \psi) \to \psi(T^*)\phi(T), \phi, \psi \in C^\infty$$

is continuous. Identifying $\phi(x)\psi(y) \in C^\infty(R^2)$ with $\phi \otimes \psi$ and using the fact that C^∞ is nuclear we deduce that 1.7 extends to a continuous map from $C^\infty(R^2)$ into $\mathcal{L}(\mathcal{H})$,

1.8
$$\phi \to \phi(T), \phi \in C^\infty(R^2).$$

We call the map defined in 1.8 the hereditary C^∞ functional calculus for T. A simple fact is that if $p \in \mathcal{P}_1$ and $\phi \in C^\infty(R^2)$ is defined by $\phi(x, y) = p(x, y), x, y \in R$, and $p(T)$ is defined by 1.2 and $\phi(T)$ is defined as in 1.8, then $\phi(T) = p(T)$.

Also, we remark that the Fourier inversion formula allows a concrete formula for the calculus defined in 1.8. Namely,

$$\phi(T) = \frac{1}{\pi} \int_{-\infty}^{\infty} \int_{-\infty}^{\infty} \hat{\phi}(s, t) e^{itT^*} e^{isT} ds\, dt$$

whenever $\phi \in C^\infty(R^2)$.

One of the great advantages of hereditary algebra is that it generalizes rather effortlessly to several variables for tuples of commuting operators. Since the geometry of the lifting and interpolation structures in several variables has a notorious reputation and since these structures have a dual encoding in hereditary algebra (as we hope these notes will make clear) the ease of this generalization is of profound significance.

3

A fundamental problem that hereditary polynomials are well equipped to handle is that of proving lifting theorems. The situation is this. One is given two collections of operators $\mathcal{B}$ and $\mathcal{F}$ with $\mathcal{B} \subseteq \mathcal{F}$. $\mathcal{F}$ is quite large (at least compared to $\mathcal{B}$) and $\mathcal{B}$ consists of operators that are *nice* (at least compared to the general operator of $\mathcal{F}$). *Nice*, which is a relative term, typically means that the elements of $\mathcal{B}$ have a model of sufficient simplicity to easily settle many of the operator theoretic questions (What is spectrum of T?, What are cyclic vectors for T?, Is T reflexive?, What is the commutant of T?, What is closure of the polynomials in T? (any topology), What is the lattice of T?, etc.) for the elements $T \in \mathcal{B}$ that might be quite difficult for the general element of $\mathcal{F}$. It should be clear in the situation just described an ideal environment would be created if the following were true.

1.9
> Given $T \in \mathcal{F}$ there exists $B \in \mathcal{B}$ and an invariant subspace
> $\mathcal{N}$ for B such that T is unitarily equivalent to $B|\mathcal{N}$.

We refer to 1.9 as a lifting theorem. The reader will find a great many examples of $\mathcal{B}$'s and $\mathcal{F}$'s that satisfy 1.9 in the subsequent text. For now we merely cite three simple examples: $\mathcal{B}$ = contractive normals and $\mathcal{F}$ = contractive subnormals; or $\mathcal{B}$ = coisometries and $\mathcal{F}$ = contractions; or $\mathcal{B}$ = unitries and $\mathcal{F}$ = isometries.

We note that in the examples just considered the set $\mathcal{B}$ actually has additional structure. In each example it is true that $\mathcal{B}$ is closed with respect to direct sums and unital representations. Thus, in each case if $T_\lambda \in \mathcal{B}$ for all $\lambda \in \Lambda$, an index set, then $\oplus_{\lambda \in \Lambda} T_\lambda \in \mathcal{B}$. Also, if $T \in \mathcal{B}, T \in \mathcal{L}(\mathcal{H})$ and $\pi : \mathcal{L}(\mathcal{H}) \to \mathcal{L}(\mathcal{K})$ is a representation with $\pi(1) = 1$, then $\pi(T) \in \mathcal{B}$. Another interesting fact is that in each case there exists a universal element in $\mathcal{B}$, i.e. there exists a C^* algebra A with unit and there exists an element $a \in A$ such that

1.10
> $T \in \mathcal{B} \cap \mathcal{L}(\mathcal{H})$ if and only if there exists a representation
> $\pi : A \to \mathcal{L}(\mathcal{H})$ with $\pi(1) = 1$ and $\pi(a) = T$.

We leave as exercises to verify that if $\mathcal{B}$ = contractive normals, then 1.10 holds with $A = C(D^-)$ and $a \in C(D^-)$ defined by $a(z) = z$ (spectral theorem), that if $\mathcal{B}$ = coisometries then 1.10 holds with $a = S^*$, the adjoint of the unilateral shift (Wold decomposition), and that if $\mathcal{B}$ = isometries, then 1.10 holds with $A = C(\partial D)$ and a defined by $a(z) = z$. Other examples of $\mathcal{B}$'s and universal elements can be found in the papers. We remark that of the thirty or so examples within classical operator theory of the situation described by 1.9, that in all of them $\mathcal{B}$ possesses the structure considered in this paragraph.

The following theorem is the most central fact about hereditary polynomials. Set $\mathcal{P}_\infty = \cup_{n=1}^\infty \mathcal{P}_n$.

Theorem 1.11. *Let A be a C^* algebra with unit, let $a \in A$, let H be a Hilbert space and assume that $T \in \mathcal{L}(\mathcal{H})$. The following are equivalent.*

(a) *T is unitarily equivalent to an operator of the form $\pi(a)|\mathcal{N}$ where π is a unital representation of A (into some $\mathcal{L}(\mathcal{K})$) and $\mathcal{N}$ is invariant for $\pi(a)$.*

(b) *If $p \in \mathcal{P}_\infty$ and $p(a) \geq 0$, then $p(T) \geq 0$.*

Evidently, if $\mathcal{B} \subseteq \mathcal{F}$ and a is a universal element for $\mathcal{B}$, then Theorem 1.11 asserts that 1.10 holds if and only if

1.12 $\qquad\qquad p(T) \geq 0$ whenever $p \in \mathcal{P}_\infty, p(a) \geq 0$, and $T \in \mathcal{F}$.

One way to get a quick gauge of the power of Theorem 1.11, is to observe that condition (a) (once the operators of the form $\pi(a)$ have been classified) is essentially a geometric condition. On the other hand, condition (b) of Theorem 1.11 is far more algebraic, and in particular applications where the specific nature of T is taken into account usually reduces to a concrete problem in analysis. We illustrate the actual performance of Theorem 1.11 by proving the Nagy dilation theorem in the following form.

1.12
$\qquad$ If T is a contraction then there exists a coisometry J and an invariant subspace $\mathcal{N}$ for J such that $T = J|\mathcal{N}$.

To construct the proof we first observe that if S denotes the unweighted unilateral shift of multiplicity one acting on H^2, then $J \in \mathcal{L}(\mathcal{K})$ is a coisometry if and only if there exists a representation $\pi : \mathcal{L}(H^2) \to \mathcal{L}(\mathcal{K})$ with $\pi(1) = 1$ and $\pi(S^*) = J$. Thus, 1.12 will follow from Theorem 1.11 if we can show that

1.13 $\qquad\qquad$ If $\|T\| \leq 1, p \in \mathcal{P}_\infty, p(S^*) \geq 0$, then $p(T) \geq 0$.

To make further progress we need to analyze the condition $p(S^*) \geq 0$. To keep things simple we shall only consider the case where $p \in \mathcal{P}_1$. Using the Szego kernel functions in H^2 one deduces that if $p \in \mathcal{P}_1$, then

1.14
$\qquad\qquad p(S^*) \geq 0$ if and only if $\left(p(z, \overline{\lambda}) \dfrac{1}{1 - \overline{\lambda}z} \right)$ is a positive

$\qquad\qquad$ semidefinite kernel on $D \times D$.

5

But a positive semidefinite kernel $g(z, \bar{\lambda})$ analytic in z and coanalytic in λ on $D \times D$ has the form

$$g(z, \lambda) = \sum_{j=1}^{\infty} \overline{g_j(\lambda)} g_j(z),$$

with subuniform convergence on $D \times D$ and g_j analytic on D. Thus, from 1.14 we see that if $p \in \mathcal{P}_1$ and $p(S^*) \geq 0$ then

1.15
$$p(z, y) = \sum_{j=1}^{\infty} \breve{g}_j(y)(1 - yx)g_j(x).$$

Regarding 1.15 as a hereditary equation, we thus see that

1.16
$$p(T) = \sum_{j=1}^{\infty} g_j(T)^*(1 - T^*T)g_j(T) \geq 0$$

which establishes 1.13. In 1.16 the right side of the equation is positive since $1 - T^*T \geq 0 (\|T\| \leq 1)$. Also, there is the small matter making sense of the sum in 1.16. This problem is handled most simply by replacing T with $rT(r < 1)$ and then letting r tend to 1. Hereditary polynomials are very stable under such tricks. For more complete details of the above argument, consult [4].

In light of Theorem 1.11, it is natural to introduce the following notation.

Definition 1.17. If A is a C^* algebra with unit and $a \in A$ then $H(a)$, the hereditary manifold of A is defined by

$$H(a) = \{p(a) | p \in \mathcal{P}_1\}.$$

We observe that if P denotes the polynomials of one variable and $P(a)$ is defined by $P(a) = \{q(a) | q \in P\}$, then $\mathcal{P}_1$ may be regarded as a bimodule over P with the left and right actions given by the formula

$$(q_1 p q_2)(x, y) = q_1(y)p(x, y)q_2(x)$$

for $q_1, q_2 \in P$ and $p \in \mathcal{P}_1$. In a similar manner $H(a)$ may be regarded as a bimodule over $P(a)$. Let us agree to say that a map $\phi : H(a_1) \to H(a_2)$ is a *hereditary homomorphism* if ϕ is a homomorphism on bimodules. Evidently, a map ϕ, defined on $H(a)$ is a hereditary homomorphism if and only if $\phi(p(a)) = p(\phi(a))$ and it is clear that condition (b) in Theorem 1.11 is equivalent to asserting that

1.18
T is the image of a (i.e. $T = \phi(a)$) under a completely positive hereditary homomorphism ϕ defined on $H(a)$.

6

The above observations should be contrasted with Arveson's approach to dilation theory. The idea is simple. Modify condition 1.18 to the form:

1.19
T is the image of a (i.e. $T = \phi(a)$) under a completely contractive algebra homomorphism ϕ defined on $P(a)$.

Arveson's basic result (from the point of view of dilation theory) is then that condition 1.19 implies condition (a) of Theorem 1.11 with the word invariant replaced with the word semi-invariant.

One might succinctly summarize our observations concerning Theorem 1.11 in the following two principles: lifts correspond to completely positive hereditary homomorphisms; dilations correspond to completely contractive algebra homomorphisms. Needless to say, these two principles generalize to more complicated algebras and their hereditary manifolds. In particular, to see an implementation of Theorem 1.11 on $R(K)$ consult [6] and to see an implementation in several variables consult [9].

The result in [6] also points out a consideration based on a theorem of Sarason (see [24]) that is of profound significance for dilation theory. Sarason's observation was that to a dilation corresponds a *difference* of invariant subspaces, a so called semi-invariant subspace. Thus for exmple, the Nagy dilation is a combination of two lifting theorems: the result that a contraction lifts to a coisometry and the fact that an isometry liefts to a unitary. Sarason's observation makes lifting theory inherently more general than dilation theory.

We close this section with a result of great technical significance when one is trying to use Theorem 1.11 in a concrete situation.

Theorem 1.20. *Let A be a C^* algebra with unit, let $\mathcal{H}$ be Hilbert space, and assume that $a \in A$. Let $\phi : H(a) \to \mathcal{L}(\mathcal{H})$ be hereditary homomorphism. If $\phi(a)$ is n-cyclic, then ϕ is completely positive if and only if ϕ is n-positive.*

2. Families

In this section we shall answer the following question. Given a collection of operators $\mathcal{F}$, when does there exist a set of hereditary polynomials α that *characterizes* $\mathcal{F}$ in the sense the 2.1 below holds?

2.1
$$T \in \mathcal{F} \text{ if and only if } p(T) \geq 0 \text{ for all } p \in \alpha?$$

For reasons that shall become clear we shall assume that $\mathcal{F}$ is bounded, i.e. that there exists a constant c such that

$$T \in \mathcal{F} \text{ implies } \|T\| \leq c.$$

Observe first that if $\mathcal{F}$ is a bounded collection of operators that satisfies 2.1 and $\{T_\lambda | \lambda \in \Lambda\}$ is a collection of operators each in F, then $\oplus_{\lambda \in \Lambda} T_\lambda$ is in $\mathcal{F}$. We summarize this property of $\mathcal{F}$ by saying $\mathcal{F}$ is *closed with respect to direct sums*. Equally trivial is the observation that if $T \in \mathcal{F} \cap \mathcal{L}(\mathcal{H})$ and π is a unital representation, $\pi : \mathcal{L}(\mathcal{H}) \to \mathcal{L}(\mathcal{K})$, then $\pi(T) \in \mathcal{F}$, i.e. $\mathcal{F}$ is closed with respect to representations. In the sequel we shall refer to a bounded collection of operators that closed with respect to direct sums and representations as a C^*-collection. We note in passing that if $\mathcal{F}$ is a C^*-collection, then $\mathcal{F}$ is also closed with respect to unitary equivalence and restriction to direct summands.

An only slightly more subtle condition that a bounded collection of operators that satisfies 2.1 must satisfy is that it be hereditary, i.e. if $T \in \mathcal{F} \cap \mathcal{L}(\mathcal{H})$ and $\mathcal{N} \subseteq \mathcal{H}$ is an invariant subspace for T, then $T|\mathcal{N} \in \mathcal{F}$. The proof that $\mathcal{F}$ is hereditary amply demonstrates the special structure of a hereditary polynomial with respect to invariant subspaces. Note that if $T \in \mathcal{L}(\mathcal{H}), \mathcal{N}$ is invariant for $\mathcal{H}$, and P denotes the orthogonal projection of $\mathcal{H}$ onto $\mathcal{N}$, then $(T|\mathcal{N})^i = (T^i)|\mathcal{N}$ and $(T|\mathcal{N})^{*j} = PT^{*j}|\mathcal{N}$. Thus, $y^j x^i(T|\mathcal{N}) = Py^j x^i(T)|\mathcal{N}$ and if $p \in \mathcal{P}$, then

$$2.2 \qquad\qquad p(T|\mathcal{N}) = Pp(T)|\mathcal{N}.$$

Thus, if $p \in \mathcal{P}_n$, then $p(T|\mathcal{N}) = P^{(n)}p(T)|\mathcal{N}^{(n)}$. In particular, if $\mathcal{F}$ is a bounded collection satisfying 2.1, we see that if $T \in \mathcal{F}, \mathcal{N}$ is invariant for T, and $p \in \alpha \cap \mathcal{P}_n$, then

$$p(T|\mathcal{N}) = P^{(n)}p(T)|\mathcal{N}^{(n)} \geq 0$$

since $p(T) \geq 0$. Thus $p(T|\mathcal{N}) \geq 0$ whenever $p \in \alpha$. But this implies $T|\mathcal{N} \in \mathcal{F}$ and establishes that $\mathcal{F}$ is hereditary.

Summarizing, we have shown that if F is a bounded collection of operators and there exists $\alpha \subseteq \mathcal{P}$ satisfying 2.1, then

$$2.3 \qquad\qquad \mathcal{F} \text{ is closed w.r.t. direct sums,}$$

$$2.4 \qquad\qquad \mathcal{F} \text{ is closed w.r.t. representations,}$$

and

2.5 $\mathcal{F}$ is hereditary.

By a *family* we shall mean any bounded collection of operators satisfying conditions 2.3–2.5.

Theorem 2.6. *Let $\mathcal{F}$ be a bounded collection of operators. There exists a set $\alpha \subseteq \mathcal{P}_\infty$ such that*

$$T \in \mathcal{F} \text{ if and only if } p(T) \geq 0 \text{ whenever } p \in \alpha$$

if and only if

$$\mathcal{F} \text{ is a bounded family.}$$

3. Some Examples

In this section we shall set down a number of examples of families that have occurred naturally in operator theory. We trust that the reader will perceive the value of the following notations and not be too critical of our use of naive set theory. For α a subset of $\mathcal{P}_\infty$ let

$$\alpha^\perp = \{T | T \text{ is an operator and } \forall_{p \in \alpha} p(T) \geq 0\}.$$

If $\mathcal{F}$ is a collection of operators, let

$$\mathcal{F}^\perp = \{p \in \mathcal{P}_\infty | \forall_{T \in \mathcal{F}} p(T) \geq 0\}.$$

Evidently, if $\mathcal{F}$ is a bounded family, then Theorem 2.6 guarantees that there exist a set $\alpha \subseteq \mathcal{P}_\infty$ such that $\mathcal{F} = \alpha^\perp$. The computation of α is a fundamental problem.

Given a family $\mathcal{F}$, it is a tautology based on the Arveson extension theorem for completely positive maps that

$$\mathcal{F} = (\mathcal{F}^\perp)^\perp.$$

Also, keep in mind that given $\mathcal{F}$ there are in general many $\alpha \in \mathcal{P}_\infty$ such that $\mathcal{F} = \alpha^\perp$. By an *economical* α, we mean an α that is small relative to $\mathcal{F}^\perp$. Admittedly, this notion is somewhat vague.

Example 3.1. $\mathcal{F} = \{T| \; \|T\| \leq c \text{ and } T \text{ is self adjoint}\}$.

In this example it is easy to see that

$$\mathcal{F} = \{c^2 - yx, y - x, x - y\}^{\perp}.$$

To compute $\mathcal{F}^{\perp}$, observe that $\mathcal{F}^{\perp} = \cup_{n=1}^{\infty} \mathcal{F}^{\perp} \cap \mathcal{P}_n$ and that $p \in \mathcal{P}_n$ gives rise canonically to an $n \times n$ matrix valued function $\widetilde{\mathbf{p}}$ defined on R by $\widetilde{\mathbf{p}}(t) = (p_{ij}(t,t))$. From the spectral theorem it follows that $p \in \mathcal{F}^{\perp} \cap \mathcal{P}_n$ if and only if $\widetilde{\mathbf{p}}(t,t) \geq 0$ whenever $t \in [-c, c]$.

Example 3.2. $\mathcal{F} = \{T|T \text{ is an isometry}\}$.

Obviously the isometries form a bounded family and $\mathcal{F} = \{1 - yx, yx - 1\}^{\perp}$. Since every isometry can be extended to a unitary, from the spectral theorem it follows that if $p \in \mathcal{P}_n$ and $(pi(e^{i\ominus}, e^{-i\ominus})) \geq 0$, then $p \in \mathcal{F}^{\perp}$. The converse follows immediately upon observing that $e^{i\ominus} \in \mathcal{L}(C)$ is an isometry.

Example 3.3. $\mathcal{F} = \{T|T \text{ is a contraction}\}$.

$\mathcal{F}$ is the object of Nagy-Foias [14] and deBranges-Rovnyak [10] theories. That $\mathcal{F} = \{1 - yx\}^{\perp}$ is not an assertion to strain the gray cells. Since every contraction has a coisometric extension, the von Neumann-Wold decomposition implies that

$$\mathcal{F}^{\perp} = \{S^*\}^{\perp},$$

where S denotes the unilateral shift of multiplicty one. $\{S^*\}^{\perp}$ is computed in [4]. Proposition 2.5 of that paper implies that $p \in \{S^*\}^{\perp} \cap \mathcal{P}_n$ if and only if $(p_{ij}(\bar{z}, \lambda) \frac{1}{1 - \bar{z}\lambda})$ is nonnegative definite $n \times n$ matrix kernel on $\mathbb{D}$, the unit disc.

Example 3.4. $\mathcal{F} = \{T|T \text{ is a subnormal contraction}\}$.

In this example it is clear that $\mathcal{F}$ is a family, but not at all clear what an economical choice for α would be. In [5] it is proved that

$$\mathcal{F} = \{(1 - yx)^n | n \geq 1\}^{\perp},$$

a result that corresponds to the operator valued Hausdorf moment theorem of Nagy and MacNerny ([20] and [19]) and Embry's characterization of subnormality [13].

It is also the case that

$$\mathcal{F} = \{p \in \mathcal{P}|p(z, \bar{z}) \geq 0, z \in \mathbb{D}\}^{\perp}$$

a result that follows from Stinespring's theory of completely positive maps. It is thus not surprising (and easy to prove using the spectral theorem) that

$$\mathcal{F}^{\perp} \cap \mathcal{P}_n = \{p| \text{ the } n \times n \text{ matrix } (p_{ij}(z, \overline{z})) \text{ is positive definite for all } z \in \mathbf{D}\}.$$

Example 3.5. $\mathcal{F} = \{T|T \text{ is a hyponormal contraction}\}$.

Recall T is said to be hyponormal if $T^*T - TT^* \geq 0$. For a lucid account of the basic theory of hyponormal operators consult [12]. In particular, a remark on page 7 of that treatise implies that F is hereditary and thus, F is a family. It appears nontrivial to analyze $\mathcal{F}^{\perp}$. John Bunce has pointed out in private correspondence that $\mathcal{F} = \alpha^{\perp}$ if

$$\alpha = \left\{ 1 - yx, \begin{bmatrix} 1 & y \\ x & yx \end{bmatrix} \right\}.$$

Example 3.6. In [15] Helton asked the following question: which operators have the form M_t on a Sobolev space? By a Sobolev space is meant the completion of $C^{\infty}(\mathbf{R})$ with respect to an inner product $[\cdot, \cdot]$ of the form

$$3.7 \qquad\qquad [\phi, \psi]_u = \sum_{i,j=0}^{n} u_{ij} \left(\phi^{(i)} \overline{\psi^{(j)}} \right)$$

where $u = (u_{ij})$ is an $(n+1) \times (n+1)$ array of compactly supported distributions on $\mathbf{R}$ that result in $[\cdot, \cdot]_u$ being positive semidefinite. If $L^2(u)$ denotes this completion (you must first factor out $\ker[\cdot, \cdot]_u$), then M_u is the operator densely defined on $L^2(u)$ by

$$(M_u \phi)(t) = t\phi(t), \phi \in C^{\infty}(\mathbf{R}).$$

Helton's characterization of the operators M_u was astonishing in that it was both simple and purely algebraic.

Theorem 3.8. Let $T \in \mathcal{L}(\mathcal{H})$ and assume that T is cyclic. T has the form "multiplication by t" on a Sobolev space of order n if and only if $(y - x)^{2n+1}(T) = 0$.

It is easy to envision that there is a multiplicity analog of Theorem 3.8 and we shall not state it here because of the somewhat lengthy technical details. Rather, let us be content to say that Helton's Theorem is the observation that if

$$\mathcal{F} = \{T| \|T\| \leq c \text{ and } T \text{ has the form } M_t \text{ on a Sobolev space of order } n\}$$

then
$$\mathcal{F} = \{c^2 - yx, (y-x)^{2n+1}, (x-y)^{2n+1}\}^\perp.$$

Example 3.9. Fix a constant $c \geq 1$. For $T \in \mathcal{L}(\mathcal{H})$ define the n-polynomial bound of $T, \mathrm{pol}_n(T)$ by the formula

3.10
$$\mathrm{pol}_n(T) = \sup\{\ \|(p_{ij}(T))\|\ \Big|\ \|\max_{|z|\leq 1}\|(p_{ij}(z))\| \leq 1\}.$$

In 3.10 the sup is over $n \times n$ matrix valued polynomials of one variable, and $(p_{ij}(T))$ acts on $\mathcal{L}(\mathcal{H}^{(n)})$. Following Halmos [18] we say that T is *polynomially bounded* if

$$\mathrm{pol}_1(T) < \infty.$$

Following Paulsen [22] we say that T is n-polynomially bounded if

$$\mathrm{pol}_n(T) < \infty,$$

and *completely polynomially bounded* if

$$\sup_{n\geq 1} \mathrm{pol}_n(T) < \infty.$$

For $p = (p_{ij}(z))$ and $n \times n$ matrix of polynomials of one variable define $\check{p} = \overline{(p_{ji}(\bar{z}))}$. One checks that if $T \in \mathcal{L}(\mathcal{H})$, then $p(T)^* = \check{p}(T^*)$ in $\mathcal{L}(\mathcal{H}^{(n)})$. Thus, if $c^2 - \check{p}(y)p(x)$ is regarded as an element of $\mathcal{P}_n$ we have that

$$\|(p_{ij}(T))\| \leq c \text{ if and only if } (c^2 - \check{p}(y)p(x))(T) \geq 0.$$

It follows that if $\mathcal{F}$ is the collection of operators s.t. $\mathrm{pol}_n(T) \leq c$, then

$$\mathcal{F} = \alpha_n^\perp,$$

where

$$\alpha_n = \{c^2 - \check{p}(y)p(x)|p \text{ is an } n \times n \text{ matrix polynomial of one variable s.t.}$$
$$\max_{|z|\leq 1}\|p(z)\| \leq 1\}.$$

Also, if $\mathcal{F}$ is the collection of operators s.t. $\sup_{n\geq 1} \mathrm{pol}_n(T) \leq c$, then

3.11
$$\mathcal{F} = (\cup_{n=0}^\infty \alpha_n)^\perp.$$

12

Example 3.12. Fix a separable but infinite dimensional Hilbert space $\mathcal{K}$ and let $c \geq 1$. Let $\mathcal{S}$ denote the set of all invertible $S \in \mathcal{L}(\mathcal{K})$ such that $\|S\| \|S^{-1}\| \leq c$ and let $\mathcal{C}$ denote the set of all $C \in \mathcal{L}(\mathcal{K})$ such that $\|C\| \leq 1$. Let $\mathcal{J}$ be the Hilbert space defined by

$$\mathcal{J} = \bigoplus_{\substack{S \in \mathcal{S} \\ C \in \mathcal{C}}} \mathcal{K}$$

and define $J \in \mathcal{L}(\mathcal{J})$ by

$$J = \bigoplus_{\substack{S \in \mathcal{S} \\ C \in \mathcal{C}}} SCS^{-1}.$$

If we let $\alpha = \{J\}^{\perp}$ and set $\mathcal{F} = \alpha^{\perp}$, then a straightforward consequence of Theorem 1.11 is that $\mathcal{F}^{\perp} = \alpha$ and that

$T \in \mathcal{F} \cap \mathcal{L}(\mathcal{H})$ if and only if there exists an invertible $S \in \mathcal{L}(\mathcal{H})$ with $\|S\| \|S^{-1}\| \leq c$ and there exists a $C \in \mathcal{L}(\mathcal{H})$ with $\|C\| \leq 1$ such that $T = SCS^{-1}$.

We remark that an unsolved problem in operator theory is to ascertain whether the following is true.

Question 3.13. *(Halmos) If $T \in \mathcal{L}(\mathcal{H})$ and $\mathrm{pol}_1(T) < \infty$, is $T \in \mathcal{F}$ for an appropriately chosen c?*

On the other hand, Paulson has shown in [22] that the family $\mathcal{F}$ defined in 3.11 is the same as the family we have just defined. In our present framework this result corresponds exactly to the explicit computation of an economical $\mathcal{B} \subseteq \mathcal{P}_{\infty}$ such that $\mathcal{F} = \mathcal{B}^{\perp}$.

Example 3.14. In this example we shall use hereditary rational functions. Let $\mathcal{K}$ be a compact subset of the plane and let $\mathcal{F}_n$ denote the collection of all operators such that $\mathcal{K}$ is an n-spectral set for T, i.e. $\sigma(T) \subseteq \mathcal{K}$ and

$$\|(f_{ij}(T))\| \leq \max_{z \in K} \|(f_{ij}(z))\|$$

for all $n \times n$ matrix-valued rational functions $f = (f_{ij})$ with poles off K. Mimicking the notation of Example 3.7, we see immediately that

$$\mathcal{F}_n = \alpha_n^{\perp}$$

where $\alpha_n \subseteq \mathcal{P}_n$ is defined by

$$\alpha_n = \{\|f\|^2 - \check{f}(y)f(x)|f \text{ is an } n \times n \text{ matrix rational function with poles of } K\}.$$

13

4. Operators as Linear Functionals

The theme of this section is an operator may be regarded as a linear functional on hereditary polynomials. To the operators that belong in a given family $\mathcal{F}$, there exists a *unified* representation of the corresponding linear functionals. When Choquet Theory is applied to the collection linear functionals that arise from $\mathcal{F}$, one obtains an integral representation of the general operator in $\mathcal{F}$. When $\mathcal{F}$ consists of self-adjoints, this integral representation is the spectral theorem for self-adjoints and when $\mathcal{F}$ is the isometries it is the spectral theorem for unitaries. When $\mathcal{F}$ is the contractions, it is Nagy Dilation Theorem, and when $\mathcal{F}$ contractive subnormals, it is the spectral theorem for normal operators (with spectrum in $\mathbf{D}^-$).

To simplify the discussion we shall consider only the cyclic case. Let $T \in \mathcal{L}(\mathcal{H})$ and assume that $\gamma \in \mathcal{H}$ is a cyclic vector for T. Define a linear functional $(T, \gamma)^{\wedge}$ on $\mathcal{P}_1$ by the formula

4.1
$$(T, \gamma)^{\wedge}(p) = < p(T)\gamma, \gamma >, \ p \in \mathcal{P}_1.$$

Our first observation is that the pair (T, γ) is completely determined up to unitary equivalence by the linear functional $(T, \gamma)^{\wedge}$.

Theorem 4.2. *Let $T_1 \in \mathcal{L}(\mathcal{H}_1), T_2 \in \mathcal{L}(\mathcal{H}_2), \gamma_1 \in \mathcal{H}_1, \gamma_2 \in \mathcal{H}_2$, and assume that γ_1 and γ_2 are cyclic for T_1 and T_2 respectively. There exists a Hilbert space isomorphism $U : \mathcal{H}_1 \to \mathcal{H}_2$ such that $T_1 = U^* T_2 U$ and $U\gamma_1 = \gamma_2$ if and only if $(T_1, \gamma_1)^{\wedge} = (T_2, \gamma_2)^{\wedge}$.*

To prove Theorem 4.2 simply consider the map $U : \mathcal{H}_1 \to \mathcal{H}_2$ defined by $U(q(T)\gamma_1) = q(T)\gamma_2, q$ a polynomial in one variable.

Now let $\mathcal{F}$ be a family. Since for simplicity we are only considering the cyclic case, we need to assume that $\mathcal{F}$ is *ancestral*, i.e. that any of the following equivalent properties hold:

4.2
$$\text{there exists } \alpha \subseteq \mathcal{P}_1 \text{ such that } \mathcal{F} = \alpha^{\perp},$$

4.3
$$\mathcal{F} = (\mathcal{F}^{\perp} \cap \mathcal{P}_1)^{\perp},$$

4.4 $T \in \mathcal{F}$ if and only if $T|\mathcal{N} \in \mathcal{F}$ whenever $\mathcal{N}$ is a cyclic invariant subspace for T.

If we let
$$F = \{(T, \gamma) | T \in \mathcal{F}, \gamma \text{ is cyclic for } T, \|\gamma\| = 1\},$$

and for $(T_1, \gamma_1), (T_2, \gamma_2) \in F$ we define $(T_1, \gamma_1) \approx (T_2, \gamma_2)$ whenever there exists a Hilbert space isomorphism U that intertwines T_1 and T_2 with $U\gamma_1 = \gamma_2$, then it is clear from 4.1 and Theorem 4.2 that we can unambigously define the following map $^\wedge$ on $F/\approx$. If $[(T, \gamma)]$ is an equivalence class in $F/\approx$, then $[(T, \gamma)]^\wedge$ is the linear functional on $\mathcal{P}_1$ defined by

4.5
$$[(T, \gamma)]^\wedge(p) = <p(T)\gamma, \gamma>, \ p \in \mathcal{P}_1.$$

Our goal now will be to concretely realize $F/\approx$ as a convex compact subset of a topological vector space. Define a Banach space B in the following way. For $p \in \mathcal{P}_1$ define $\|p\|$ by,

$$\|p\| = \sup_{T \in \mathcal{F}} w(p(T)) = \sup_{T \in \mathcal{F}} \sup_{\|\gamma\|=1} |<p(T)\gamma, \gamma>|.$$

Evidently, $\| \cdot \|$ is a seminorm. Set $B_0 = \mathcal{P}_1 \backslash N$ where $N \subseteq \mathcal{P}_1$ is the subspace $\{p| \, \|p\| = 0\}$. Then $\| \cdot \|$ is a norm on B_0. Let B be the completion of B_0 with respect to $\| \cdot \|$. Clearly, B is a Banach space and $\mathcal{P}_1$ may be regarded as a dense submanifold of B. In particular, if α is chosen so that 4.2 is satisfied, then $\alpha \subseteq B$ and we set $\mathcal{B} = \{\breve{q}(y)pq(x)|q$ is a poly of one variable and $p \in \alpha\}$. The *state space* of $\mathcal{F}, \Delta$, is defined as a subset of B^*, the dual of B.

Definition 4.6. Let Δ be the set of all $\mu \in B^*$ such that the following two conditions hold.

(a) $\mu(p) \geq 0$ whenever $p \in \mathcal{B}$.

(b) $\mu(1) = 1$.

We begin with the observation that Δ does not depend upon the choice of α (this is not obvious). Also, we note that Δ is both convex and compact in the w^* topology (this is obvious). The remaining point is that 4.5 defines a natural 1-1 correspondence between Δ and $F/\approx$. Thus, 4.5 defines a linear functional $\mu = [T, \gamma]^\wedge$ on $\mathcal{P}_1$. In fact, μ extends to be continuous on B and thus, $\mu \in B^*$. That $\mu(p) \geq 0$ whenever $p \in \mathcal{B}$ is a tautology based on the facts that $T \in \mathcal{F}$ and 4.2 holds. $\mu(1) = 1$ because $\|\gamma\| = 1$. Thus, we have shown that if $[(T, \gamma)] \in F$, then $[(T, \gamma)]^\wedge$ is in Δ.

How, given $\mu \in \Delta$, do we construct $T \in \mathcal{F} \cap \mathcal{L}(\mathcal{H})$ and $\gamma \in \mathcal{H}$, cyclic for T, such that $[(T, \gamma)]^\wedge = \mu$? Assume $\mu \in \Delta$. Define an inner product on polynomials of one variable by setting

$$[q_1, q_2]_\mu = \mu(q_2(y)q_1(x)).$$

Let $H^2(\mu)$ denote the completion of the polynomials of one variable with respect to $[\cdot,\cdot]_\mu$ and let M_μ dente the operator on $H^2(\mu)$ densely defined by

$$(M_\mu q)(z) = zq(z).$$

It follows from Definition 4.6 that M_μ is in $\mathcal{F}$ (not obvious). By construction the polynomial 1 is a cyclic vector for M_μ and condition (b) in Definition 4.6 implies that $\|1\| = 1$. Thus, $(M_\mu, 1) \in F$. It is a tautology that $[(M_\mu, 1)]^\wedge = \mu$.

We summarize the preceding constructions in the following theorem.

Theorem 4.7. *The map $\wedge$ is a 1-1 transformation from $F/\approx$ onto Δ, the state space of $\mathcal{F}$. If $\mu \in \Delta$, then $(M_\mu, 1) \in F$ and $[(M_\mu, 1)]^\wedge = \mu$. In particular, if $(T, \gamma) \in F$ and $\mu = [(T, \gamma)]^\wedge$, then $(T, \gamma) \approx (M_\mu, 1)$.*

A number of interesting ideas and powerful proof techniques flow naturally from Theorem 4.7. We illustrate with the following (by no means exhaustive) list of examples.

Example 4.8. The construction of extrinsic models through the representation of linear functionals.

Let $\mathcal{F}$ be the famiy defined in Example 3.1. A simple argument shows that the map $\Phi : B_0 \to C([-c, c])$ defined by

$$\Phi(p)(t) = p(t, t), \ t \in [-c, c]$$

is an isometry. Φ thus extends to an isometric isomorphism that identifies B and $C([-c, c])$. In particular, the state space of $\mathcal{F}$ is a subset of $C([-c, c])^*$, the finite regular Borel measures on $[-c, c]$. Needless to say, Δ is the probability measures on $[-c, c]$. If $\mu \in \Delta$, then

$$H^2(\mu) \ni f(z) \to f(t) \in L^2(\mu)$$

is an isometric isomorphism, M_μ is unitarily equivalent to M_t on $L^2(\mu)$, and the final statement of Theorem 4.7 (that $(T, \gamma) \approx (M_\mu, 1)$) is simply the assertion that a cyclic self-adjoint operator with spectrum in $[-c, c]$ is unitarily equivalent to M_t on $L^2(\mu)$ for some positive measure μ supported in $[-c, c]$.

Thus the spectral theorem (for cyclic self-adjoint operators) is a combination of two ideas. The first idea is to establish that if $T \in \mathcal{F}$, then

4.8 $\qquad (T, \gamma)^\wedge(p) \geq 0$ whenever $p \in \mathcal{P}_1$ and $p(t, t) \geq 0$ for all $t \in [-c, c]$.

As the reader certainly knows, the proof (see [23] for example) of 4.8 is almost purely algebraic, relying as it does on manipulating the defining polynomials for $\mathcal{F}$ and the spectral mapping theorem. The second idea in the proof of the spectral theorem is to represent the liner function in 4.8 using a measure.

There exists a probability measure μ supported on $[-c, c]$ such that

4.9
$$(T, \gamma)^\wedge(p) = \int p(t, t) d\mu \text{ whenever } p \in \mathcal{P}_1.$$

Example 4.10. Let $\mathcal{F}$ be the family defined in Example 3.6. Consider the continuous sesquilinear functional defined on $C^\infty(\mathbb{R})$ by 3.7. By the nuclearity of $C^\infty(\mathbb{R})'$ we deduce there exists a $u \in C^\infty(\mathbb{R}^2)'$ such that

4.11
$$[\varphi, \psi]_u = u(\varphi(x)\overline{\psi(y)})$$

whenever $\varphi, \psi \in C^\infty(\mathbb{R})$. In particular, if φ and ψ are polynomials q_1 and q_2 of one variable we have

$$[(q_2(y)q_1(x))(M_u)1, 1]_u = [q_2(M_u^*)q_1(M_u)1, 1]_u$$
$$= [q_1(M_u)1, \check{q}_2(M_u)1]_u$$
$$= [q_1, \check{q}_2]_u$$
$$= u(q_2(y)q_1(x)).$$

By linearity, we see that in fact,

$$[p(M_u)1, 1]_u = u(p)$$

for all $p \in \mathcal{P}_1$. Since $\mathcal{P}_1$ is dense in $C^\infty(\mathbb{R}^2)$ and since $[p(M_u)1, 1]_u = (M_u, 1)^\wedge(p)$ we see that Theorem 3.8 implies the following assertion.

4.12
If $(T, \gamma) \in F, 4$ then there exists $u \in C^\infty(\mathbb{R}^2)$ such that $(T, \gamma)^\wedge(p) = u(p)$ for all $p \in \mathcal{P}_1$.

Thus, Helton's theorem may be regarded as a representation theorem for the elements in the state space of $\mathcal{F}$. Observe also that the Sobolev space of Theorem 3.8 is $H^2(u)$. To precisely nail down the state space of $\mathcal{F}$, observe that if $(T, \gamma) \in F$ and u is defined by 4.12, then

4.13
$$u \text{ is pos. def.,}$$
$$(y - x)^{2n+1}u = 0 \text{ (since } T \text{ is } 2n + 1 - \text{symmetric)},$$
$$(c^2 - yx)u \text{ is pos. def. (since } \|T\| \leq c\text{), and}$$
$$u(1) = 1 \text{ (since } \|\gamma\| = 1\text{)}.$$

Conversely, it is a tautology that any $u \in C^\infty(\mathbb{R}^2)'$ satisfying conditions 4.3 is in Δ, the state space of $\mathcal{F}$. Summarizing,

Theorem 4.14. *The state space of $\mathcal{F}, \Delta$, is the collection of all $u \in C^{\infty}(\mathbb{R}^2)'$ that satisfy conditions 4.13.*

On the one hand Theorem 4.14 is adequate (Theorem 3.8 is an immediate corollary), but on the other hand it drives home an important point. At the analogous point in the theory of self adjoints, having just identified the state space as the probability measures, we interface for free, with the vast theory of measures, and thus the concrete representation of the state space leads to a rather complete theory of self adjoint operators. In the present context, however, it is clear that the theory of local forms is not sufficiently developed to allow the model implied by Theorem 4.14 to solve many of the operator theoretic questions one might have about $2n + 1$-symmetric operators.

Example 4.15. In this example we investigate the implications of Choquet Theory when applied to the state space of a family.

First we make some general observations. If $\mathcal{F}$ is a family (satisfying 4.2-4.3) and Δ is a state space, then we know from the convexity and compactness of Δ that Δ possesses extreme points. Furthermore, we know that if E denotes the set extreme points and $\mu \in \Delta$ then there exists a probability measure m on E^- such that

4.16
$$\mu = \int_{E^-} v \, dm(v).$$

In operator theory language 4.16 is a lifting theorem. Let

$$\mathcal{J} = \int_{\oplus} H^2(v) \, dm(v),$$

$$J = \int_{\oplus} M_v \, dm(v),$$

and define $V : H^2(\mu) \to \mathcal{J}$ by

$$V(q) = \int_{\oplus} q \, dm(v).$$

Then 4.16 asserts that V is an isometry and $V M_\mu = JV$, i.e., that T has an extension to J. Since any cyclic operator in F is unitarily equivalent to M_μ for some $\mu \in \Delta$ we obtain the following result.

Theorem 4.17. *If $T \in \mathcal{F}$ and T is cyclic, then T has an extension to a direct integral of the form*

$$\int_{\oplus} M_v \, dm(v)$$

18

where m is supported on the closure of the extreme points of Δ.

Theorem 4.17 has far reaching implications for lifting theory. Suppose one is in the presence of a family $\mathcal{F}$ for which one wants to develop a model theory. One way to proceed is to compute the extreme points of Δ, and then introduce as the model operators of $\mathcal{F}$ the direct integrals of the extreme points of $\mathcal{F}$. Theorem 4.17 then guarantees that every cyclic $T \in \mathcal{F}$ has an extension to a model operator of $\mathcal{F}$.

In the first example we consider $\mathcal{F}$ is the isometries. It is easy to calculate that if $\mu \in \Delta$ and μ is extreme, then there exists a point $e^{i\theta} \in \partial \mathbb{D}$ such that

$$4.18 \qquad\qquad \mu(p) = p(e^{i\theta}, e^{-i\theta}), \quad q \in \mathcal{P}_1.$$

We deduce from 4.18 that M_μ is a rank 1 unitary. Theorem 4.17 implies that every cyclic isometry has an extension to a direct integral of rank 1 unitaries. This latter fact, often referred to as Bochner's theorem, is a combination of two facts: the spectral theorem for cyclic unitaries, and the fact that a cyclic isometry lifts to a cyclic unitary.

Let $\mathcal{F}$ be the contractive subnormals. If $\mu \in \Delta$ and μ is extreme, then there exists a point $z \in \mathbb{D}^-$ such that

$$4.19 \qquad\qquad \mu(p) = p(z, z), \quad p \in \mathcal{P}_1,$$

If 4.19 holds, then M_μ is the rank one operator multiplication by z on $\mathbb{C}$. Theorem 4.17 asserts that a cyclic subnormal has an extension to a direct integral of scalar operators. This is the spectral theorem.

Let $\mathcal{F}$ be the contractions. If $p \in \Delta$ is extreme, then there exists $e^{i\theta} \in \partial \mathbb{D}$ such that 4.18 holds or there exists a unit vector $f \in H^2$ such that

$$4.20 \qquad\qquad \mu(p) = <p(S^*)f, f>, \quad p \in \mathcal{P}_1.$$

In 4.20, S is the shift operator $(S\varphi(z) = z\varphi(z))$ acting on H^2, the classical Hardy space. From 4.20 one sees that M_μ is unitarily equivalent to S^* restricted to a cyclic invariant subspace. In particular, every extreme point of Δ has an extension to a coisometry of multiplicity one. Since direct integrals of coisometries are coisometries, Theorem 4.17 implies that every cyclic contraction has an extension to a coisometry.

We now consider a new implication of extreme point analysis. Let $\mathcal{F}$ be the family defined in Example 3.6 and then further considered in Example 4.10. In this case Theorem 4.14 guarantees that the cyclic state space of $\mathcal{F}, \Delta$ can be identified with the set of $u \in C^{\infty}(\mathbb{R}^2)'$ such that conditions 4.13 hold. The following result then obtains. For simplicity we take $c = 1$. Let the $(n+1) \times (n+1)$ matrix J_0 be the jordan cell with 1's on the superdiagonal and 0's elsewhere. For $t \in [-1,1]$ define $\varphi_t : \mathbf{D} \to \mathbf{D}$ by $\varphi_t(z) = \frac{z+t}{1+tz}$. Let $J_t = \varphi_t(J_0)$.

Theorem 4.21. $v \in \Delta$ is an extreme point of Δ if and only if $v \approx (J_t, \gamma)$ for some $t \in [-1,1]$ and some $\gamma \in \mathbb{C}^{n+1}$ with $\|\gamma\| = 1$.

Since direct integrals of operators of the J_t are of the form $A + Q$ where $A = A^*$, $QA = AQ$, and $Q^{n+1} = 0$ we see that Theorem 4.17 and Theorem 4.21 imply the following result originally conjectured by Helton in [17].

Theorem 4.22. Let T be a cyclic operator. $(y-x)^{2n+1}(T) = 0$ if and only if T has an extension to an operator of the form $A + Q$ where $A = A^*$, $QA = AQ$, and $Q^{n+1} = 0$.

5. The Boundary

In this section we shall prove a theorem which asserts that if $\mathcal{F}$ is a family then there exists an *optimal* model theory for $\mathcal{F}$. If $\mathcal{F}$ is a family we say $\mathcal{B}$ is a model for $\mathcal{F}$ if the following two conditions hold.

5.1 $\qquad \mathcal{B}$ is closed with respect to unital representations and direct sums.

5.2 $\qquad$ If $T \in \mathcal{F} \cap \mathcal{L}(\mathcal{H})$, then there exists a $B \in \mathcal{B}$ such that $\mathcal{H}$ is invariant for B and $T = B|\mathcal{H}$.

A family $\mathcal{F}$ has in general many models. For example, $\mathcal{F}$ is a model for $\mathcal{F}$. We, however, shall be interested in models that are as small as possible. After all, what good is a model unless its elements are highly distinguished?

Theorem 5.3. Let $\mathcal{F}$ be a family. There exists a model $\mathcal{B}_0$ for $\mathcal{F}$ with the property that $\mathcal{B}_0 \subseteq \mathcal{B}$ whenever $\mathcal{B}$ is a model for $\mathcal{F}$.

Before continuing, we observe that the $\mathcal{B}_0$ which Theorem 5.3 guarantees exists is obviously unique. According, we refer to $\mathcal{B}_0$ as the *boundary of $\mathcal{F}$* and write $\mathcal{B}_0 = \partial \mathcal{F}$. We consider the following examples.

20

Example 5.4. Let $\mathcal{F}$ = isometries. In this case $\partial\mathcal{F}$ = unitaries. Obviously, the unitaries are closed with respect to unital representations and direct sums. Also, any isometry can be extended to a unitary. Thus, the unitaries are a model for the isometries. To see that any model for the isometries must contain the unitaries, observe that any model for the isometries must contain the rank one unitaries and then note, as a consequence of 5.1 and the spectral theorem, it must contain the unitaries as well.

Example 5.5. Using arguments similar to those in Example 5.4, it can be deduced that if $\mathcal{F}$ is the contractive subnormals, then $\partial\mathcal{F}$ is the contractive normals.

Example 5.6. If $\mathcal{F}$ is the contractive self adjoints, then the fact that every invariant subspace of a self adjoint operator is reducing implies that $\partial\mathcal{F} = \mathcal{F}$.

Example 5.7. Let $\mathcal{F}$ be the contractions. The Nagy dilation theorem implies that the coisometries are a model for $\mathcal{F}$. That in fact $\partial\mathcal{F}$ is the coisometries follows from Proposition 5.9 below.

To construct a proof of Theorem 5.3 we need the notion of an *extremal operator*.

Definition 5.8. Let $\mathcal{F}$ be a family. We say $T \in \mathcal{F}$ is *extremal for $\mathcal{F}$* if $\mathcal{N}$ is reducing for S whenever $S \in \mathcal{F}, \mathcal{N}$ is invariant for S, and $T = S|\mathcal{N}$.

The significance of extremal operators for model theory lies in the following two propositions.

Proposition 5.9. Let $\mathcal{F}$ be a family. If $\mathcal{B}$ is a model for $\mathcal{F}$ and S is extremal for $\mathcal{F}$, then $S \in \mathcal{B}$.

Proposition 5.10. Let $\mathcal{F}$ be a family. If $T \in \mathcal{F}$, then there exists an $S \in \mathcal{F}$ and an invariant subspace $\mathcal{N}$ for S such that S is extremal for $\mathcal{F}$ and $T = S|\mathcal{N}$.

The proof of Proposition 5.9 is easy. Observe that if $\mathcal{B}$ is a model for $\mathcal{F}$ and S is extremal for $\mathcal{F}$, then there exists $F \in \mathcal{B}$ and $\mathcal{N}$ invariant for R such that $S = R|\mathcal{N}$. Since S is extremal, $\mathcal{N}$ is reducing for R. Since $\mathcal{B}$ is a model for $\mathcal{F}$, 5.1 holds and thus $S \in \mathcal{B}$.

The proof of Proposition 5.10 is much more involved and here we merely indicate the idea of the proof. Let $T \in \mathcal{F}$. If T is external we are done. Thus, T may be assumed not extremal and there exist $T_1 \in \mathcal{F}$ and $\mathcal{N}_1$ invariant for T_1 such that $T = T_1|\mathcal{N}$, and $\mathcal{N}_1$ is not reducing for T_1. If T_1 is extermal, we are done. Thus, as before, T_1 may be assumed not extremal and there exist $T_2 \in \mathcal{F}$ and $\mathcal{N}_2$ invariant for T_2 such that $T_1 = T_2|\mathcal{N}_2$ and $\mathcal{N}_2$ is not reducing for T_2. Continuing in this manner by transfinite induction one obtains a sequence of operators T_α and invariant subspaces $\mathcal{N}_\alpha$ indexed by the ordinals. By careful bookkeeping on the condition

$$\mathcal{N}_\alpha \text{ is not reducing for } T_\alpha,$$

one deduces the existence of an ordinal α_0 such that T_α is extremal for $\mathcal{F}$. Since $T = T_\alpha | \mathcal{N}_1$ that completes the proof.

From Propositions 5.9 and 5.10 it is easy to deduce Theorem 5.3. Let Ω denote the collection of all subsets $\mathcal{B} \subseteq \mathcal{F}$ that contain the extremal operators for $\mathcal{F}$ and satisfy 5.1. Zorn's lemma guarantees that Ω contains a minimal element $\mathcal{B}_0$. Proposition 5.10 implies that $\mathcal{B}_0$ is a model for $\mathcal{F}$ and Proposition 5.9 guarantees that $\mathcal{B}_0$ is a subset of any model for $\mathcal{F}$.

References

1. Abrahamse, M. B. and R. G. Douglas, A class of operators related to multiply-connected domains, Adv. in Math. **19** (1976), 1-43.

2. Agler, J., Subjordan operators, Indiana University doctoral dissertation, 1980.

3. ———, Sub-jordan operators: Bishop's theorem, spectral inclusion, and spectral sets, Jour. of Op. Th. (1980), 373-395.

4. ———, The Arveson extension theorem and coanalytic models, Int. Eq. and Op. Th. **5** (1982), 608-631.

5. ———, Hypercontractions and subnormality, Jour. of Op. Th.

6. ———, Rational dilation on an annulus, Ann. of Math. **121** (1985), 537-563.

7. Arveson, W. B., Subalgebras of C^*-algebras, Acta Mth. **123** (1969), 141-224.

8. ———, Subalgebras of C^*-algebras II, Acta Math. **128** (1972), 271-308.

9. Athavale, A., Holomorphic kernels and commuting operators, (preprint).

10. deBranges, L. and J. Rovnyak, The existence of invariant subspaces, B.A.M.S. **70** (1964), 718-721.

11. ———, Canonical models in quantum scattering theory, *Perturbation Theory and Its Applications in Quantum Mechanics*, Ed. by C. H. Wilcox (New York-London-Sidney, 1966), 295-392.

12. Clancey, K., Seminormal operators (*Lecture Notes in Mathematics*), Vol. **742**, Springer-Verlag, Berlin, Heidelberg, and New York.

13. Embry, M., A generalization of the Halmos-Braun criterion for subnormality, Acta Sci. Math. (Szeged) **35** (1973), 61-64.

14. Sz-Nagy, B. and C. Foias, *Harmonic Analysis of Operators on Hilbert Space*, North-Holland, Amsterdam-London.

15. Helton, J. W., Operators with a representation as multiplication by x on a Sobolev space, Colloquia Math. Soc. Janos Bolyai **5**, Hilbert Space Operators, Tihany, Hungary (1970), 279-287.

16. ———, Jordan operators in infinite dimensions and Sturm-Liouville conjugate point theory, B.A.M.S. **78** (1972), 57-62.

17. ———, Infinite dimensional Jordan operators and Sturm-Liouville conjugate point theory, T.A.M.S. **170** (1972), 305-331.

18. Halmos, P. R., Ten problems in Hilbert space, B.A.M.S. **76** (1970), 887-933.

19. MacNerney, J., Hermitian moment sequences, T.A.M.S. **103** (1962), 45-81.

20. Sz.-Nagy, B., A moment problem for self adjoint operators, Acta Sci. Hungar. **3** (1952A), 285-292.

21. ———, Sur les contractions de l'espace de Hilbert, Acta Sci. Math. **15** (1953), 87-92.

22. Paulsen, V. I., Every completely polynomially bounded operator is similar to a contraction, Jour. Func. Anal. **55** (1984), 1-17.

23. Radjavi, J. and P. Rosenthal, Invariant Subspaces, Springer-Verlag, New York, Heidelberg, and Berlin.

24. Sarason, D., On spectral sets having connected complement, Acta Sci. Math. **26** (1965), 289-299.

25. ———, The H^p spaces of an annulus, M.A.M.S. **56** (1965).

Jim Agler

Department of Mathematics

University of California, San Diego

LaJolla, California 92093

U.S.A.

Applications of Several Complex Variables to Multiparameter Spectral Theory

by

Raúl E. Curto

0. Introduction

Let $\mathcal{B}$ be a commutative Banach algebra with identity 1, and let $M_{\mathcal{B}}$ be the maximal ideal space of $\mathcal{B}$. The Gelfand transform $\widehat{\ } : \mathcal{B} \to C(M_{\mathcal{B}})$ is defined by $\widehat{a}(\varphi) = \varphi(a), a \in \mathcal{B}, \varphi \in M_{\mathcal{B}}$. The spectrum of an element $a \in \mathcal{B}$, $\sigma_{\mathcal{B}}(a)$, defined as $\{\lambda \in \mathbb{C} : (a-\lambda)\mathcal{B} \neq \mathcal{B}\}$, can then be described by the formula

$$\sigma_{\mathcal{B}}(a) = \widehat{a}(M_{\mathcal{B}}).$$

In other words, a is invertible in $\mathcal{B}$ if and only if $\widehat{a} \neq 0$. Moreover, there exists a homomorphism $f \mapsto f(a)$ from $A(\sigma_{\mathcal{B}}(a))$, the algebra of germs of functions analytic in a neighborhood of $\sigma_{\mathcal{B}}(a)$, into $\mathcal{B}$ such that

(i) $1(a) = 1$;

(ii) $z(a) = a$; and

(iii) $\widehat{f(a)} = f \circ \widehat{a}$, for all $f \in A(\sigma_{\mathcal{B}}(a))$.

If Γ is a smooth rectifiable curve and $\Gamma = \partial U$, where U is an open set containing $\sigma_{\mathcal{B}}(a)$, then

$$f(a) := \frac{1}{2\pi i} \int_{\Gamma} f(z)(z-a)^{-1}\, dz.$$

As a consequence, one gets

$$\sigma_{\mathcal{B}}(f(a)) = \widehat{f(a)}(M_{\mathcal{B}}) = f[\widehat{a}(M_{\mathcal{B}})] = f(\sigma_{\mathcal{B}}(a)).$$

When $\mathcal{B}$ is no longer commutative, say $\mathcal{B} = \mathcal{L}(\mathcal{X})$, $\mathcal{X}$ a Banach space, one looks for a commutative subalgebra $\mathcal{A}$ containing a. In general, $\sigma_{\mathcal{A}}(a)$ will depend on $\mathcal{A}$, but if $\mathcal{A}$ is maximal (as an abelian subalgebra), then $\sigma_{\mathcal{A}}(a)$ is independent of $\mathcal{A}$. Therefore, we can define $\sigma_{\mathcal{B}}(a) := \sigma_{\mathcal{A}}(a)$, where $\mathcal{A}$ is a maximal abelian subalgebra of $\mathcal{B}$ containing a. It turns out that if $\mathcal{R}$ is the smallest inverse-closed closed subalgebra of $\mathcal{B}$ containing a, then $\sigma_{\mathcal{R}}(a) = \sigma_{\mathcal{B}}(a)$, so that $\sigma_{\mathcal{R}}(a)$ is already the smallest spectrum for a in the collection $\{\sigma_{\mathcal{A}}(a) : a \in \mathcal{A} \text{ (abelian) } \subseteq \mathcal{B}\}$.

In the case of several Banach algebra elements, the (joint) spectral theory is more complicated. Not only is there a difference between the study of commutative and noncommutative n-tuples, but also there is a clear distinction between algebraic and spatial joint spectra. In these lectures we shall concentrate on the elementary aspects of the commutative theory. We shall first consider the algebraic viewpoint, initiated by R. Arens, A. Calderón, and L. Waelbroeck in the 1950's, and then turn to the spatial approach, with emphasis in the notion developed by J. L. Taylor in 1970. Here we shall restrict attention mainly to the Hilbert space situation, and describe the projection property and functional calculus following F. -H. Vasilescu's program, with a number of simplifications and improvements due to us that make, in our opinion, the subject matter more understandable and accessible. We also approach the various notions of joint spectra from an axiomatic viewpoint, due to W. Żelazko, which highlights the importance of the projection property in the spectral mapping theorem; at the same time we can see how spatial notions of nonsingularity really amount to the nonvanishing of the Gelfand transform on suitable subsets of the maximal ideal space of a certain commutative Banach algebra.

All through these notes we keep an eye on the applications of the theory and its connections with several complex variables. We also devote Sections 6, 7, and 8 to the Fredholm theory and the study of Bergman n-tuples for various domains in $\mathbb{C}^n$. Many more interrelations with complex analysis, sheaf theory, complex geometry, and geometric measure theory take place as we go deeper into multiparameter operator theory, but we do not deal with them here. We have chosen to keep the level elementary so that the basic ideas and main examples get proper attention. The spatial theory is already 15-20 years old, but only relatively recently has it been simplified enough (at least in the Hilbert space case) to make it accessible to someone with little or no knowledge of homological algebra; only basic notions of Banach algebras, single operator theory, and several complex variables are required to follow these notes.

Throughout the paper, we shall consistently use the following notation:

$\mathcal{X}$: a vector space or Banach space,

$\mathcal{H}$: a Hilbert space,

$\mathcal{L}(\mathcal{X})$: the algebra of linear transformations (respectively bounded operators) on $\mathcal{X}$, where $\mathcal{X}$ is a vector space (respectively a Banach space),

$\mathcal{B}$: a Banach algebra or C^*-algebra,

$M_{\mathcal{B}}$: the maximal ideal space of a commutative Banach algebra,

$\mathbb{C}^n$: the complex n-dimensional space,

$\mathbb{C}[x]$: the algebra of polynomials in the indeterminate x

Ω : a bounded open subset of $\mathbb{C}^n$,

$: =$: the symbol for definition (see, for instance, Definition 1.1)

$R(T)$: the range of the linear transformation T,

$N(T)$: the kernel of the linear transformation T,

$P_{\mathcal{M}}$: the orthogonal projection of $\mathcal{H}$ onto the subspace $\mathcal{M} \subset \mathcal{H}$,

$\Lambda[e], \Lambda, \Lambda_n$: the exterior algebra on n generators $e_1, \dots, e_n$.

In general, I have tried and made an effort to give proper credit to the various mathematicians involved with the development of the theory; however, if a mathematical result appears to have no reference attached, it should not be assumed that it is due to the author.

1. Algebraic Joint Spectra

Let $\mathcal{B}, M_{\mathcal{B}}, \widehat{}$ be as before ($\mathcal{B}$ commutative), and let $a = (a_1, \dots, a_n) \in \mathcal{B}^n$ ($n \geq 1$).

Definition 1.1. We say that a is *invertible with respect to* $\mathcal{B}$ if there exist $b_1, \dots, b_n \in \mathcal{B}$ such that $a \circ b := \sum_{i=1}^{n} a_i b_i = 1$. The (algebraic) *spectrum* of a in $\mathcal{B}$ is

$$\sigma_{\mathcal{B}}(a) := \{\lambda \in \mathbb{C}^n : a - \lambda \text{ is not invertible in } \mathcal{B}\}.$$

The following is a well-known fact. For a general account of algebraic joint spectra, see [59, Chapter III] and [69, Chapter III].

Proposition 1.2.

$$\sigma_{\mathcal{B}}(a) = \{\lambda \in \mathbb{C}^n : (a - \lambda) \circ \mathcal{B}^n \neq \mathcal{B}\}$$
$$= \widehat{a}(M_{\mathcal{B}}) := \{(\varphi(a_1), \dots, \varphi(a_n)) : \varphi \in M_{\mathcal{B}}\}.$$

Corollary 1.3. Let $p : \mathbb{C}^n \to \mathbb{C}^k$ be a *polynomial mapping* and let $a \in \mathcal{B}^n$. Then $\sigma_{\mathcal{B}}(p(a)) = p(\sigma_{\mathcal{B}}(a))$. In particular, $\sigma_{\mathcal{B}}(a_i) = \{\lambda_i : \lambda \in \sigma_{\mathcal{B}}(a)\}$ ($i = 1, \dots, n$).

Proof:

$$\sigma_{\mathcal{B}}(p(a)) = \widehat{p(a)}(M_{\mathcal{B}}) = p[\widehat{a}(M_{\mathcal{B}})] = p(\sigma_{\mathcal{B}}(a)). \qquad \blacksquare$$

Definition 1.4. Let K be a bounded subset of $\mathbf{C}^n$. The *polynomially convex hull* of K is

$$\widehat{K} := \{\lambda \in \mathbf{C}^n : |p(\lambda)| \leq \|p\|_K \text{ for all } p \in \mathbf{C}[z]\}$$

(here $\mathbf{C}[z]$ stands for the algebra of polynomials in z). Clearly, $K \subseteq \widehat{K}$. When $K = \widehat{K}$, we say that K is *polynomially convex*.

Examples 1.5.

(i) When $n = 1$, K is polynomially convex if and only if $\pi^1(K) = 0$, i.e., if and only if $\mathbf{C} \setminus K$ has no bounded components. For $n > 1$, there exist *contractible* K's which are not polynomially convex (see Example 8.5 below).

(ii) If $n, k \geq 1$ and $K \ (\subseteq \mathbf{C}^n)$ and $L \ (\subseteq \mathbf{C}^k)$ are polynomially convex, then $K \times L \ (\subseteq \mathbf{C}^{n+k})$ is polynomially convex. In particular, the unit polydisk $\overline{\mathbf{D}}^n$ is polynomially convex. More generally, $(K \times L)\hat{} = \widehat{K} \times \widehat{L}$.

(iii)

$$\mathcal{B}_n := \left\{ \lambda \in \mathbf{C}^n : \|\lambda\| := \left(\sum_{i=1}^{n} |\lambda_i|^2 \right)^{1/2} \leq 1 \right\}$$

is polynomially convex.

(iv) The union of two disjoint compact convex subsets of $\mathbf{C}^n$ is polynomially convex, and so is the disjoint union of three closed balls. The union of three disjoint polydisks may not be polynomially convex, while it is unknown whether that is the case for four disjoint closed balls.

Theorem 1.6. (Oka-Weil) ([59, Theorem 5.1]) *Let K be polynomially convex, and let Ω be an open subset of $\mathbf{C}^n$ containing K. If $f \in A(\Omega)$, the algebra of functions analytic in Ω, then f is the uniform limit on K of a sequence of polynomials.*

Polynomially convex sets can always be regarded as maximal ideal spaces of polynomial algebras, as the following result, itself a particular case of Theorem 1.8, shows.

Theorem 1.7. *Let K be a compact subset of $\mathbf{C}^n$ and let $P(K)$ be the closure in $C(K)$ of $\mathbf{C}[z]$. Then $M_{P(K)}$ is canonically homeomorphic to $\widehat{K}$ via the map $\varphi \mapsto \varphi(z)$. (Strictly speaking, we should write $\varphi(\xi)$ instead of $\varphi(z)$, where $\xi(z) := z \ (z \in \mathbf{C}^n)$; unless confusion could arise, we shall usually avoid the use of ξ.)*

Theorem 1.8. *Let $a = (a_1, \ldots, a_n) \in \mathcal{B}^n$ and assume that a generates $\mathcal{B}$, i.e., $\mathbf{C}[a]$ is dense in $\mathcal{B}$. Then $M_{\mathcal{B}}$ can be identified with $\sigma_{\mathcal{B}}(a)$ via the map $\varphi \mapsto \varphi(a)$. Moreover, $\sigma_{\mathcal{B}}(a)$ is polynomially convex.*

Proof: Define $\Phi : M_{\mathcal{B}} \to \mathbb{C}^n$ by $\Phi(\varphi) = \varphi(a)$, $\varphi \in M_{\mathcal{B}}$. Clearly Φ is continuous, one-to-one, and $\Phi(M_{\mathcal{B}}) = \hat{a}(M_{\mathcal{B}}) = \sigma_{\mathcal{B}}(a)$. To prove that $\sigma_{\mathcal{B}}(a)$ is polynomially convex, let $\lambda \in \sigma_{\mathcal{B}}(\hat{a})$, i.e., $|p(\lambda)| \le \|p\|_{\sigma_{\mathcal{B}}(a)}$, for all $p \in \mathbb{C}[z]$. Now

$$\|p\|_{\sigma_{\mathcal{B}}(a)} = \|p\|_{\widehat{a}(M_{\mathcal{B}})} = \|\widehat{p(a)}\|_{M_{\mathcal{B}}} \le \|p(a)\|.$$

Therefore, the map $p(a) \mapsto p(\lambda)$ extends to a multiplicative linear functional φ on $\mathcal{B}$. Moreover, $\varphi(a) = \lambda$, so that $\lambda \in \hat{a}(M_{\mathcal{B}}) = \sigma_{\mathcal{B}}(a)$. ∎

Our next goal is to construct an analytic functional calculus for $\sigma_{\mathcal{B}}$. There are at least two instances where the construction can be carried out explicitly:

(A) Since $\sigma_{\mathcal{B}}(a_i) = P_i(\sigma_{\mathcal{B}}(a))$ $(P_i(z) = z_i$, $i = 1, \ldots, n)$, if D is an open polynomial containing $\sigma_{\mathcal{B}}(a)$, and $f \in A(D)$, then

$$f(z) = \frac{1}{(2\pi i)^n} \int_{\partial D} f(w) \prod_{i=1}^{n} (w_i - z_i)^{-1} \, dw_1 \cdot \ldots \cdot dw_n (z \in \sigma_{\mathcal{B}}(a))$$

(∂D is the distinguished boundary of D), so that we can define

$$f(a) := \frac{1}{(2\pi i)^n} \int_{\partial D} f(w) \prod_{i=1}^{n} (w_i - a_i)^{-1} \, dw_1 \cdot \ldots \cdot dw_n.$$

In particular, if

$$f(z) = \sum_{|\alpha| \le N} c_\alpha z^\alpha$$

is a polynomial, then

$$f(a) = \sum_{|\alpha| \le N} c_\alpha \, a^\alpha.$$

(B) If $\sigma_{\mathcal{B}}(a)$ is polynomially convex and $f \in A(\Omega)$, where Ω is an open subset of $\mathbb{C}^n$ containing $\sigma_{\mathcal{B}}(a)$, pick K polynomially convex such that $\sigma_{\mathcal{B}}(a) \subseteq \text{int.} (K) \subseteq K \subseteq \Omega$. Let $\{p_m\}_{m=1}^\infty$ be a sequence of polynomials converging uniformly to f on K (such a sequence exists by the Oka-Weil Theorem). Then one defines $f(a) := \lim_m p_m(a)$ (of course, one needs to prove that $\{p_m(a)\}_{m=1}^\infty$ is a Cauchy sequence in $\mathcal{B}$ and that the limit is independent of $\{p_m\}$).

We shall now describe the general construction of the functional calculus, following the line discovered by L. Waelbroeck. First, we would like to recall what a functional calculus for $\sigma_{\mathcal{B}}$ is. By a functional calculus for $\sigma_{\mathcal{B}}$, we mean a family of continuous homomorphisms $\Phi_a : A(\sigma_{\mathcal{B}}(a)) \to \mathcal{B}$, indexed by tuples a of elements in $\mathcal{B}$. Therefore, if $a \in \mathcal{B}^n$ and $f \in A(\sigma_{\mathcal{B}}(a))$, we shall be looking for an element $f(a) \in \mathcal{B}$.

Definition 1.9. (Waelbroeck ([119]) For $a \in \mathcal{B}^n$, the *rational spectrum* of a in $\mathcal{B}$ is $\sigma_R(a)$, where R is the smallest inverse-closed closed subalgebra of $\mathcal{B}$ that contains $a_1, \ldots, a_n$.

In general, $\sigma_B(a) \subseteq \sigma_R(a)$. Also, $\sigma_R(a)$ is *rationally convex*, i.e., if $\lambda \in \mathbb{C}^n$ and $|q(\lambda)| \leq \|q\|_{\sigma_R(a)}$ for all rational functions q (with singularities off $\sigma_R(a)$), then $\lambda \in \sigma_R(a)$.

Definition 1.10. Let $p : \mathbb{C}^n \to \mathbb{C}^k$ be a polynomial mapping, let $D \subseteq \mathbb{C}^n$ and $\Delta \subseteq \mathbb{C}^k$ be open polydomains, and let $\Omega(p, D, \Delta) := \{z \in D : p(z) \in \Delta\}$. When D and Δ are polydiscs with equal polyradii, $\overline{\Omega(p, D, \Delta)}$ is called a polynomial polyhedron; polynomial polyhedra are polynomially convex ([59, III.1]). The map $\iota : \Omega(p, D, \Delta) \to D \times \Delta$ given by $\iota(z) := (z, p(z))$ induces by duality the restriction map $\rho := \iota^* : A(D \times \Delta) \to A(\Omega(p, D, \Delta))$.

Theorem 1.11. (Cartan) ([119, I.3.2]) ρ is onto.

Observe that if $g(z, w) = p(z) - w, z \in D, w \in \Delta$, then the ideal $I(g)$ generated by g satisfies the property

$$f \in I(g) \Rightarrow \rho(f) \equiv 0.$$

The following theorem asserts that the null space of ρ is precisely $I(g)$.

Theorem 1.12. (Oka-Cartan-Waelbroeck) ([119, I.3.2])

$$A(D \times \Delta)/I(g) \simeq A(\Omega(p, D, \Delta)).$$

We can now give Waelbroeck's construction of the functional calculus. First, we need the following key result.

Theorem 1.13. (Waelbroeck) ([119, II.3.3]) *Let $a \in \mathcal{B}^n$ and let Ω be an open subset of $\mathbb{C}^n$ containing $\sigma_R(a)$. Then there exist polydomains D and Δ and a polynomial mapping $p : D \to \Delta$ such that $\sigma_R(a) \subseteq \Omega(p, D, \Delta) \subseteq \Omega$.*

The Construction of the Functional Calculus.

Let $a \in \mathcal{B}^n$, let $\Omega \supseteq \sigma_R(a)$, and let $f \in A(\Omega)$. By Theorem 1.13 there exist p, D and Δ such that $\sigma_R(a) \subseteq \Omega(p, D, \Delta) \subseteq \Omega$. Since $f \in A(\Omega)$, we have $f \in A(\Omega(p, D, \Delta))$, so that $f = \rho(F)$ for some $F \in A(D \times \Delta)$ (Theorem 1.11). Also, it is not hard to see that $\sigma_R(a, p(a)) \subseteq D \times \Delta$. Define

$$f(a) := F(a, p(a)),$$

where

$$F(a, p(a)) := \frac{1}{(2\pi i)^{n+k}} \int_{\partial D} \int_{\partial \Delta} F(z, w) \prod_{i=1}^{n} \prod_{j=1}^{k} (z_i - a_i)^{-1} (w_j - p_j(a))^{-1} \, dz \, dw$$

(see (A) on page 173 above, after Theorem 1.8). One needs to check that $f(a)$ is independent of $\Omega(p, D, \Delta)$ and of F. Also, if $f_i \in A(\Omega_i)$ $(i = 1, 2)$ and $f_1 = f_2$ on $\Omega_1 \cap \Omega_2$, then $f_1(a) = f_2(a)$ (see [119], [16]). When f is a polynomial

$$\left[f(z) = \sum_{|a| \le N} c_\alpha z^\alpha \right],$$

one can take $\Omega = \mathbb{C}^n$, $D \supseteq \sigma_R(a)$, $\Delta = \mathbb{D}$ and $p \equiv 0$ to get

$$f(a) = F(a, 0) = \frac{1}{(2\pi i)^n} \int_{\partial D} F(z, 0) \prod_{i=1}^{n} (z_i - a_i)^{-1} \, dz,$$

so that $f(a)$ is again given by

$$\sum_{|\alpha| \le N} c_\alpha a^\alpha.$$

To get the functional calculus for σ_B, one needs the following clever observation.

Lemma 1.14. (The Arens-Calderón trick) Let $a \in B^n$ and let $\Omega \supseteq \sigma_B(a)$. Then there exists $b = (b_1, \ldots, b_k) \in B^k$ such that $\Omega \supseteq P_a(\widehat{\sigma}(a, b))$, where $P_a : \mathbb{C}^{n+k} \to \mathbb{C}^n$ takes (z, w) to z, and where $\widehat{\sigma}(a, b)$ is the spectrum of (a, b) in the Banach subalgebra of B generated by a and b.

Proof: If $\lambda \notin \sigma_B(a)$, there exists $b = (b_1, \ldots, b_n) \in B^n$ such that $(a - \lambda) \circ b = 1$. Let $\varphi \in M_{((a,b))}$, where $((a,b))$ is the Banach subalgebra of B generated by (a, b). Then $(\varphi(a) - \lambda) \circ \varphi(b) = 1$, so that $\varphi(a) \ne \lambda$. Therefore, $\lambda \notin P_a(\widehat{\sigma}(a, b))$. (Recall that $\widehat{\sigma}(a, b)$ can be identified with $M_{((a,b))}$ (Theorem 1.8).) Then there exists an open neighborhood $\Omega(\lambda)$ of λ such that

$$\Omega(\lambda) \cap P_a(\widehat{\sigma}(a, b)) = \phi.$$

Since $\widehat{\sigma}(a) \backslash \Omega$ is compact, it can be covered by finitely many $\Omega(\lambda)$'s, say $\widehat{\sigma}(a) \backslash \Omega \subseteq \cup_{i=1}^{\ell} \Omega(\lambda_i)$. Thus, we can now conclude that

$$(\mathbb{C}^n \backslash \Omega) \cap P_a(\widehat{\sigma}(a, b^{(1)}, \ldots, b^{(\ell)}) = \phi,$$

for some tuples $b^{(1)}, \ldots, b^{(\ell)}$. To finish the proof, take $b = (b^{(1)}, \ldots, b^{(\ell)})$. (Recall that $P_a \widehat{\sigma}(a, b) \subseteq P_a \widehat{\sigma}(a, b^{(i)})$ for all $i = 1, \ldots, \ell$.) ∎

31

Theorem 1.15. (Shilov, Arens-Calderón, Waelbroeck) *There exists a continuous homomorphism $f \mapsto f(a)$ from $A(\sigma_B(a))$ into B such that*

(i) $1(a) = 1$;

(ii) $z_i(a) = a_i$ $(i = 1, \ldots, n)$; and

(iii) $\widehat{f(a)} = f \circ \hat{a}$, for all $f \in A(\sigma_B(a))$.

Consequently, $\sigma_B(f(a)) = f(\sigma_B(a))$.

Proof: Let $f \in A(\sigma_B(a))$. By Lemma 1.14, there exists $b \subset B$ such that $f \circ P_a \in A(\hat{\sigma}(a, b)) \subseteq A(\sigma_R(a,b))$. Define $f(a) := (f \circ P_a)(a,b)$. To prove (iii), observe that $f(a)^\wedge = f \circ \hat{a}$ whenever $f \in A(D)$, where D is a polydomain containing $\sigma_B(a)$. For a general $f \in A(\sigma_B(a))$, write $f(a) = (f \circ P_a)(a, b)$ for some $b \subset B$. Now

$$(f \circ P_a)(a, b) = F((a, b), p(a, b))$$

for some $F \in A(D \times \Delta)$ and p a polynomial. Therefore,

$$f(a)^\wedge = (f \circ P_a)(a, b)^\wedge = F((a, b)^\wedge, p(a, b)^\wedge)$$
$$= F((\hat{a}, \hat{b}), p(\hat{a}, \hat{b})) = f \circ \hat{a}. \quad \blacksquare$$

Remark 1.16.

(i) For finitely generated Banach algebras, Theorem 1.15 was proved by G. Shilov ([101]).

(ii) For an extension to locally convex algebras, see [98].

(iii) Waelbroeck proved that $f \mapsto f(a)$ is a homomorphism, a fact not obvious from [11] and [101]. R. Arens later showed in [10] that it was indeed possible to prove that fact with the techniques of [11] and [101]. Arens also extended the functional calculus to certain inverse limits of Banach algebras, and to vector-valued functions.

Theorem 1.17. (Uniqueness of the Functional Calculus [59, III.4.1]) *Suppose that $f \mapsto \tilde{f}(a)$ is another functional calculus satisfying (i), (ii), and*

(iv) $$\tilde{f}(a) = f \circ P_a(a, b) \text{ for all } a, b \text{ and } f \in A(\sigma_B(a)).$$

Then $\tilde{f}(a) = f(a)$ for all $f \in A(\sigma_B(a))$.

Proof: Let $a \in B^n$ and $f \in A(\sigma_B(a))$. By Lemma 1.14 there exists $b \in B^k$ such that $f \circ P_a \in A(\hat{\sigma}(a, b))$. Since $\hat{\sigma}(a, b)$ is polynomially convex, it follows from the Oka-Weil

Theorem, conditions (i) and (ii), and the continuity of both functional calculi, that $\tilde{f} \circ P_a(a, b) = f \circ P_a(a, b)$. Therefore, $\tilde{f}(a) = \tilde{f} \circ P_a(a, b) = f \circ P_a(a, b) = f(a)$, using (iv). ■

Remarks 1.18.

(i) Observe that if a functional calculus satisfies (i), (ii), and (iv), then it also satisfies (i), (ii), and (iii). W. Zame has shown in [122] that the Shilov-Arens-Calderón-Waelbroeck functional calculus is also unique subject to (i), (ii), and (iii) (see Corollary 5.21 below).

(ii) A somewhat different approach to the functional calculus can be found in [120].

2. Spatial Joint Spectra: The Axiomatic Approach

In this section we shall present W. Żelazko's axiomatic approach to joint spectra. Instead of dealing with Żelazko's spectra and subspectra, we shall work with the more general notion of a spectral system. A good portion of the section is devoted to examples, with special emphasis on the Taylor spectrum. The functional representation of spectral systems with the projection property will be considered in Section 3.

Definition 2.1. Let $\mathcal{B}$ be a Banach algebra with identity 1 and let $n \geq 1$. We shall let

$$\mathcal{B}^n_{\text{com}} := \{a = (a_1, \ldots, a_n) \in \mathcal{B}^n : a_i \, a_j = a_j \, a_i \text{ for all } i, j\}.$$

We shall write $a \subset \mathcal{B}$ if $a \in \mathcal{B}^n_{\text{com}}$ for some $n \geq 1$.

When $\mathcal{B}$ is a noncommutative Banach algebra with identity, say $\mathcal{B} = \mathcal{L}(\mathcal{X})$, where $\mathcal{X}$ is a Banach space, one could try to define the joint spectrum of an n-tuple $a = (a_1, \ldots, a_n)$ of elements in $\mathcal{B}$ as the algebraic joint spectrum of a relative to a maximal abelian subalgebra of $\mathcal{B}$ containing $a_1, \ldots, a_n$. Unlike the case $n = 1$, this spectrum *does* depend on the maximal abelian subalgebra, as the following example shows.

Example 2.2. (Albrecht [2]) There exist a Banach space $\mathcal{X}$, an n-tuple $a \subset \mathcal{L}(\mathcal{X})$, and maximal abelian subalgebras $\mathcal{A}_1$ and $\mathcal{A}_2$ containing a such that $\sigma_{\mathcal{A}_1}(a) \neq \sigma_{\mathcal{A}_2}(a)$.

Given a set X, we shall let $\mathcal{P}(X)$ denote the power set of X.

Definition 2.3. Let $\mathcal{B}$ be a Banach algebra. A *spectral system* for $\mathcal{B}$ is a map

$$\widetilde{\sigma} : \bigcup_{n=1}^{\infty} \mathcal{B}_{\mathrm{com}}^n \longrightarrow \mathcal{P}(\mathbf{C}^\omega)$$

($\mathbf{C}^\omega$ is the cartesian product of denumerably many copies of $\mathbf{C}$) such that

(i) $a \subset \mathcal{B} \Rightarrow \widetilde{\sigma}(a) \neq \phi$,

(ii) $a \in \mathcal{B}_{\mathrm{com}}^n \Rightarrow \widetilde{\sigma}(a) \subseteq \mathbf{C}^n \subseteq \mathbf{C}^\omega$, and

(iii) $a \subset \mathcal{B} \Rightarrow \widetilde{\sigma}(a)$ is compact.

A spectral system *on* a Banach space $\mathcal{X}$ is a spectral system *for* $\mathcal{L}(\mathcal{X})$.

Definition 2.4. Let $\mathcal{B}$ be a Banach algebra. A spectral system $\widetilde{\sigma}$ for $\mathcal{B}$ possesses the *projection property* if

$$P_a \widetilde{\sigma}(a,b) = \widetilde{\sigma}(a) \text{ and } P_b \widetilde{\sigma}(a,b) = \widetilde{\sigma}(b)$$

for all $a,b \subset \mathcal{B}$ ($P_a : \mathbf{C}^n \times \mathbf{C}^k \longrightarrow \mathbf{C}^n$, $(z,w) \mapsto z$, and similarly for P_b). σ possesses the *spectral mapping property for polynomials* if $\widetilde{\sigma}(p(a)) = p(\widetilde{\sigma}(a))$ for every polynomial map $p : \mathbf{C}^n \longrightarrow \mathbf{C}^k$ and for every n-tuple $a \in \mathcal{B}_{\mathrm{com}}^n$.

Examples 2.5.

(i) If $\mathcal{B}$ is commutative, $\sigma_\mathcal{B}, \sigma_R$ and $\widehat{\sigma}$ are all spectral systems. However, only $\sigma_\mathcal{B}$ has the projection property.

(ii) For $a \subset \mathcal{B}$ we let (a), $(a)'$, and $(a)''$ denote the Banach subalgebras of $\mathcal{B}$ *generated* by $a_1, \ldots, a_n$ and 1, by its *commutant* (relative to $\mathcal{B}$), and by its *double commutant*, respectively. We define

$$\widehat{\sigma}(a) := \sigma_{(a)}(a), \qquad \text{(cf. Lemma 1.14)}$$

$$\sigma''(a) := \sigma_{(a)''}(a),$$

and

$$\sigma'(a) := \{\lambda \in \mathbf{C}^n : (a-\lambda) \circ (a)' \neq (a)'\}.$$

More generally, if $\mathcal{A}$ is any (closed) subalgebra of $\mathcal{B}$ containing a in its center, we let

$$\sigma_\mathcal{A}(a) := \{\lambda \in \mathbf{C}^n : (a-\lambda) \circ \mathcal{A} \neq \mathcal{A}\}.$$

$\widehat{\sigma}$, σ' and σ'' are spectral systems *without* the projection property ([103]). Also $\sigma' \subseteq \sigma'' \subseteq \widehat{\sigma}$.

In general, when we refer to an *algebraic spectrum*, we mean $\sigma_\mathcal{A}$ for some $\mathcal{A}$. For general facts about $\sigma_\mathcal{A}$, see [29], [48], [49], [50], [104], [105], [116], [123].

(iii) The *left* spectrum is defined by

$$\sigma_\ell(a) := \{\lambda \in \mathbb{C}^n : \mathcal{B} \circ (a - \lambda) \neq \mathcal{B}\} \ (a \subset \mathcal{B}).$$

Similarly, the *right* spectrum is given by

$$\sigma_r(a) := \{\lambda \in \mathbb{C}^n : (a - \lambda) \circ \mathcal{B} \neq \mathcal{B}\} \ (a \subset \mathcal{B}).$$

The *Harte spectrum* is $\sigma_H := \sigma_\ell \cup \sigma_r$. $\sigma_\ell, \sigma_r \subseteq \sigma_H \subseteq \sigma'$; moreover, σ_ℓ, σ_r and σ_H all possess the projection property ([19], [63], [64]).

(iv) If $\mathcal{B} = \mathcal{L}(\mathcal{X})$, where $\mathcal{X}$ is a Banach space, the *approximate point spectrum* and *defect spectrum* are defined by

$$\sigma_\pi(a) := \{\lambda \in \mathbb{C}^n : a - \lambda \text{ is not jointly bounded below}\}$$

and

$$\sigma_\delta(a) := \{\lambda \in \mathbb{C}^n : a - \lambda \text{ is not jointly onto}\}.$$

($a \subset \mathcal{L}(\mathcal{X})$ is *jointly bounded below* if there exists a positive constant $\varepsilon > 0$ such that

$$\sum_{i=1}^n \|a_i x\| \geq \varepsilon \|x\| \qquad (\text{all } x \in \mathcal{X});$$

a is *jointly onto* if

$$\sum_{i=1}^n a_i \mathcal{X} = \mathcal{X}.)$$

σ_π and σ_δ are spectral systems *with* the projection property ([26]).

(v) Let $\sigma_{\pi,k}$ ($k = 0, \ldots, n-1$) and $\sigma_{\delta,k}$ ($k = 1, \ldots, n$) be the generalized approximate point and defect spectra introduced by Z. Slodkowski. As shown in [102], $\sigma_{\pi,k}$ and $\sigma_{\delta,k}$ are spectral systems with the projection property.

(vi) For $a \subset \mathcal{L}(\mathcal{X})$ let $\sigma_\Pi(a, \mathcal{X}) := \sigma(a_1, \mathcal{X}) \times \cdots \times \sigma(a_n, \mathcal{X})$, where $\sigma(a_i, \mathcal{X})$ denotes the ordinary spectrum of a_i as an element of $\mathcal{L}(\mathcal{X})$. σ_Π is a spectral system with the projection property; however, σ_Π does not have the spectral mapping property for polynomial mappings. (Example: Let p be a nontrivial idempotent in $\mathcal{L}(\mathcal{X})$. Then $\sigma_\Pi(p, p) = \{(0,0), (0,1), (1,0), (1,1)\}$, so that $\{\lambda_1 + \lambda_2 : (\lambda_1, \lambda_2) \in \sigma_\Pi(p, p)\} = \{0, 1, 2\}$, while $\sigma(2p) = \{0, 2\}$.) We shall see in Section 3 that if a spectral system $\tilde{\sigma}$ satisfies locally an inclusion of the form $\tilde{\sigma} \subseteq \sigma_\mathcal{B}$, where $\mathcal{B}$ is commutative, then the spectral mapping property for polynomial mappings follows from the projection property.

35

(vii) (Taylor's spectrum) Let $\Lambda = \Lambda[e] = \Lambda_n[e]$ be the exterior algebra on n genera-tors $e_1, \ldots, e_n$, with identity $e_0 = 1$. Λ is the algebra of forms in $e_1, \ldots, e_n$ with complex coefficients, subject to the collapsing property $e_i\, e_j + e_j\, e_i = 0$ $(1 \leq i,j \leq n)$. Let $E_i : \Lambda \longrightarrow \Lambda$ be given by $E_i \xi = e_i \xi$ $(i = 1, \ldots, n)$. $E_1, \ldots, E_n$ are the *creation operators*. Clearly $E_i\, E_j + E_j\, E_i = 0$ $(1 \leq i,j \leq n)$. Λ can be regarded as a Hilbert space if we declare $\{e_{i_1} \cdot \ldots \cdot e_{i_k} : 1 \leq i_1 < \ldots < i_k \leq n\}$ as an orthonormal basis. Then $E_i^* \xi = \xi'$, where $\xi = e_i \xi' + \xi''$ is the unique decomposition of a form ξ as the sum of an element in the range of E_i and an element in the kernel of E_i^*. Actually, each E_i is a partial isometry, and $E_i^* E_j + E_j\, E_i^* = \delta_{ij}$ $(1 \leq i,j \leq n)$. For $\mathcal{X}$ a vector space and $A \subset \mathcal{L}(\mathcal{X})$, we define $D_A : \Lambda(\mathcal{X}) \longrightarrow \Lambda(\mathcal{X})$ $(\Lambda(\mathcal{X}) := \mathcal{X} \otimes_{\mathbf{C}} \Lambda)$ by

$$D_A := \sum_{i=1}^{n} A_i \otimes E_i.$$

Then

$$D_A^2 (x \otimes \xi) = \sum_{i,j=1}^{n} A_j\, A_i x \otimes E_j\, E_i \xi = \sum_{i<j} A_i\, A_j x \otimes (E_i\, E_j + E_j\, E_i)\xi = 0,$$

so that $R(D_A) \subseteq N(D_A)$ (R and N denote range and kernel). We say that A is nonsingular on $\mathcal{X}$ if $R(D_A) = N(D_A)$. When $n = 1$, for instance, A is nonsingular if and only if A is one-to-one and onto. The Taylor spectrum of A on $\mathcal{X}$ is

$$\sigma_T(A, \mathcal{X}) := \{\lambda \in \mathbf{C}^n : R(D_{A-\lambda}) \neq N(D_{A-\lambda})\}.$$

One can also construct a cochain complex $K(A, \mathcal{X})$, called the *Koszul complex* associated to A on $\mathcal{X}$, as follows:

$$K(A, \mathcal{X}): \; 0 \longrightarrow \Lambda^0(\mathcal{X}) \xrightarrow{D_A^0} \Lambda^1(\mathcal{X}) \xrightarrow{D_A^1} \cdots \xrightarrow{D_A^{n-1}} \Lambda^n(\mathcal{X}) \longrightarrow 0$$

(here $\Lambda^k(\mathcal{X})$ is the collection of k forms and $D_A^k := D_A\big|_{\Lambda^k(\mathcal{X})}$). Then

$$\sigma_T(A, \mathcal{X}) = \{\lambda \in \mathbf{C}^n : K(A - \lambda, \mathcal{X}) \quad \text{is not exact}\}.$$

J. L. Taylor showed that if $\mathcal{X}$ is a Banach space, then $\sigma_T(A, \mathcal{X})$ is compact and nonempty, and that $\sigma_T(A, \mathcal{X}) \subseteq \sigma'(A)$ ([104]). Moreover, σ_T carries an analytic functional calculus (so that in particular σ_T has the projection property) ([105]). We shall take a close look at σ_T in Section 4. For now, we would like to indicate how D_A reflects the joint spatial behavior of $A_1, \ldots A_n$. For $n = 1$, D_A admits the following 2×2-matrix relative to the direct sum decomposition $(\mathcal{X} \otimes e_0) \oplus (\mathcal{X} \otimes e_1)$:

$$D_A = \begin{pmatrix} 0 & 0 \\ A & 0 \end{pmatrix} \begin{matrix} e_0 \\ e_1 \end{matrix} ;$$

then

$$N(D_A) = N(A) \oplus \mathcal{X}$$

$$R(D_A) = 0 \oplus R(A)$$

$$\overline{N(D_A)/R(D_A)} = N(A) \oplus \mathcal{X}/R(A).$$

For $n = 2$,

$$D_A = \begin{pmatrix} 0 & 0 & 0 & 0 \\ A_1 & 0 & 0 & 0 \\ A_2 & 0 & 0 & 0 \\ 0 & -A_2 & A_1 & 0 \end{pmatrix} \begin{matrix} e_0 \\ e_1 \\ e_2 \\ e_1 e_2 \end{matrix} ,$$

and

$$N(D_A) = [N(A_1) \cap N(A_2) \oplus \{(x_1, x_2) : A_2 x_1 = A_1 x_2\} \oplus \mathcal{X}$$

$$R(D_A) = O \oplus \{(A_1 x_0, A_2 x_0) : x_0 \in \mathcal{X}\} \oplus [R(A_1) + R(A_2)].$$

Observe that

$$\sigma_\pi(A, \mathcal{X}) = \{\lambda \in \mathbb{C}^n : D_{A-\lambda}^0 \quad \text{is not bounded below}\}$$

and

$$\sigma_\delta(A, \mathcal{X}) = \{\lambda \in \mathbb{C}^n : D_{A-\lambda}^{n-1} \quad \text{is not onto}\},$$

so that $\sigma_\pi,\ \sigma_\delta \subseteq \sigma_T$. The set

$$\sigma_p(A, \mathcal{X}) := \{\lambda \in \mathbb{C}^n : D_{A-\lambda}^0 \quad \text{is not one-to-one}\}$$

is called the point spectrum of A on $\mathcal{X}$; $\sigma_p(A, \mathcal{X})$ is the collection of joint eigenvalues for A on $\mathcal{X}$ (λ is a *joint eigenvalue* for A on $\mathcal{X}$ if there exists a nonzero vector $x \in \mathcal{X}$ such that $(A_i - \lambda_i)x = 0$ for all $i = 1, \ldots, n$). Unlike σ_δ, σ_p is not in general a spectral system.

If $\mathcal{B}$ is a commutative Banach algebra and $a \subset \mathcal{B}$, and if we let L_a denote the n-tuple of left multiplications by the a_i's, then $\sigma_T(L_a, \mathcal{B}) = \sigma_\mathcal{B}(a)$ ([104, p. 191]). This can be shown as follows: If L_a is nonsingular on $\mathcal{B}$, then every n-form must be in the range of D_A, so that there exist elements $b_1, \ldots, b_n$ in $\mathcal{B}$ such that

$$D_A \left[\sum_{i=1}^{n} b_i \otimes e_1 \, e_2 \cdot \ldots \cdot \widehat{e_i} \cdot \ldots \cdot e_n \right] = 1 \otimes e_1 \cdot \ldots \cdot e_n$$

(where $\widehat{}$ means deletion). Then

$$\sum_{i=1}^{n} \sum_{j=1}^{n} a_j b_i \otimes e_j \, e_1 \, e_2 \cdot \ldots \cdot \widehat{e_i} \cdot \ldots \cdot e_n = 1 \otimes e_1 \cdot \ldots \cdot e_n,$$

so that

$$\sum_{i=1}^{n} a_i b_i (-1)^{i-1} = 1,$$

and therefore a is invertible with respect to $\mathcal{B}$. Conversely, if $a \circ b = 1$ for some $b \in \mathcal{B}^n$, we can let

$$\tilde{D}_b := \sum_{i=1}^{n} b_i E_i^*$$

and then establish the formula

$$D_a \tilde{D}_b + \tilde{D}_b D_a = \left[\sum_{i=1}^{n} a_i b_i \right] \otimes I = 1 \otimes I,$$

from which it easily follows that $N(D_a) = R(D_a)$. The same result, with identical proof, also holds if $\mathcal{B}$ is non-commutative, provided that the commuting n-tuple $a \subset \mathcal{B}$ is taken from the center of $\mathcal{B}$. Thus, every algebraic joint spectra $\sigma_A(a)$ can be thought of as $\sigma_T(L_a, \mathcal{A})$ ([104, p. 191]).

When $\mathcal{B}$ is a C^*-algebra, say $\mathcal{B} \subset \mathcal{L}(\mathcal{H})$, and $a \subset \mathcal{B}$, then $\sigma_T(L_a, \mathcal{B}) = \sigma_T(a, \mathcal{H})$ ([37]). This fact, known as *spectral permanence* for the Taylor spectrum, is a consequence of the following result: L_a is nonsingular on $\mathcal{B}$ if and only if $D_a + D_{a^*}^t$ is invertible (as a matrix over $\mathcal{B}$), where $a^* = (a_1^*, \ldots, a_n^*)$ and ^{t} denotes the transpose of a matrix. As a consequence we also see that $D_a + D_{a^*}^t$ determines the nonsingularity of a on $\mathcal{H}$. It is because of this that we shall usually refer to nonsingularity as *invertibility* when a is an n-tuple of C^*-algebra elements.

When $\mathcal{X}$ is a finite-dimensional space and $A \subset \mathcal{L}(\mathcal{X})$, then

$$\sigma_p(A, \mathcal{X}) = \sigma_\ell(A, \mathcal{X}) = \sigma_\pi(A, \mathcal{X}) = \sigma_r(A, \mathcal{X}) = \sigma_\delta(A, \mathcal{X})$$
$$= \sigma_T(A, \mathcal{X}) = \sigma'(A) = \sigma''(A) = \widehat{\sigma}(A)$$

([35, p. 401]). The reason for this lies in the fact that $A_1, \ldots, A_n$ can be simultaneously triangularized as

$$A_i = \begin{pmatrix} \lambda_i^{(1)} & & * \\ & \ddots & \\ 0 & & \lambda_i^{(k)} \end{pmatrix} \quad (i = 1, \ldots, n),$$

so that all of the spectra above equal $\{\lambda^{(1)}, \ldots, \lambda^{(k)}\}$.

To conclude this preliminary discussion of σ_T, let us indicate one more important property: If $A \subset \mathcal{L}(\mathcal{X})$, $\mathcal{X}$ a Banach space, and if $\mathcal{Y}$ is a subspace of $\mathcal{X}$ invariant under A (i.e., $A_i \mathcal{Y} \subseteq \mathcal{Y}$, $i = 1, \ldots, n$), then the union of any two of $\sigma_T(A, \mathcal{X}), \sigma_T(A, \mathcal{Y})$ and $\sigma_T(A, \mathcal{X}/\mathcal{Y})$ contains the third ([104, Lemma 1.2]). In particular, if $A \subset \mathcal{L}(\mathcal{X}), B \subset \mathcal{L}(\mathcal{Y})$ and

$$\begin{pmatrix} A & C \\ 0 & B \end{pmatrix} \subset \mathcal{L}(\mathcal{X} \oplus \mathcal{Y}),$$

then

$$\sigma_T\left(\begin{pmatrix} A & C \\ 0 & B \end{pmatrix}, \mathcal{X} \oplus \mathcal{Y} \right) \subseteq \sigma_T(A, \mathcal{X}) \cup \sigma_T(B, \mathcal{Y})$$

([35, Lemmas 4.4 and 4.5]). (If $C = 0$, this last inclusion becomes an equality.)

To end this section, we shall present a table summarizing the various notions of joint spectra and their properties.

TABLE 2.6*

		Spectral systems on $\mathcal{X}$								
	σ_B	σ_π σ_δ	σ_ℓ σ_r	σ_H	σ_T	σ'	σ''	σ_R	$\hat{\sigma}$	σ_Π
projection property	✓	✓	✓	✓	✓					✓
spectral mapping property for polynomial mappings	✓	✓	✓	✓	✓					
functional calculus	✓				✓					
spectral mapping property for analytic functions	✓				✓					

*($\mathcal{B}$ commutative, $\mathcal{X}$ Banach space)

Note: $\begin{smallmatrix}\sigma_\pi\\\sigma_\delta\end{smallmatrix} \subseteq \begin{smallmatrix}\sigma_\ell\\\sigma_r\end{smallmatrix} \subseteq \sigma_H \subseteq \sigma_T \subseteq \sigma' \subseteq \sigma'' \subseteq \sigma_R \subseteq \hat{\sigma}$

3. The Spectral Mapping Theorem

We shall see now the functional representation of spectral systems with the projection property. In [123], W. Żelazko defines a subspectrum on a Banach algebra $\mathcal{B}$ as a spectral system $\widetilde{\sigma}$ on $\mathcal{B}$ possessing the spectral mapping property for polynomials and such that $\widetilde{\sigma}(a) \subseteq \Pi_{i=1}^{n}\sigma_{\mathcal{B}}(a_i)$ for all $a \subset \mathcal{B}$. Żelazko then shows that subspectra admit a functional representation in terms of the Gelfand transform, as follows: If $\mathcal{A}$ is a maximal abelian subalgebra of $\mathcal{B}$, then there exists a compact nonempty subset $M_{\widetilde{\sigma}}(\mathcal{A})$ of $M_{\mathcal{A}}$ such that $\widetilde{\sigma}(a) = \widehat{a}(M_{\widetilde{\sigma}}(\mathcal{A}))$ for all $a \subset \mathcal{A}$. As a consequence, $\widetilde{\sigma}(a) \subseteq \widehat{a}(M_a) = \sigma_{\mathcal{A}}(a)$ (Żelazko proves this inclusion first, and then uses it in the proof of the functional representation).

Our approach to this circle of ideas will be the following: We shall start with a spectral system $\widetilde{\sigma}$ on a Banach space $\mathcal{X}$ possessing the projection property and such that $\widetilde{\sigma} \subseteq \sigma_{\mathcal{B}}$ for some commutative Banach algebra $\mathcal{B}$ (see Definition 3.1). We shall then find the functional representation for $\widetilde{\sigma}$ and also show the uniqueness of $M_{\widetilde{\sigma}}(\mathcal{B})$. Our hypotheses are weaker than those in [123], but we shall see that we can still recover the same results.

Definition 3.1. Let $\mathcal{X}$ be a Banach space and let $\widetilde{\sigma}$ be a spectral system on $\mathcal{X}$ possessing the projection property. Let $\mathcal{B}$ be a commutative Banach algebra acting on $\mathcal{X}$, i.e., there exists a map $\Phi : \mathcal{B} \longrightarrow \mathcal{L}(\mathcal{X})$ (alternatively, suppose that $\mathcal{X}$ is a Banach $\mathcal{B}$-module). Assume that

$$\widetilde{\sigma}\big(\Phi(a)\big) \subseteq \sigma_{\mathcal{B}}(a)$$

for all $a \subset \mathcal{B}$. Then

$$M_{\widetilde{\sigma}} = M_{\widetilde{\sigma}}(\mathcal{B}, \mathcal{X}, \Phi) := \bigcap_{a \subset \mathcal{B}} \widehat{a}^{-1}\big[\widetilde{\sigma}\big(\Phi(a)\big)\big] \subseteq M_{\mathcal{B}}.$$

Lemma 3.2. $M_{\widetilde{\sigma}}$ *is a compact and nonempty subset of* $M_{\mathcal{B}}$.

Proof: For $a \subset \mathcal{B}$ let $M_a := \widehat{a}^{-1}\big[\widetilde{\sigma}\big(\Phi(a)\big)\big]$. Since $\widetilde{\sigma}\big(\Phi(a)\big) \subseteq \sigma_{\mathcal{B}}(a)$ $(= \widehat{a}(M_{\mathcal{B}}))$, it follows that $M_a \neq \phi$. If $a^{(1)}, \ldots, a^{(k)} \subset \mathcal{B}$ and $\varphi \in M_{(a^{(1)}, \ldots, a^{(k)})}$, then $\varphi\big(a^{(1)}, \ldots, a^{(k)}\big) \in \widetilde{\sigma}\big[\big(\Phi(a^{(1)}), \ldots, \Phi(a^{(k)})\big)\big]$ and, therefore, since $\widetilde{\sigma}$ has the projection property,

$$\varphi\big(a^{(i)}\big) \in \widetilde{\sigma}\left[\Phi\big(a^{(i)}\big)\right] \qquad (i = 1, \ldots, k).$$

It follows that $\varphi \in \cap_{i=1}^{k} M_{a^{(i)}}$; thus,

$$M_{(a^{(1)}, \ldots, a^{(k)})} \subseteq \bigcap_{i=1}^{k} M_{a^{(i)}},$$

which shows that the family $\{M_a\}_{a \subset B}$ has the finite intersection property. Since M_B is compact, we must have $M_{\widetilde{\sigma}} = \cap_{a \subset B} M_a \neq \phi$. The compactness of $M_{\widetilde{\sigma}}$ follows from that of M_a, which in turn is a consequence of the continuity of $\widehat{a}$ and the fact that $\widetilde{\sigma}$ is a spectral system. ∎

Theorem 3.3. *Let $\mathcal{X}, \mathcal{B}, \Phi$ and $\widetilde{\sigma}$ be as before. If $a \subset \mathcal{B}$, then*

$$\widetilde{\sigma}\big(\Phi(a)\big) = \widehat{a}(M_{\widetilde{\sigma}}).$$

Moreover, $M_{\widetilde{\sigma}}$ is the only compact nonempty subset of M_B having this property.

Remarks 3.4.

(i) The importance of the functional representation of $\widetilde{\sigma}$ is that one recovers the formula "$\sigma(a) = \text{range } (\widehat{a})$"; in the noncommutative case, however, one must restrict $\widehat{a}$ to a certain *subset* of a maximal ideal space. The uniqueness of $M_{\widetilde{\sigma}}$ is of importance in some applications (see Corollary 3.12 below).

(ii) For the special case of σ_T, J. L. Taylor gave in [104, Theorem 3.3] a corresponding version of Theorem 3.3. We basically follow his outline for our proof below.

Proof of Theorem 3.3.

$$\widehat{a}(M_{\widetilde{\sigma}}) \subseteq \widehat{a}\big(\widehat{a}^{-1}\big[\widetilde{\sigma}(\Phi(a))\big]\big) \subseteq \widetilde{\sigma}\big(\Phi(a)\big).$$

Conversely, let $\lambda \in \widetilde{\sigma}\big(\Phi(a)\big)$. We claim that there exists $\varphi \in M_{\widetilde{\sigma}}$ such that $\varphi(a) = \lambda$. Consider the family $\{M_b \cap \widehat{a}^{-1}(\lambda)\}_{b \subset B}$. As in the proof of Lemma 3.2, one shows that

$$\bigcap_{b \subset B} (M_b \cap \widehat{a}^{-1}(\lambda)) \neq \phi$$

provided $M_b \cap \widehat{a}^{-1}(\lambda) \neq \phi$ for all $b \subset B$. To prove the latter fact, consider (a, b). Then

$$P_a\big[\widetilde{\sigma}\big(\Phi(a), \Phi(b)\big)\big] = \widetilde{\sigma}\big(\Phi(a)\big),$$

so that for some $\mu \in \widetilde{\sigma}(\Phi(b))$, $(\lambda, \mu) \in \widetilde{\sigma}(\Phi(a), \Phi(b)) \subseteq \sigma_B(a, b)$, and therefore there exists $\varphi \in M_B$ such that $\varphi(a) = \lambda$, $\varphi(b) = \mu$. In particular, $\varphi \in M_b \cap \widehat{a}^{-1}(\lambda)$. We have thus shown that $M_b \cap \widehat{a}^{-1}(\lambda) \neq \phi$ and therefore,

$$\widetilde{\sigma}\big(\Phi(a)\big) \subseteq \widehat{a}(M_{\widetilde{\sigma}}).$$

To prove that $M_{\widetilde{\sigma}}$ is unique, one uses the fact that $\{\widehat{a} : a \subset \mathcal{B}\}$ separates points in $M_{\mathcal{B}}$ as follows. Let $M \subseteq M_{\mathcal{B}}$ be such that $\widetilde{\sigma}(\Phi(a)) = \widehat{a}(M)$ for all $a \subset \mathcal{B}$. Then

$$M \subseteq \widehat{a}^{-1}\widehat{a}(M) = \widehat{a}^{-1}\,\widetilde{\sigma}\big(\Phi(a)\big) = M_a,$$

for all $a \subset \mathcal{B}$, so that $M \subseteq M_{\widetilde{\sigma}}$. Suppose there exists $\varphi \in M_{\widetilde{\sigma}} \setminus M$. Then for every $\psi \in M$ there exists a neighborhood U_ψ of ψ such that $\{\varphi\} \cap U_\psi = \phi$. We can assume that U_ψ is a basic neighborhood, i.e.,

$$U_\psi = U_\psi(a_\psi,\varepsilon) := \{\eta : |\eta(a_\psi) - \psi(a_\psi)| < \varepsilon\},$$

for $a_\psi \subset \mathcal{B}$, $\varepsilon > 0$. Then $\{\widehat{a}_\psi(\varphi)\} \cap \widehat{a}_\psi(U_\psi) = \phi$. By the compactness of M, we can get $\psi_1, \ldots, \psi_k \in M$ such that $\cup_{i=1}^k U_{\psi_i} \supseteq M$. Form $a := (a_{\psi_1}, \ldots, a_{\psi_k})$. Then $\widehat{a}(\varphi) \notin \widehat{a}(M)$. For, if $\psi \in M$, then $\psi \in U_{\psi_i}$ for some i ($1 \leq i \leq k$). If $\widehat{a}(\varphi) = \widehat{a}(\psi)$, then $\widehat{a}(\varphi) \cap \widehat{a}(U_{\psi_i}) \neq \phi$, a contradiction. Thus

$$\widehat{a}(\varphi) \notin \widehat{a}(M) = \widehat{a}(M_{\widetilde{\sigma}}).$$

But $\varphi \in M_{\widetilde{\sigma}}$, so that $\widehat{a}(\varphi) \in \widehat{a}(M_{\widetilde{\sigma}})$. We therefore obtain $M_{\widetilde{\sigma}} = M$. ∎

Corollary 3.5. (Spectral Mapping Theorem for Polynomial Mappings) *Let $\mathcal{X}, \mathcal{B}, \Phi$ and $\widetilde{\sigma}$ be as before. Let $a \subset \mathcal{B}$ and let $p : \mathbf{C}^n \longrightarrow \mathbf{C}^k$ be a polynomial mapping. Then*

$$\widetilde{\sigma}\big[\Phi(p(a))\big] = p\big[\widetilde{\sigma}(\Phi(a))\big].$$

Proof:

$$\widetilde{\sigma}\big[\Phi(p(a))\big] = \big[p(a)\big]\widehat{}(M_{\widetilde{\sigma}}) = (p \circ \widehat{a})(M_{\widetilde{\sigma}}) = p\big[\widetilde{\sigma}(\Phi(a))\big]. \quad ∎$$

Remarks 3.6.

(i) What the corollary really shows is that whenever the formula $\widehat{f(a)} = f \circ \widehat{a}$ holds, then the spectral mapping theorem also holds. For instance, that is the case for $f \in A(\sigma_{\mathcal{B}}(a))$ (see also [87]).

(ii) More on the spectral mapping theorem can be found in [68], [94].

The following special case of Corollary 3.5 encompasses the basic ideas of this section.

Corollary 3.7. *Let $\tilde{\sigma}$ be a spectral system on a Banach space $\mathcal{X}$. Assume that $\tilde{\sigma}$ possesses the projection property and that $\tilde{\sigma}(a) \subseteq \hat{\sigma}(a)$ for all $a \subset \mathcal{L}(\mathcal{X})$. Then $\tilde{\sigma}$ possesses the spectral mapping property for polynomial mappings.*

We shall give now a lower bound for $M_{\tilde{\sigma}}$.

Proposition 3.8. *Let $\mathcal{X}$, $\mathcal{B}$, Φ and $\tilde{\sigma}$ be as before. Assume that Φ satisfies the equality*

$$r_{\tilde{\sigma}}(\Phi(a)) = r_{\mathcal{B}}(a) \text{ (for all } a \in \mathcal{B}),$$

where r denotes the spectral radius, e.g.,

$$r_{\mathcal{B}}(a) := \sup\{|\lambda| : \lambda \in \sigma_{\mathcal{B}}(a)\}.$$

Then the Shilov boundary of $\mathcal{B}$, $\partial_{\mathcal{B}}$, is contained in $M_{\tilde{\sigma}}$.

Proof: We shall see that $M_{\tilde{\sigma}}$ is a closed boundary for $\mathcal{B}$. Let $a \in \mathcal{B}$. Then $\|\hat{a}\|_{M_{\tilde{\sigma}}} = r_{\tilde{\sigma}}(\Phi(a)) = r_{\mathcal{B}}(a) = \|\hat{a}\|_{M_{\mathcal{B}}}$, as desired. ∎

We shall now see some applications of Theorem 3.3 and Corollary 3.5.

Application 3.9. *Let $\tilde{\sigma}$ be a spectral system on $\mathcal{X}$ possessing the projection property and such that $\tilde{\sigma} \subseteq \hat{\sigma}$. Assume that $\tilde{\sigma}(A) = \sigma(A)$ for every $A \in \mathcal{L}(\mathcal{X})$. Then $\tilde{\sigma} \circ \tilde{\sigma} = \sigma_T \circ \sigma_T$; in other words, $\tilde{\sigma}(A) \circ \tilde{\sigma}(B) = \sigma_T(A) \circ \sigma_T(B)$ for every $A, B \in \mathcal{L}(\mathcal{X})_{\text{com}}^{(n)}$. Moreover, $\tilde{\sigma}(A)$ and $\sigma_T(A)$ have identical rationally convex hulls, namely $\sigma_R(A)$, for all $A \subset \mathcal{L}(\mathcal{X})$.*

Proof: Let $\lambda \in \tilde{\sigma}(A)$, $\mu \in \tilde{\sigma}(B)$, and let $p_\lambda(z) := \lambda \circ z, z \in \mathbf{C}^k$. Then $\lambda \circ \mu = p_\lambda(\mu) \in p_\lambda(\tilde{\sigma}(B)) = \tilde{\sigma}(p_\lambda(B)) = \sigma(p_\lambda(B)) = \sigma_T(p_\lambda(B)) = p_\lambda(\sigma_T(B)) = \{\lambda\} \circ \sigma_T(B)$. Therefore, $\lambda \circ \mu \in \tilde{\sigma}(A) \circ \sigma_T(B)$. A similar argument shows that $\tilde{\sigma}(A) \circ \sigma_T(B) = \sigma_T(A) \circ \sigma_T(B)$. To see that $\tilde{\sigma}(A)$ and $\sigma_T(A)$ have identical rationally convex hulls, observe that for every polynomial p,

$$p(\tilde{\sigma}(A)) = p(\sigma_T(A)),$$

so that

$$\tilde{\sigma}(A)^{\widehat{}\text{rat.}} = \sigma_T(A)^{\widehat{}\text{rat.}}.$$

Now, $\sigma_T(A)^{\widehat{}\text{rat.}} = \sigma_R(A)$, as it is easily seen from the spectral mapping theorem for rational functions with singularities off $\sigma_T(A)$. ∎

Application 3.10. *Let T be a completely non-unitary contraction on a Hilbert space $\mathcal{H}$. Let $\Phi : H^{\infty}(\mathbf{D}) \longrightarrow \mathcal{L}(\mathcal{H})$ be the Sz. Nagy-Foias functional calculus for T ($\Phi(u) = u(T)$). In [57], C. Foias and W. Mlak studied a spectral mapping theorem for Φ, and showed that*

$$\sigma\big(u(T)\big) = \widehat{u}\big(\sigma_{\text{ext}}(T)\big),$$

where $\sigma_{\text{ext}}(T)$, the extended spectrum of T, is a certain compact nonempty subset of $M_{H^{\infty}}$, the maximal ideal space of $H^{\infty}(\mathbf{D})$; namely,

$$\sigma_{\text{ext}}(T) := \bigcap_{u \in H^{\infty}} \widehat{u}^{-1}\big[\sigma\big(u(T)\big)\big].$$

If $u \subset H^{\infty}$, and if $\widetilde{\sigma}$ is a spectral system with the projection property such that $\widetilde{\sigma} \subseteq \sigma_{H^{\infty}}$, we can apply Theorem 3.3 with $\mathcal{X} = \mathcal{H}$, $B = H^{\infty}(\mathbf{D})$ and Φ as above, and then get

$$\widetilde{\sigma}\big(u(T)\big) = \widehat{u}\big(M_{\widetilde{\sigma}}(H^{\infty}(\mathbf{D}), \mathcal{H}, \Phi)\big).$$

As in [57], the basic question now is to find a geometric representation of $\widehat{u}(M_{\widetilde{\sigma}})$. The following theorem and its corollaries, obtained in [43], give a partial answer.

Theorem 3.11. *([43]) Let T be a c.n.u. contraction and let $u \subset H^{\infty}$. Let $\widetilde{\sigma}$ be a spectral system with the projection property such that $\widetilde{\sigma} \subseteq \sigma_{H^{\infty}}$ and $\sigma_H \subseteq \widetilde{\sigma}$. Then*

$$u\big(\sigma(T) \cap \mathbf{D}\big)^{-} \subseteq \sigma_H\big(u(T)\big) \subseteq \widetilde{\sigma}\big(u(T)\big) \subseteq u(\mathbf{D})^{-}.$$

Corollary 3.12. *If $\sigma(T) = \overline{\mathbf{D}}$, then*

$$M_{\sigma_H} = M_{\widetilde{\sigma}} = M_{H^{\infty}}.$$

Proof: Clearly $M_{\sigma_H} \subseteq M_{\widetilde{\sigma}} \subseteq M_{H^{\infty}}$, so it suffices to prove that $M_{\sigma_H} = M_{H^{\infty}}$. By the uniqueness of M_{σ_H}, this amounts to showing that $\widehat{u}(M_{H^{\infty}}) = \sigma_H(u(T))$ for every $u \subset H^{\infty}$. Since $\widehat{u}(M_{H^{\infty}}) = u(\mathbf{D})^{-}$ always, and $\sigma_H\big(u(T)\big) = u(\mathbf{D})^{-}$ by Theorem 3.11, the proof of the corollary is complete. ∎

Corollary 3.13. *Let $u \subset H^{\infty}$ and let $T_u = (T_{u_1}, \ldots, T_{u_n})$ be the n-tuple of Toeplitz operators on the Hardy space $H^2(\mathbf{T})$. Then $\sigma_T(T_u) = u(\mathbf{D})^{-}$.*

Proof: $T_u = u(T)$, where T is the unilateral shift. ∎

44

Application 3.14. *Let $\mathcal{X}, \mathcal{B}, \phi$ and $\widetilde{\sigma}$ be as in Definition 3.1, and let $GL(n, \mathcal{B})$ denote the collection of invertible $n \times n$ matrices over $\mathcal{B}$. For $a \in \mathcal{B}^n$ and $m \in GL(n, \mathcal{B})$ let ma be the product of the matrix m and the column matrix determined by a. Assume that $\Phi(a)$ is $\widetilde{\sigma}$-invertible. Then $\Phi(ma)$ is $\widetilde{\sigma}$-invertible for all $m \in GL(n, \mathcal{B})$.*

Proof: From Theorem 3.3,

$$\widetilde{\sigma}(\Phi(ma)) = \widehat{ma}(M_{\widetilde{\sigma}}) = \{(ma)\widehat{}(\varphi) : \varphi \in M_{\widetilde{\sigma}}\}$$
$$= \{\varphi(m)\varphi(a) : \varphi \in M_{\widetilde{\sigma}}\}.$$

If $m \in GL(n, \mathcal{B})$ then $\varphi(m) \in GL(n, \mathbb{C})$; moreover, $\varphi(a) \neq 0$ $(\varphi \in M_{\widetilde{\sigma}})$. Therefore, $\varphi(m)\varphi(a) \neq 0$ for all $\varphi \in M_{\widetilde{\sigma}}$, so that $\Phi(ma)$ is $\widetilde{\sigma}$-invertible. ∎

To conclude this section, we shall establish a functorial property of $M_{\widetilde{\sigma}}$.

Proposition 3.15. *Let $\mathcal{X}$ be a Banach space, let $\widetilde{\sigma}$ be a spectral system on $\mathcal{X}$ with the projection property, and let $\mathcal{B}_1$ and $\mathcal{B}_2$ be commutative Banach algebras with identity, acting on $\mathcal{X}$ via maps Φ_1 and Φ_2 so that $\widetilde{\sigma}[\Phi_i(a^{(i)})] \subseteq \sigma_{\mathcal{B}_i}(a^{(i)})$ $(i = 1, 2)$. In addition, let $\alpha : \mathcal{B}_1 \longrightarrow \mathcal{B}_2$ be a unital algebra homomorphism such that the following diagram is commutative:*

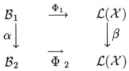

where β is an algebra automorphism of $\mathcal{L}(\mathcal{X})$. Assume also that $\widetilde{\sigma}(\beta(A)) = \widetilde{\sigma}(A)$ for all $A \subset \mathcal{L}(\mathcal{X})$. Then

$$\alpha^*\left[M_{\widetilde{\sigma}}(\mathcal{B}_2, \mathcal{X}, \Phi_2)\right] \subseteq M_{\widetilde{\sigma}}(\mathcal{B}_1, \mathcal{X}, \Phi_1),$$

where $\alpha^*(\psi) := \psi \circ \alpha$, $\psi \in M_{\mathcal{B}_2}$. If α is an isomorphism, then the inclusion above becomes an equality.

4. The Projection Property for σ_T

We shall devote this section to a proof of the projection property for the Taylor spectrum ([104, Lemma 3.1]). We shall present a simplification of F.-H. Vasilescu's Proof in [115], making more evident the use of nilpotent operators of index 2 and pointing out the relevance of Voiculescu's Theorem in this context. Vasilescu's Proof, which basically follows the outline of J. Bunce's Proof of the projection property for the left spectrum ([19, Proposition 1]), makes implicit use of Voiculescu's Theorem. Our proof shows explicitly the role played by that theorem. We shall begin with a series of lemmas.

Lemma 4.1. *Let T be an operator on a Hilbert space $\mathcal{H}$, and assume that $T^2 = 0$. The following statements are equivalent:*

(i) $R(T) = N(T)$.

(ii) $R_T := T + T^*$ *is invertible.*

(iii) $\square_T := T^*T + TT^*$ *is invertible.*

$\square_T$ will be called the *Laplacian* of T. When (i) holds, we shall say that T is *exact*.

Proof: (Cf. [34, Proposition II.3.1], [35, Proposition 3.1], [36, Proposition 3.4], [110].)

(i) $\Rightarrow$ (ii) Let $y \in \mathcal{H}$. Since $R(T) = N(T)$, we can write $y = Tx + T^*z$, where $x \perp N(T)$ and $z \perp N(T^*)$. Then $y = (T + T^*)(x + z)$, since $x \in R(T^*)$ and $z \in R(T)$. Thus $T + T^*$ is onto, i.e., $T + T^*$ is invertible.

(ii) $\Rightarrow$ (iii) $R_T^2 = (T + T^*)^2 = T^*T + TT^* = \square_T$.

(iii) $\Rightarrow$ (i) Let $x \in N(T)$. Then $\square_T x = T^*Tx + TT^*x = TT^*x$, so that $x = \square_T^{-1}TT^*x$. Since $\square_T$ commutes with TT^*, we have $x = TT^*\square_T^{-1}x \in R(T)$. Therefore $N(T) \subseteq R(T)$, The other inclusion follows from the condition "$T^2 = 0$." ∎

Corollary 4.2. *Let $A \subset \mathcal{L}(\mathcal{H})$. Then $\sigma_T(A, \mathcal{H})$ is a closed subset of $\mathbf{C}^n$.*

Proof: The map $A \mapsto R_{D_A}$ is real analytic in the norm topology. Since the set of invertible operators in $\mathcal{L}(\Lambda(\mathcal{H}))$ is open, it follows that

$$\{\lambda \in \mathbf{C}^n : R(D_{A-\lambda}) = N(D_{A-\lambda})\}$$

is an open subset of $\mathbf{C}^n$. ∎

The following result contains a new formula for the orthogonal projection onto the range of T in case T is exact.

Lemma 4.3. *Let $T \in \mathcal{L}(\mathcal{H})$ with $T^2 = 0$. Then:*

(i) $\overline{R(T)} = N(T) \Leftrightarrow N(\square_T) = 0$.

(ii) R_T invertible $\Rightarrow P_{R(T)} = TR_T^{-1}$ (here $P_{R(T)}$ is the orthogonal projection of $\mathcal{H}$ onto $R(T)$).

Proof:

(i) $N(\Box_T) = N(T) \cap N(T^*) = N(T) \Theta R(T)$.

(ii) Let $P := TR_T^{-1}$. Then $P^* = R_T^{-1}T^* = TR_T^{-1} = P$. Also, $R_T P^2 R_T = R_T T R_T^{-1} T R_T^{-1} R_T = T^*T = R_T P R_T$, so that $P^2 = P$. Thus P is a projection. Finally,

$$R(T) = R(TR_T^{-1}) = R(P). \quad \blacksquare$$

Lemma 4.4. Let $T \in \mathcal{L}(\mathcal{X})$ with $T^2 = 0$. If there exists $S \in \mathcal{L}(\mathcal{X})$ such that $TS + ST = I$, then $N(T) = R(T)$.

Proof: Let $x \in N(T)$. Then $TSx + STx = x$, or $TSx = x$, so that $x \in R(T)$. $\quad \blacksquare$

Lemma 4.5. ([104, Lemma 1.1]) Let $A \subset \mathcal{L}(\mathcal{X})$. Then $\sigma_T(A, \mathcal{X}) \subseteq \sigma'(A)$.

Proof: ([115, Lemma 2.5]) Assume A is invertible relative to $(A)'$, i.e., there exists $B \subset (A)'$ such that $A \circ B = I$. Then $S := \sum_{i=1}^{n} B_i E_i^*$ satisfies the equation

$$D_A S + S D_A = \sum_{i,j} (A_i B_j E_i E_j^* + B_j A_i E_j^* E_i)$$

$$= \sum_{i,j} A_i B_j (E_i E_j^* + E_j^* E_i)$$

$$= \sum_{i=1}^{n} A_i B_i = I.$$

By the previous lemma, $N(D_A) = R(D_A)$, so that A is Taylor invertible. $\quad \blacksquare$

In the next lemma, we establish that for $(A, B) \subset \mathcal{L}(\mathcal{H})$, $\sigma_T((A, B), \mathcal{H}) \subseteq \sigma_T(A, \mathcal{H}) \times \sigma_T(B, \mathcal{H})$.

Lemma 4.6. Let $A, B \subset \mathcal{L}(\mathcal{H})$ and assume that $N(D_A) = R(D_A)$ and that $(A, B) \subset \mathcal{L}(\mathcal{H})$. Then $N(D_{(A,B)}) = R(D_{(A,B)})$.

Proof: Observe that $D_{(A,B)} = D_A + \tilde{D}_B$, where $\tilde{D}_B := \sum_{j=1}^{m} B_j E_{n+j}$. We shall show that if $\eta \in \Lambda_{n+m}(\mathcal{H})$ and $(D_A + \tilde{D}_B)\eta = 0$, then $\eta = (D_A + \tilde{D}_B)\xi$, for some $\xi \in \Lambda_{n+m}(\mathcal{H})$ (we shall actually give a *formula* for ξ in terms of η). It suffices to assume that η is homogeneous of degree p $(0 \le p \le n + m)$, i.e., η is a linear combination of forms of degree p in $e_1, \ldots, e_{n+m}$; this is so because $D_A + \widetilde{D}_B$ maps homogenous p-forms to homogeneous $(p+1)$-forms. Now write $\eta = \eta_0 + \cdots + \eta_p$, where η_i has degree i in $e_{n+1}, \ldots, e_{n+m}$. We get:

$$
\begin{cases}
D_A\, \eta_0 && = 0 \\
D_A\, \eta_1 + \widetilde{D}_B \eta_0 = 0 \\
\quad \cdots \\
D_A\, \eta_p + \widetilde{D}_B\, \eta_{p-1} = 0 \\
\widetilde{D}_B\, \eta_p = 0.
\end{cases}
$$

Then $\eta_0 \in N(D_A) = R(D_A)$, so that $\eta_0 = D_A\, R_A^{-1}\, \eta_0$, by Lemma 4.3 (for brevity, we write R_A for R_{D_A}). Let $\xi_0 := R_A^{-1}\xi_0$. Then

$$
D_A\eta_1 + \widetilde{D}_B\, D_A\, \xi_0 = 0
$$

or

$$
D_A(\eta_1 - \widetilde{D}_B\, \xi_0) = 0 \,,
$$

so that $\eta_1 - \widetilde{D}_B\xi_0 = D_A\xi_1$, where $\xi_1 = R_A^{-1}(\eta_1 - \widetilde{D}_B\, \xi_0)$. Similarly, $\eta_i = \widetilde{D}_B\xi_{i-1} + D_A\xi_i (i = 1, \ldots, p)$. Now, $\deg(\eta_p\,;\ e_1, \ldots, e_n) = \deg(\widetilde{D}_B\xi_{p-1}\,;\ e_1, \ldots, e_n) = 0$, so that $D_A\xi_p = 0$, or $\xi_p = 0$. Thus

$$
\eta_p = \widetilde{D}_B\xi_{p-1}.
$$

Moreover,

$$
(*) \qquad \xi_{p-1} = \sum_{k=0}^{p-1} (-1)^k R_A^{-1} \left(\widetilde{D}_B R_A^{-1} \right)^k \eta_{p-1-k}.
$$

Then

$$
(D_A + \widetilde{D}_B)(\xi_0 + \cdots + \xi_{p-1}) = \eta,
$$

as desired. ∎

To prove the remaining half of the projection property (i.e., that $\sigma_T(A, \mathcal{H}) \subseteq P_A\sigma_T((A, B), \mathcal{H}))$, one must show that whenever $R(D_A) \neq N(D_A)$, then $R(D_{(A, B-\mu)}) \neq N(D_{(A, B-\mu)})$ for at least one $\mu \in \mathbb{C}^m$. Suppose for simplicity that $\mathcal{H}$ is finite-dimensional, $A = (A_1)$ and $B = (B_1)$. If A is not invertible, then $N(A) \neq 0$; since B maps $N(A)$ into $N(A)$, we can look at $B|_{N(A)}$ and find an eigenvalue $\mu \in \mathbb{C}$. Then $(0, \mu)$ is a joint eigenvalue for (A, B), which shows that $(0, \mu) \in \sigma_T((A, B), \mathcal{H})$. The general

48

case is of course much more complicated than this simple example, but the idea of the proof is basically the same. One finds, however, that there are two distinctive cases whenever A is not invertible on $\mathcal{H}$: either $\overline{R(D_A)} \neq \overline{N(D_A)}$ or $\overline{R(D_A)} = N(D_A)$. In the former, one can imitate the construction above with $N(A)$ replaced by $N(D_A) \ominus R(D_A)$. In the latter case, however, there is not enough room between $N(D_A)$ and $R(D_A)$! It is to handle this case that one needs Voiculescu's Theorem. First, we need a definition.

Definition 4.7. Let $A, \widetilde{A} \subset \mathcal{L}(\mathcal{H})$. We say that A is approximately unitarily equivalent to $\widetilde{A}$ (in symbols, $A \simeq_a \widetilde{A}$) if there exists a sequence $\{U_k\}_{k=1}^\infty$ of unitary operators on $\mathcal{H}$ such that $U_k A U_k^* := (U_k A_1 U_k^*, \cdots, U_k A_n U_k^*) \longrightarrow \widetilde{A}$, as $k \to \infty$ (naturally, this implies that $\widetilde{A}$ must also be an n-tuple). $\simeq_a$ is an equivalence relation on commuting n-tuples of operators on $\mathcal{H}$, which preserves all spectral properties, as the reader can easily verify (see for instance [44, Corollary 4.4]). In particular, $\sigma_T(A, \mathcal{H}) = \sigma_T(\widetilde{A}, \mathcal{H})$ whenever $A \simeq_a \widetilde{A}$.

The next lemma, which slightly generalizes the Gap Theorem of [42], states that if A is not invertible on $\mathcal{H}$, one can always assume $\overline{R(D_A)} \neq N(D_A)$, up to approximate unitary equivalence (for $n = 1$, this fact was proved by C. Apostol in [8]).

Lemma 4.8. Let $(A, B) \subset \mathcal{L}(\mathcal{H})$. If $R(D_A) \neq \overline{R(D_A)} = N(D_A)$, then there exists $(\widetilde{A}, \widetilde{B}) \subset \mathcal{L}(\mathcal{H}), (\widetilde{A}, \widetilde{B}) \simeq_a (A, B)$, such that $\overline{R(D_{\widetilde{A}})} \neq N(D_{\widetilde{A}})$.

Proof: Recall that $N(\Box_A) = N(D_A) \cap N(D_A^*) = \{0\}$, so that $\Box_A$ cannot have closed range (being a self-adjoint operator). Therefore, there exists a projection $Q \in \mathcal{L}(\Lambda_n(\mathcal{H}))$ such that Q is not compact but $\Box_A Q$ is a compact operator. Let $\mathcal{A}$ be the C^*-algebra generated by $A_1, \ldots, A_n, B_1, \ldots, B_m, I$ and $\{Q_{ij}\}_{i,j=1}^{2^n}$, and let $p : \pi(\mathcal{A}) \to \mathcal{L}(\mathcal{K})$ be a faithful $*$-representation of $\pi(\mathcal{A})$ on a separable Hilbert space $\mathcal{K}$, where π is the Calkin map. By Voiculescu's Theorem [118],

$$\mathrm{id}_{\mathcal{A}} \simeq_a \mathrm{id}_{\mathcal{A}} \oplus (\rho \circ \pi),$$

so that $(A, B) \simeq_a (A', B') \subset \mathcal{L}(\mathcal{H} \oplus \mathcal{K})$. Let

$$Q' = 0 \oplus (\rho \circ \pi)(Q).$$

Then $\Box_{A'} Q' = 0$ and $Q' \neq 0$, so that $\overline{R(D_{A'})} \neq N(D_{A'})$. Since $\mathcal{H} \oplus \mathcal{K}$ is isometrically isomorphic to $\mathcal{H}$, we obtain $(\widetilde{A}, \widetilde{B}) \subset \mathcal{L}(\mathcal{H})$, $(\widetilde{A}, \widetilde{B}) \simeq_a (A, B)$ and $N(D_{\widetilde{A}}) \neq \overline{R(D_{\widetilde{A}})}$. ∎

We can now state the main result of this section.

Theorem 4.9. ([104, Lemma 3.1]) *Let* $(A, B) \subset \mathcal{L}(\mathcal{H})$. *Then* $P_A \sigma_T((A, B), \mathcal{H}) = \sigma_T(A, \mathcal{H})$.

Proof: The inclusion "$\subseteq$" follows from Lemma 4.6. For the reverse inclusion, let $\lambda \in \sigma_T(A)$. By Lemma 4.8, we can assume that $Q := P_{N(\square_{A-\lambda})} \neq 0$. It is also clear that we can assume $m = 1$, i.e., $B \in \mathcal{L}(\mathcal{H})$. Consider $\tilde{B} = B \otimes I_\Lambda$ on $\Lambda_n(\mathcal{H})$. Since $Q \neq 0$ we must have $\sigma_r(Q\tilde{B}Q) \neq \phi$, so that for some $\mu \in \mathbb{C}$, $R(Q(\tilde{B} - \mu)Q) \neq R(Q)$. We now claim that $(A - \lambda, B - \mu)$ cannot be invertible on $\mathcal{H}$. The verity of our claim will be a consequence of the following series of lemmas.

Lemma 4.10. ([115, Proof of Lemma 2.12, formula (2.2)]) *Let* $A \subset \mathcal{L}(\mathcal{X})$ *and let* $B \in \mathcal{L}(\mathcal{X})$ *be such that* $(A, B) \subset \mathcal{L}(\mathcal{X})$. *Assume that* $N(D_{(A,B)}) = R(D_{(A,B)})$. *Then*

$$N(D_A) = R(D_A) + \tilde{B}N(D_A),$$

where $\tilde{B} = B \otimes I_\Lambda \in \mathcal{L}(\mathcal{X} \otimes \Lambda)$.

Proof: Since $\tilde{B}$ commutes with D_A, it follows that $R(D_A) + \tilde{B}N(D_A) \subseteq N(D_A)$. Let $\xi \in N(D_A)$, i.e., $D_A\xi = 0$. Consider $\eta := E_{n+1}\xi \in \Lambda_{n+1}(\mathcal{X})$. Then

$$D_{(A,B)}\eta = D_A\eta + (B \otimes E_{n+1})\eta = D_A E_{n+1}\xi = -E_{n+1}D_A\xi = 0,$$

so that $\eta \in N(D_{(A,B)})$. Then $\eta = D_{(A,B)}\theta$, where $\theta \in \Lambda_{n+1}(\mathcal{X})$. Write $\theta = \theta' + E_{n+1}\theta''$, where θ', $\theta'' \in \Lambda_n(\mathcal{X})$. We thus have:

$$E_{n+1}\xi = \eta = D_A\theta' + D_A E_{n+1}\theta'' + \tilde{B}E_{n+1}\theta'$$
$$= D_A\theta' + E_{n+1}(\tilde{B}\theta' - D_A\theta'').$$

It follows that $\xi = \tilde{B}\theta' - D_A\theta''$, with $D_A\theta' = 0$, which shows that $\xi \in R(D_A) + \tilde{B}N(D_A)$, as desired. ∎

Lemma 4.11. *Let* $T \in \mathcal{L}(\mathcal{H})$ *with* $T^2 = 0$, *and let* $S \in \mathcal{L}(\mathcal{H})$ *be such that* $ST = TS$ *and* $N(T) = R(T) + SN(T)$. *Then* $R(C) = R(P)$, *where* $P := P_{N(T)}$ *and* $C := PT + PSP$.

Proof: Clearly, $R(C) \subseteq R(P)$. Conversely, let $y \in R(P)$, i.e., $y \in N(T)$. Then $y = Tx + Sz$, with $Tz = 0$ and $x \perp N(T)$. Then

$$(PT + PSP)(x + z) = PTx + PSPx + PTz + PSPz$$
$$= Tx + Sz = y$$

(observe that $PSPz = Sz$, since $SN(T) \subseteq N(T)$), so that $y = C(x + z) \in R(C)$. ∎

Lemma 4.12. Let $T \in \mathcal{L}(\mathcal{H})$ with $T^2 = 0$, let $S \in \mathcal{L}(\mathcal{H})$ be such that $ST = TS$, let $P = P_{N(T)}, Q = P_{N(T) \cap N(T^*)}$ and let $C = PT + PSP$. Then

(i) $QSQ = QSP$;
(ii) $QCQ = QC = QSQ$; and
(iii) $R(C) = R(P) \Rightarrow R(QSQ) = R(Q)$.

Proof:

(i) Let $R := P_{\overline{R(T)}} = P - Q$. Then $PSP = SP$ and $RSR = SR$, so that

$$QSQ = (P - R)S(P - R) = PSP + RSR - PSR - RSP$$
$$= PSP + SR - PSPR - RSP \qquad \text{(since } R \le P)$$
$$= PSP - RSP = (P - R)SP = QSP.$$

(ii) $QCQ = Q(PT + PSP)Q = QPSPQ = QSQ$, since $Q \le P$. Also,

$$QC = QPSP + QPT = QSP + QT$$
$$= QSP \qquad \text{(since } T^*Q = 0)$$
$$= QSQ \qquad \text{(by (i))}.$$

(iii) Since $R(C) = R(P)$, we can find an operator $D \in \mathcal{L}(\mathcal{H})$ such that $CD = P$ (by Douglas's factorisation Theorem, for example [51]). Then

$$QCDQ = Q,$$

so that

$$(QCQ)(QDQ) = QCQDQ = QCDQ \qquad \text{(by (ii))}$$
$$= Q,$$

and therefore

$$(QSQ)(QDQ) = Q,$$

which implies $R(QSQ) = R(Q)$. ∎

Proof of Theorem 4.9. (Conclusion) Since $R(Q(\widetilde{B} - \mu)Q) \ne R(Q)$, Lemma 4.12 (iii) says that $R(C) \ne R(P)$, where $P = P_{N(D_{A-\lambda})}$ and $C = PD_{A-\lambda} + P(\widetilde{B} - \mu)P$. By Lemma 4.11, this then implies that $N(D_{A-\lambda}) \ne R(D_{A-\lambda}) + (\widetilde{B} - \mu)N(D_{A-\lambda})$, and by Lemma 4.10, we must therefore have $N(D_{A-\lambda,B-\mu}) \ne R(D_{A-\lambda,B-\mu})$, so that $(\lambda, \mu) \in \sigma_T((A, B), \mathcal{H})$. This completes the proof. ∎

5. The Analytic Functional Calculus for σ_T

J. L. Taylor developed in [105] the analytic functional calculus for σ_T. He based the construction on a Cauchy-Weil integral formula in homology; Taylor later simplified the construction ([106]), but it was F.-H. Vasilescu in [112], [115] who found a much simpler version (for the case of n-tuples of elements in a C^*-algebra) (see also [116]). We shall present here an outline of Vasilescu's construction of the functional calculus. The main problem is to make sense out of $f(a)$, where a is a commuting n-tuple and f is analytic in a neighborhood of $\sigma_T(a)$. Of course, if f is analytic in a neighborhood of $\sigma_B(a)$, where B is some commutative Banach algebra containing a, $f(a)$ should agree with Waelbroeck's and Arens and Calderón's "$f(a)$." In particular, if f is analytic in a polydisc D containing $\sigma_T(a)$, then $f(a)$ must be given by the formula

$$\frac{1}{(2\pi i)^n} \int_{\partial D} f(w) \prod_{i=1}^{n} (w_i - a_i)^{-1} \, dw_i,$$

where ∂D denotes the distinguished boundary of D.

We shall explain Vasilescu's construction in the case of Hilbert spaces. By spectral permanence ([37]) (see Example 2.5 (vii)), there is no loss of generality in doing this. We first need to recall some facts about the Martinelli kernel.

Definition 5.1. Let Ω be a bounded and open subset of $\mathbf{C}^n$ and assume that the boundary of $\Omega, \partial\Omega$, is C^1. Let $f \in C^1(\Omega)$. We define

$$\partial f := \frac{\partial f}{\partial z_1} \, dz_1 + \cdots + \frac{\partial f}{\partial z_n} \, dz_n$$

$$\bar{\partial} f := \frac{\partial f}{\partial \bar{z}_1} \, d\bar{z}_1 + \cdots + \frac{\partial f}{\partial \bar{z}_n} \, d\bar{z}_n$$

and

$$df := \partial f + \bar{\partial} f.$$

More generally, if ξ is a $C^\infty(\Omega)$-valued, (k,ℓ)-form in $dz_1, \ldots, dz_n, d\bar{z}_1, \ldots, d\bar{z}_n$, say,

$$\xi = f dz_{i_1} \wedge \cdots \wedge dz_{i_k} \wedge d\bar{z}_{j_1} \wedge \cdots \wedge d\bar{z}_{j_\ell},$$

we let

$$\partial\xi := \partial f \wedge dz_{i_1} \wedge \cdots \wedge dz_{i_k} \wedge d\bar{z}_{j_1} \wedge \cdots \wedge d\bar{z}_{j_\ell},$$

and similarly for $\bar{\partial}$ and d. It follows that each of ∂, $\bar{\partial}$ and d gives rise to operators on $C^\infty(\Omega) \otimes \Lambda[dz, d\bar{z}]$ such that $\partial^2 = \bar{\partial}^2 = d^2 = 0$. The Koszul complex associated

52

with d is called the de Rham complex over Ω; the one associated with $\bar{\partial}$ is called the $\bar{\partial}$-complex. For us, the $\bar{\partial}$-complex will be the restriction of the classical $\bar{\partial}$-complex to $C^\infty(\Omega) \otimes \Lambda[d\bar{z}]$, that is, $\bar{\partial}$ acting on $(0,\ell)$ forms (this complex is sometimes called the Dolbeault complex (see [74, p. 231]). Also $A(\Omega) := \{f \in \Lambda^0(C^\infty(\Omega)) : \bar{\partial}f = 0\}$. (For this and other basic facts about several complex variables, the reader is referred to [60], [62], [69], [73], [74].)

Definition 5.2. For $z \neq 0$, $z \in \mathbb{C}^n$, let

$$M(z) := (n-1)! \sum_{j=1}^{n} (-1)^{j-1} \frac{\bar{z}_j}{\|z\|^{2n}} \bigwedge_{\substack{1 \le k \le n \\ k \neq j}} d\bar{z}_k,$$

where $\|z\|^2 := \sum_{i=1}^{n} |z_i|^2$. M is the Martinelli kernel.

Theorem 5.3. (Bochner-Martinelli Formula) ([74, Theorem 1.14]) *Let* $f \in C^1(\bar{\Omega})$. *Then*

$$f(z) = \frac{1}{(2\pi i)^n} \int_{\partial\Omega} f(w)\, M(w-z) \wedge dw$$
$$- \frac{1}{(2\pi i)^n} \int_{\Omega} \bar{\partial}f(w) \wedge M(w-z) \wedge dw \qquad (z \in \Omega).$$

Proof: (Sketch) Apply Stokes's Theorem to

$$\omega(w) := f(w)M(w-z) \wedge dw$$

on $\Omega_{z,\varepsilon} := \Omega \setminus \bar{B}(z,\varepsilon)$, and then let $\varepsilon \to 0$. ∎

Corollary 5.4. *If* $f \in A(\Omega) \cap C^1(\bar{\Omega})$, *then*

$$f(z) = \frac{1}{(2\pi i)^n} \int_{\partial\Omega} f(w)M(w-z) \wedge dw \qquad (z \in \Omega).$$

Remarks 5.5.

(i) M is an $(n-1)$-differential form in $d\bar{z}_1, \ldots, d\bar{z}_n$ such that $\bar{\partial}M = 0$.

(ii) If $n = 1$, $M(z) = \frac{1}{z}$, the Cauchy kernel.

(iii) If $n \geq 2$, M is not analytic (in the sense that its C^∞-coefficients are not analytic functions) and, therefore, M does not *create* analytic functions (as the Cauchy kernel does), but it nevertheless reproduces analytic functions.

Definition 5.6. Let $\beta \subset A\big(\Omega, \mathcal{L}(\mathcal{H})\big)$, the space of $\mathcal{L}(\mathcal{H})$-valued analytic functions in Ω. We define

$$D_\beta : C^\infty\big(\Omega, \Lambda(\mathcal{H})\big) \longrightarrow C^\infty\big(\Omega, \Lambda(\mathcal{H})\big)$$

by

$$(D_\beta f)(z) := D_{\beta(z)} f(z), \qquad z \in \Omega.$$

We now identify $C^\infty(\Omega, \mathcal{H}) \otimes \Lambda[e]$ and $C^\infty(\Omega, \mathcal{H}) \otimes \Lambda[e] \otimes \Lambda[d\bar{z}]$ with $C^\infty(\Omega, \mathcal{H} \otimes \Lambda[e])$ and $C^\infty(\Omega, \mathcal{H} \otimes \Lambda[e, d\bar{z}])$, respectively. We extend D_β and $\bar{\partial}$ to $C^\infty(\Omega, \mathcal{H} \otimes \Lambda[e, d\bar{z}])$ as follows:

$$D_\beta(f \otimes \xi \otimes \eta) := D_\beta(f \otimes \xi) \otimes \eta$$

and

$$\bar{\partial}(f \otimes \xi \otimes \eta) := \sum_{i=1}^{n} \frac{\partial f}{\partial \bar{z}_i} \otimes \xi \otimes (d\bar{z}_i \wedge \eta),$$

where $f \in C^\infty(\Omega, \mathcal{H})$, $\xi \in \Lambda[e]$ and $\eta \in \Lambda[d\bar{z}]$. Clearly, $\bar{\partial} D_\beta + D_\beta \bar{\partial} = 0$.

Lemma 5.7. Let $D_\beta, \bar{\partial}$ be as above. If $N(D_\beta) = R(D_\beta)$, then $N(D_\beta + \bar{\partial}) = R(D_\beta + \bar{\partial})$.

Proof: Imitate the proof of Lemma 4.6. ∎

Later on we shall specialize to the case $\beta(z) = z - A$, where $A \subset \mathcal{L}(\mathcal{H})$, so the reader should keep this example in mind for the construction that follows.

Let $\beta \subset A\big(\Omega, \mathcal{L}(\mathcal{H})\big)$ and let $x \in \mathcal{H}$. Define

$$\eta = \eta_0 \equiv \mathrm{E}x := x \otimes e_1 \wedge \cdots \wedge e_n.$$

Then $D_\beta \eta = 0$ (since η is an n-form in $e_1, \ldots, e_n$) and $\bar{\partial} \eta = 0$ (since η is constant in z), so that $\eta \in N(D_\beta + \bar{\partial})$. Assume for a moment that $N(D_\beta) = R(D_\beta)$. By Lemma 5.7, we must then have

$$\eta = (D_\beta + \bar{\partial})\xi,$$

where

$$\xi_{n-1} = \sum_{k=0}^{n-1} (-1)^k R_\beta^{-1} \big(\bar{\partial} R_\beta^{-1}\big)^k \eta_{n-1-k}$$

$$= (-1)^{n-1} R_\beta^{-1} \big(\bar{\partial} R_\beta^{-1}\big)^{n-1} \mathrm{E}x.$$

(recall the formula for ξ_{n-1} that we obtained in the proof of Lemma 4.6!). Moreover, $\bar{\partial}\xi_{n-1} = \eta_n = 0$.

54

Definition 5.8. Let $\beta \subset A(\Omega, \mathcal{L}(\mathcal{H}))$. The spectrum of β in Ω is

$$\sigma_T^\Omega(\beta, \mathcal{H}) := \left\{ \lambda \in \Omega : N(D_{\beta(\lambda)}) \neq R(D_{\beta(\lambda)}) \right\}.$$

The set $\sigma_T^\Omega(\beta, \mathcal{H})$ is closed in Ω, since $\lambda \notin \sigma_T^\Omega(\beta, \mathcal{H})$ if and only if $D_{\beta(\lambda)}$ is exact on $\Lambda(\mathcal{H})$ if and only if $R_{\beta(\lambda)}$ is invertible on $\Lambda(\mathcal{H})$, and the function $\lambda \mapsto R_{\beta(\lambda)}$ is continuous in the norm topology.

We can now give the definition of the Martinelli kernel, which generalizes the resolvent $(z - A)^{-1}$, used when $n = 1$.

Definition 5.9. Let $\beta \subset A(\Omega, \mathcal{L}(\mathcal{H}))$ and let $z \notin \sigma_T^\Omega(\beta, \mathcal{H})$. We let

$$M(\beta)(z) := R_{\beta(z)}^{-1} \left(\bar{\partial}_z R_{\beta(z)}^{-1} \right)^{n-1} E \Big|_{\mathcal{H} \otimes \Lambda^\circ}.$$

$M(\beta)$ is the Martinelli kernel associated with β.

Remarks 5.10.

(i) $\deg(M(\beta); d\bar{z}_1, \ldots, d\bar{z}_n) = n - 1$.

(ii) $\bar{\partial} M(\beta) = 0$.

Lemma 5.11. If $\dim \mathcal{H} = 1$ and $\beta(z) = z$, then $M(\beta)$ is the ordinary Martinelli kernel

$$M(\beta)(z) = (n-1)! \sum_{j=1}^n (-1)^{j-1} \frac{\bar{z}_j}{\|z\|^{2n}} \bigwedge_{\substack{1 \leq k \leq n \\ k \neq j}} d\bar{z}_k \qquad (z \neq 0).$$

Proof: This is a straightforward computation (see [115, Theorem 3.5]). ■

Definition 5.12. Let $\beta \subset A(\Omega, \Lambda(\mathcal{H}))$ and let $f \in A(\Omega, \mathcal{H})$. Define

$$\nu_\beta(f) := \frac{1}{(2\pi i)^n} \int_{\partial \Omega'} M(\beta)(z) f(z) \, dz,$$

where $\sigma_T^\Omega(\beta, \mathcal{H}) \subseteq \Omega' \subseteq \overline{\Omega'} \subseteq \Omega$ and Ω' has C^1-boundary.

Lemma 5.13. ν_β is a well-defined, bounded operator on $\mathcal{H}$.

Proof: See [115, Lemma 3.6]. ■

Lemma 5.14. Let $\beta \subset A(\Omega, \mathcal{L}(\mathcal{H}))$ and let $B \in \mathcal{L}(\mathcal{H})$ be an operator commuting with β, i.e., $B\beta(z) = \beta(z)B$ for all $z \in \Omega$. Then $\nu_\beta(f)B = B\nu_\beta(f)$ for all $f \in A(\Omega)$.

Proof: See [115, Proposition 3.9]. ■

Lemma 5.15. ([115, Lemma 3.11]) Let $\beta \subset A(\Omega, \mathcal{L}(\mathcal{H}))$, let $f \in A(\Omega, \mathcal{H})$, and let $n = 1$. Then

$$\nu_\beta(f) = \frac{1}{2\pi i} \int_\Gamma \beta(z)^{-1} f(z) \, dz \,,$$

where $\Gamma = \partial\Omega'$ and $\sigma_T^\Omega(\beta, \mathcal{H}) \subseteq \Omega' \subseteq \overline{\Omega}' \subseteq \Omega$ and Γ is C^1.

Proof: Since $n = 1$,

$$M(\beta) = R_\beta^{-1} E \mid_{\mathcal{H} \otimes \Lambda^\circ} = (\beta E + \beta^* E^*)^{-1} E \mid_{\mathcal{H} \otimes \Lambda^\circ}$$
$$= (\beta^{-1} E^* + \beta^{*-1} E) E \mid_{\mathcal{H} \otimes \Lambda^\circ} = \beta^{-1}\,. \qquad \blacksquare$$

Lemma 5.16. Let Ω be a bounded open subset of $\mathbb{C}^{n+m}$, let Ω_m be the projection of Ω onto the last m coordinates, let $\beta = (\beta_1, \ldots, \beta_n) \subset A(\Omega, \mathcal{L}(\mathcal{H}))$ and let $\gamma = (\gamma_1, \ldots, \gamma_m) \subset A(\Omega_m, \mathcal{L}(\mathcal{H}))$ be such that $(\beta, \gamma) \subset A(\Omega, \mathcal{L}(\mathcal{H}))$ (for $z \in \mathbb{C}^n$, $w \in \mathbb{C}^m$ and $(z, w) \in \Omega$, extend γ to Ω by $\gamma(z, w) := \gamma(w)$). If $f \in A(\Omega)$, then

$$\nu_{(\beta, \gamma)}(f) = \nu_\gamma(\nu_\beta(f))\,.$$

Proof: See [115, Theorem 3.8]. $\blacksquare$

Lemma 5.17. Let $\Omega = \Omega_1 \times \cdots \times \Omega_n$ (where each Ω_i is a subset of $\mathbb{C}$), let $\beta \subset A(\Omega, \mathcal{L}(\mathcal{H}))$ and let $f \in A(\Omega)$. Then

$$\nu_\beta(f) = \frac{1}{(2\pi i)^n} \int_{\Gamma_1} \cdots \int_{\Gamma_n} \beta_1(z_1)^{-1} \cdots \beta_n(z_n)^{-1} f(z) \, dz \,,$$

where Γ_i surrounds $\sigma_T^{\Omega_i}(\beta_i, \mathcal{H})$ $(i = 1, \ldots, n)$.

Proof: Use Lemma 5.15 together with Lemma 5.16 to give an induction argument. $\blacksquare$

Theorem 5.18. (Taylor)([105], [116]) Let $A \subset \mathcal{L}(\mathcal{H})$ and let $\Omega \supseteq \sigma_T(A, \mathcal{H})$. Let Ω' be such that $\sigma_T(A, \mathcal{H}) \subseteq \Omega' \subseteq \overline{\Omega}' \subseteq \Omega$ and Ω' has C^1-boundary. Then the map

$$A(\Omega) \to \mathcal{L}(\mathcal{H})$$
$$f \mapsto f(A) := \nu_\beta(f)$$

(where $\beta(z) = z - A$) is a unital continuous homomorphism such that:

(i) $z_i(A) = A_i$ $(i = 1, \ldots, n)$.

(ii) $f(A) \in (A)''$.

(iii) If $f = 0$ on a neighborhood of $\sigma_T(A,\mathcal{H})$, then $f(A) = 0$.

(iv) For every relatively compact open subset Φ of Ω containing $\sigma_T(A,\mathcal{H})$, there exists a constant $C_\Phi > 0$ such that

$$\|f(A)\| \leq C_\Phi \sup\{|f(z)| : z \in \Phi\}$$

for all $f \in A(\Omega)$.

Proof:

(i) Let f be a polynomial. Then $f \in A(\mathbb{C}^n)$, and by Lemma 5.17,

$$f(A) = \frac{1}{(2\pi i)^n} \int_{\Gamma_1} \cdots \cdot \int_{\Gamma_n} f(z) \prod_{i=1}^{n} (z_i - A_i)^{-1} \, dz,$$

where Γ_i *surrounds* $\sigma(A_i)$ $(i = 1, \ldots, n)$. As we mentioned in Section 1 (Construction of the Functional Calculus), the expression above equals A_i whenever $f(z) = z_i$ $(i = 1, \ldots, n)$.

(ii) follows from Lemma 5.14.

(iv) C_Φ can be taken as the L^1-norm of $M(\beta)/(2\pi)^n$ on $\partial\Omega'$ provided $\Omega' \subset \Phi$. ∎

Theorem 5.19. (Taylor ([105, Theorem 4.8])) *Let* $A \subset \mathcal{L}(\mathcal{H})$, $\Omega \supseteq \sigma_T(A,\mathcal{H})$, *and let* $f \subset A(\Omega)$. *Then*
$$\sigma_T(f(A),\mathcal{H}) = f(\sigma_T(A,\mathcal{H})).$$

Proof: By the results of Section 3, it suffices to see that f commutes with the Gelfand transform of $(A)''$ whenever $f \in A(\Omega)$. We use an induction argument (see [115, Theorem 4.3]). Since we know that the statement is true for $n = 1$, assume it is true for n-tuples, and let $\widetilde{A} = (A, A_{n+1})$ and $\widetilde{f} \in A(\widetilde{\Omega})$, where $\widetilde{\Omega} \supseteq \sigma_T(\widetilde{A},\mathcal{H})$. Then

$$\widetilde{f}(\widetilde{A}) = \frac{1}{2\pi i} \int_{\Gamma_{n+1}} (w - A_{n+1})^{-1} \widetilde{f}_w(A) \, dw,$$

where $\widetilde{f}_w = f(\cdot, w)$, and Γ_{n+1} *surrounds* $\sigma(A_{n+1})$, by Lemmas 5.15 and 5.16. Therefore, if $\varphi \in M_{(\widetilde{A})''}$, we have

$$\varphi\left(\widetilde{f}(\widetilde{A})\right) = \frac{1}{2\pi i} \int_{\Gamma_{n+1}} (w - \varphi(A_{n+1}))^{-1} \varphi\left(\widetilde{f}_w(A)\right) \, dw.$$

57

Since $\varphi \mid_{(A)''} \in M_{(A)''}$, we have $\varphi\left(\tilde{f}_w(A)\right) = \tilde{f}_w(\varphi(A))$, by the induction hypothesis. Thus,

$$\varphi\left(\tilde{f}(\tilde{A})\right) = \frac{1}{2\pi i} \int_{\Gamma_{n+1}} (w - \varphi(A_{n+1}))^{-1} \tilde{f}_w(\varphi(A)) \, dw$$
$$= \tilde{f}\left(\varphi(\tilde{A})\right) ,$$

so that $\left(\tilde{f}(\tilde{A})\right)^{\wedge} = \tilde{f} \circ \tilde{\tilde{A}}$, as desired. ∎

We shall now state the uniqueness of the analytic functional calculus for σ_T.

Theorem 5.20. (Putinar ([84])) *Let $A \subset \mathcal{L}(\mathcal{H})$ and let $\Omega \supseteq \sigma_T(A, \mathcal{H})$. Then there exists a unique continuous unital $A(\Omega)$-functional calculus extending the polynomial calculus and satisfying the spectral mapping theorem.*

Putinar actually proved a more general version of Theorem 5.20, allowing $\mathcal{H}$ to be any Banach space. One can then get the uniqueness of the algebraic functional calculus as a consequence.

Corollary 5.21. (Zame ([122, Theorem 1])) *For $\mathcal{B}$ a commutative B-algebra with identity, the Shilov-Arens-Calderón-Waelbroeck functional calculus is unique (subject to the conditions (i), (ii), and (iii) of Theorem 1.15).*

Proof: Recall that $\sigma_{\mathcal{B}}(a) = \sigma_T(L_a, \mathcal{B})$ for all $a \subset \mathcal{B}$ and observe that if $f \mapsto f(a)$ is a functional calculus for a, then $f \mapsto f(L_a) := L_{f(a)}$ is a functional calculus for L_a. Now use Putinar's result for L_a acting on $\mathcal{B}$. ∎

Remark 5.22. The importance of Putinar's uniqueness result lies in the fact that no matter which construction we use for $f(A)$ we are bound to obtain the same element of $\mathcal{L}(\mathcal{H})$. For instance, if $f \in A(\sigma_R(A))$ (recall $\sigma_R(A)$ is the rational spectrum of A), then $f(A)$ has three meanings, a priori, one in the sense of Arens-Calderón-Waelbroeck, one in the sense of Taylor, and one in the sense of Vasilescu. As it turns out, they all agree and the value $f(A)$ can then be computed in any of the three ways. The reader will notice that, strictly speaking, we cannot credit Theorem 5.18 to Taylor until we show that Vasilescu's "$f(A)$" is the same as Taylor's "$f(A)$;" after Theorem 5.20, however, we know that both "$f(A)$'s" are the same.

Theorem 5.23. (The Substitution Theorem) ([83]; cf. [59, III.4.6] for algebraic spectra) *Let $A \subset \mathcal{L}(\mathcal{H})$, $\Omega \supseteq \sigma_T(A, \mathcal{H})$, $f \subset A(\Omega)$, and let $g \subset A(\tilde{\Omega})$, where $\tilde{\Omega} \supseteq f(\Omega)$.*

Then:

$$(g \circ f)(A) = g[f(A)].$$

We shall now give some applications of the results in this section.

Application 5.24. (The Shilov Idempotent Theorem for σ_T) ([105, Theorem 4.9])
Let $A \subset \mathcal{L}(\mathcal{H})$ and assume that $\sigma_T(A,\mathcal{H}) = K_1 \cup K_2$, where K_1 and K_2 are two nonempty disjoint compact subsets of $\sigma_T(A,\mathcal{H})$. Then there exist subspaces $\mathcal{M}_1$ and $\mathcal{M}_2$ of $\mathcal{H}$ such that $\mathcal{M}_1 \cap \mathcal{M}_2 = (0)$, $\mathcal{M}_1 + \mathcal{M}_2 = \mathcal{H}$, $(A)'\mathcal{M}_i \subseteq \mathcal{M}_i$ $(i = 1, 2)$ and $\sigma_T\left(A \mid_{\mathcal{M}_i}, \mathcal{M}_i\right) = K_i$ $(i = 1, 2)$.

Proof: Consider an analytic function f which is constantly one on a neighborhood of K_1 and constantly zero on a neighborhood of K_2. Then $f(A)$ is an idempotent in the bicommutant of A. Now take $\mathcal{M}_1 = R(f(A))$. ∎

Application 5.24. (continued) ([44], [56]) *A is jointly similar to*

$$A = \begin{pmatrix} A^{(1)} & 0 \\ 0 & A^{(2)} \end{pmatrix} \begin{matrix} \mathcal{M}_1 \\ \mathcal{M}_1^{\perp}, \end{matrix}$$

where $\sigma_T\left(A^{(i)}\right) = K_i$ $(i = 1, 2)$.

Proof: The map $V : \mathcal{M}_2 \longrightarrow \mathcal{M}_1^{\perp}$, given by $Vx = P_{\mathcal{M}_1^{\perp}} x$, is one-to-one and onto. Moreover,

$$P_{\mathcal{M}_1^{\perp}} A \mid_{\mathcal{M}_1^{\perp}} = VAV^{-1},$$

as one easily verifies. Thus,

$$A = \begin{pmatrix} A^{(1)} & * \\ 0 & A^{(2)} \end{pmatrix} \begin{matrix} \mathcal{M}_1 \\ (\mathcal{M}_1)^{\perp}, \end{matrix}$$

where $A^{(1)} := A \mid_{\mathcal{M}_1}$ and $A^{(2)} := VAV^{-1}$. Since

$$\sigma_T\left(A^{(2)}, \mathcal{M}_1^{\perp}\right) = \sigma_T\left(A \mid_{\mathcal{M}_2}, \mathcal{M}_2\right) = K_2$$

(see, for instance, basic fact (iv) after Remarks 6.7 on page 67 below), we have

$$\sigma_T\left(A^{(1)}, \mathcal{M}_1\right) \cap \sigma_T\left(A^{(2)}, \mathcal{M}_1^{\perp}\right) = \phi.$$

By the generalization of Rosenblum's Corollary shown in [44, Theorem 2.4], it follows that A is jointly similar to $A^{(1)} \oplus A^{(2)}$. (For a different proof, obtained independently by L. Fialkow, see [56, Lemma 2.1].) ∎

Application 5.25. (Taylor ([105, Theorem 4.5])) *Let $A, B \in \mathcal{L}(\mathcal{H})^{(n)}_{com}$ and assume that $T \in \mathcal{L}(\mathcal{H})$ intertwines A and B, i.e., $A_i T = T B_i$ $(i = 1, \ldots, n)$. Let $f \subset A\big(\sigma_T(A, \mathcal{H}) \cup \sigma_T(B, \mathcal{H})\big)$. Then T also intertwines $f(A)$ and $f(B)$.*

Proof: Consider
$$
C = \begin{pmatrix} A & 0 \\ 0 & B \end{pmatrix} \quad \text{and} \quad S = \begin{pmatrix} 0 & T \\ 0 & 0 \end{pmatrix}.
$$
Then $C \subset \mathcal{L}(\mathcal{H} \oplus \mathcal{H})$, $S \in \mathcal{L}(\mathcal{H} \oplus \mathcal{H})$ and $S \in (C)'$. Since $\sigma_T(C, \mathcal{H} \oplus \mathcal{H}) = \sigma_T(A, \mathcal{H}) \cup \sigma_T(B, \mathcal{H})$ (see the last paragraph in Section 2), we know that $f \subset A\big(\sigma_T(C, \mathcal{H} \oplus \mathcal{H})\big)$. Therefore, $f(C)S = Sf(C)$, because $f(C) \in (C)''$. Now, it is straightforward to check that
$$
f(C) = \begin{pmatrix} f(A) & 0 \\ 0 & f(B) \end{pmatrix}
$$
(clearly $M(z - C) = M(z - A) \oplus M(z - B)$), and therefore $f(A)T = Tf(B)$. $\blacksquare$

Application 5.26. (Paulsen ([79])) *Let $A \subset \mathcal{L}(\mathcal{H})$ and let $\Omega \supseteq \sigma_T(A, \mathcal{H})$. Then A is jointly similar to A', where A' is the compression of a normal n-tuple N on a Hilbert space $\mathcal{K} \supseteq \mathcal{H}$ with $\sigma_T(N, \mathcal{K}) = \partial\Omega$, to a semi-invariant subspace, i.e.,*
$$
N = \begin{pmatrix} * & * & * \\ 0 & A' & * \\ 0 & 0 & * \end{pmatrix} \mathcal{H}.
$$

Proof: The map $f \mapsto f(A)$ from the algebra $\text{Rat}(\overline{\Omega})$ of rational functions with singularities off $\overline{\Omega}$ to $\mathcal{L}(\mathcal{H})$ is completely bounded (see Theorem 5.18 (iv)) above, and the proof of [78, Theorem 8.13]). Then the result follows from standard facts from the theory of completely bounded maps ([78]). $\blacksquare$

Application 5.27. ([44, Theorem 2.1]) *Let $\Omega \subset \mathbb{C}^n$ and let $\mathcal{R}$ be a functional Hilbert space on Ω, i.e., $\mathcal{R} \subseteq A(\Omega)$, $f \mapsto f(\lambda)$ is bounded on $\mathcal{R}$ for all $\lambda \in \Omega$, $1 \in \mathcal{R}$, and M_{z_j} is bounded in $\mathcal{R}$ for all $j = 1, \ldots, n$. Let $A \subset \mathcal{L}(\mathcal{H})$ be such that $\sigma_T(A, \mathcal{H}) \subseteq \Omega$. Then there exist a subspace $\mathcal{M} \subseteq \mathcal{R} \otimes \mathcal{H}$ such that $(M_z \otimes I)\mathcal{M} \subseteq \mathcal{M}$, and an operator $W : \mathcal{H} \longrightarrow (\mathcal{R} \otimes \mathcal{H})\ominus\mathcal{M}$ such that W is one-to-one and onto and*
$$
M_z \otimes I = \begin{pmatrix} * & * \\ 0 & A' \end{pmatrix},
$$
where $A' = WAW^{-1}$.

Proof: We shall give a brief outline; for details the reader is referred to [44].

(i) Let k be a reproducing kernel for $\mathcal{R}$, i.e., k is a sesquianalytic function on Ω such that $f(\mu) = \langle f, k(\bar{\mu}, \cdot) \rangle_{\mathcal{R}}$, for all $\mu \in \Omega$.

(ii) Define $\Gamma : \mathcal{H} \to \mathcal{R} \otimes \mathcal{H}$ by

$$(\Gamma x)(\lambda) := k(A^*, \lambda) x$$
$$= \frac{1}{(2\pi i)^n} \int_{\Sigma^*} k(\mu, \lambda) M(\mu - A^*) x \, d\mu,$$

where Σ^* *surrounds* $\sigma_T(A^*, \mathcal{H})$, which equals $\overline{\sigma_T(A, \mathcal{H})} := \{\bar{\lambda} : \lambda \in \sigma_T(A, \mathcal{H})\}$, by [35, Lemma 4.3] or [36, Corollary 3.14].

(iii) W is bounded and bounded below, and $\Gamma A^* = (M_z^* \otimes I) \Gamma$, so that

$$\mathcal{M} := (\mathcal{R} \otimes \mathcal{H}) \ominus R(\Gamma)$$

and $W : (\Gamma^*|_{R(\Gamma)})^{-1}$ have the required properties. ∎

Remarks 5.28.

(i) The model described above is the extension to several variables of the functional model in [13] (see also [27]).

(ii) For arbitrary Ω, it is true that $\Omega \subseteq \sigma_T(M_z, \mathcal{R})$ (indeed, if $\lambda \in \Omega$, $M_z^* k(\bar{\lambda}, \cdot) = \bar{\lambda} k(\bar{\lambda}, \cdot)$, so that $\lambda \in \sigma_r(M_z, \mathcal{R})$). Moreover, if Ω has an envelope of holomorphy $\tilde{\Omega} \subseteq \mathbb{C}^n$, it is also the case that $\tilde{\Omega} \subset \sigma_T(M_z, \mathcal{R})$. Thus, if $\sigma_T(A, \mathcal{H}) = \partial \mathbb{B}_n$ ($n \geq 2$), then $\overline{\mathbb{B}}_n \subseteq \sigma_T(M_z, \mathcal{R})$. Therefore, unlike the case $n = 1$, the spectrum of the model M_z can be much larger than the spectrum of A. In the theory of closures of joint similarity orbits, for instance, it is of interest to find other models where the two spectra (that of the model and that of the modelled n-tuple) have comparable sizes.

Application 5.29. Upper semicontinuity of separate parts of the Taylor spectrum ([44, Lemma 4.3]) *Let $A \subset \mathcal{L}(\mathcal{H})$ and assume that $\sigma_T(A, \mathcal{H}) = K_1 \cup K_2$, where K_1 and K_2 are compact and $K_1 \neq \phi$. Let $\Omega \supseteq K_1$ be such that $\overline{\Omega} \cap K_2 = \phi$. Then there exists $\delta > 0$ such that $\sigma_T(B, \mathcal{H}) \cap \Omega \neq \phi$, whenever $B \subset \mathcal{L}(\mathcal{H})$ and $\|B - A\| := \max_{1 \leq i \leq n} \|B_i - A_i\| < \delta$.*

Application 5.30. (Fialkow ([55, Example 3.5]) *The Harte spectrum σ_H does not carry an analytic functional calculus. Consider the following pair of operators on $\mathcal{H} = \ell^2(\mathbb{Z}_+) \otimes \ell^2(\mathbb{Z}_+)$, where $\ell^2(\mathbb{Z}_+)$ denotes the Hilbert space of square summable complex* *sequences:*

$$A_1 := \begin{pmatrix} U_+^* & & & \\ & U_+^* & & \\ & & \ddots & \end{pmatrix}, \qquad A_2 := \begin{pmatrix} 0 & & & \\ U_+^* & 0 & & \\ & I & & \\ & & \ddots & \end{pmatrix},$$

where U_+ is the unilateral shift on $\ell^2(\mathbb{Z}_+)$. Suppose $P \in \mathcal{L}(\mathcal{H})$, $P^2 = P$ and $PA_i = A_iP$ $(i = 1, 2)$. A matrix computation shows that P must be 0 or I. Therefore, $\sigma_T(A,\mathcal{H})$ is connected (by Application 5.24); indeed, $\sigma_T(A,\mathcal{H}) = \overline{\mathbb{D}}^2$, as shown in [39, Example 4.1]. However, $\sigma_H(A,\mathcal{H}) = \{(0,0)\} \cup (\mathbb{T} \times \mathbb{D}) \cup (\mathbb{D} \times \mathbb{T})$, so that $(0,0)$ is isolated in $\sigma_H(A,\mathcal{H})$ (see [39] and [55]). Therefore, $(0,0)$ cannot be excised by means of an analytic functional calculus.

Application 5.31. ([41, Theorem 2.11]) Let $A \subset \mathcal{L}(\mathcal{H})$, let $\mathcal{B}$ be a commutative Banach algebra containing A and let λ be an isolated point of $\sigma_\mathcal{B}(A)$. Then λ is isolated in $\sigma_T(A, \mathcal{H})$ and, a fortiori, λ is isolated in $\sigma_\ell(A) \cap \sigma_r(A)$.

Proof: First we establish that if λ is isolated in $\sigma_\mathcal{B}(A)$ then λ must belong to $\sigma_T(A, \mathcal{H})$. Since $\sigma_\mathcal{B}(A) = \{\lambda\} \dot\cup K$ and $\sigma_T(A, \mathcal{H}) \subseteq \sigma_\mathcal{B}(A)$, we see that $(\lambda \notin \sigma_T(A, \mathcal{H}) \Rightarrow \sigma_T(A, \mathcal{H}) \subseteq K)$. Let $P = f(A) \in \mathcal{B}$ be the idempotent associated with λ constructed via the Arens-Calderón-Waelbroeck functional calculus. Then $\sigma_\mathcal{B}(P) = f(\sigma_\mathcal{B}(A)) = \{0,1\}$, while $\sigma_T(P, \mathcal{H}) = f(\sigma_T(A, \mathcal{H})) = \{0\}$. Since $\sigma_\mathcal{B}(P) \subseteq \widehat{\sigma}(P)$ and $\widehat{\sigma}(P) = [\sigma_T(P, \mathcal{H})]^{\widehat{}}$, we get a contradiction. Thus, $\lambda \in \sigma_T(A, \mathcal{H})$. Now use Application 5.24 to show that $\lambda \in \sigma_\ell(A) \cap \sigma_r(A)$. ■

6. Fredholm Theory

In this section we shall investigate the notion of Fredholm-ness for commuting n-tuples of operators on a Hilbert space $\mathcal{H}$.

Definition 6.1. ([34], [36]) Let $A \subset \mathcal{L}(\mathcal{H})$. We say that A is Fredholm (in symbols, $A \in \mathcal{F}(\mathcal{H})$) if $N(D_A)/R(D_A)$ is finite-dimensional (this implies that $R(D_A)$ is closed). The Taylor essential spectrum of A on $\mathcal{H}$ is

$$\sigma_{Te}(A,\mathcal{H}) := \{\lambda \in \mathbb{C}^n : A - \lambda \notin \mathcal{F}(\mathcal{H})\}.$$

By analogy with Lemma 4.1 and with Atkinson's Theorem, we have the following result.

Theorem 6.2. (cf. [34] and [36]) *Let* $A \subset \mathcal{L}(\mathcal{H})$. *The following statements are equivalent:*

(i) $A \in \mathcal{F}(\mathcal{H})$.

(ii) $R_A = D_A + D_A^*$ *is Fredholm.*

(iii) $\square_A = D_A^* D_A + D_A D_A^*$ *is Fredholm.*

(iv) $L_a = (L_{a_1}, \ldots, L_{a_n})$ *is invertible on* $\mathcal{Q}(\mathcal{H})$ $(a := \pi(A))$.

(v) $R_a := D_a + D_{a*}^t \in M_{2^n}(\mathcal{Q}(\mathcal{H}))^{-1}$.

(vi) $\square_a := D_{a*}^t D_a + D_a D_{a*}^t \in M_{2^n}(\mathcal{Q}(\mathcal{H}))^{-1}$.

Sketch of proof. The equivalence of (i), (ii), and (iii) can be obtained by adapting the proof of Lemma 4.1. The equivalence of (iv), (v), and (vi) is contained in the spectral permanence theorem [37]. To prove (i) $\Leftrightarrow$ (iv) (Atkinson's Theorem), one tries and follows the proof for $n = 1$. For instance, if $x \in \mathcal{Q}(\mathcal{H}) \otimes \Lambda[e]$ and $D_{L_a}(x) = 0$, then $D_A(X)$ is compact, where $X \in \mathcal{L}(\mathcal{H}) \otimes \Lambda[e]$ and $\pi(X) = x$. Therefore,

$$\square_A(X) - D_A D_A^*(X)$$

is compact. Since $\square_A$ is Fredholm (by (iii)), we can find a pseudoinverse $\tilde{\square}_A$ such that $\square_A \tilde{\square}_A - I$ and $\tilde{\square}_A \square_A - I$ are both compact. Since $\tilde{\square}_A$ will commute (up to compacts) with $D_A D_A^*$, we then have

$$X - D_A D_A^* \tilde{\square}_A(X)$$

compact, so that $x \in R(D_{L_a})$. ∎

Example 6.3. ([36, Theorem 5]) Let U_+ be the unilateral shift on $\ell^2(\mathbb{Z}_+)$ and let $\mathcal{H} = \ell^2(\mathbb{Z}_+) \otimes \ell^2(\mathbb{Z}_+)$. Then $\sigma_{Te}(U_+ \otimes I, I \otimes U_+) = (\overline{\mathbf{D}} \times \mathbf{T}) \cup (\mathbf{T} \times \overline{\mathbf{D}})$. (There is actually a much more general fact which gives the essential spectrum of arbitrary tensor products on Hilbert space ([61]).)

The next result says that Fredholm n-tuples are associated with Koszul complexes made up of a finite-dimensional piece and an exact piece.

Proposition 6.4. *Let* $A \subset \mathcal{L}(\mathcal{H})$ *be Fredholm. Then the Koszul complex for* A *can be written as*

$$K(A, \mathcal{H}) = L(A, \mathcal{H}) \oplus M(A, \mathcal{H}),$$

where L *is exact and* M *is finite-dimensional. (Caution:* L *and* M *will not be Koszul, in general.)*

Proof: Define L and M as follows:

$$M^k := N\left(D_A^k\right) \cap N\left((D_A^{k-1})^*\right) \quad (= N(\square_A^k)),$$

and

$$L^k := \Lambda^k(\mathcal{H}) \ominus M^k \quad (k = 1, \ldots, n).$$

Since A is Fredholm, M^k is finite-dimensional for all k. Also,

$$D_A^k\big|_{M^k} = 0 \quad \text{and} \quad R\left(D_A^k\right) \subseteq L^{k+1},$$

so that

$$D_A^k L^k \subseteq L^{k+1}.$$

Then,

$$\{M^k, 0\} \quad \text{and} \quad \{L^k, D_A^k\big|_{L^k}\}$$

are two cochain complexes such that $K = L \oplus M$. We must show that L is exact. Observe that

$$N\left(D_A^k\right) = R\left(D_A^{k-1}\right) \oplus M^k,$$

so that

$$L^k = R\left(D_A^{k-1}\right) \oplus \left(N(D_A^k)\right)^\perp.$$

Let $x \in L^k$ be such that $D_A^k x = 0$. Then $x \in N(D_A^k)$ and also $x = y + z$, where $y \in R(D_A^{k-1})$ and $z \perp N(D_A^k)$. Since $D_A^k(x) = D_A^k(y) = 0$, we must have $z \in N(D_A^k)$, and therefore $z = 0$. Then $x = y \in R(D_A^{k-1})$, say $x = D_A^{k-1}(u)$, where $u \in N(D_A^{k-1})^\perp \subseteq L^{k-1}$. Therefore,

$$N\left(D_A^k\big|_{L^k}\right) \subseteq R\left(D_A^{k-1}\big|_{L^{k-1}}\right),$$

showing that L is exact. ∎

Definition 6.5. Let $A \in \mathcal{F}(\mathcal{H})$ and let $K(A, \mathcal{H}) = L(A, \mathcal{H}) \oplus M(A, \mathcal{H})$. We define the index of A by

$$\text{ind}(A) = \chi(M) := \sum_{k=0}^{n} (-1)^k \dim(M^k)$$

(χ is the Euler characteristic of the finite-dimensional cochain complex M).

Theorem 6.6. ([36, Theorem 3]) ind $: \mathcal{F}(\mathcal{H}) \to \mathbb{Z}$ *is continuous, invariant under compact perturbations, and onto* $\mathbb{Z}$. *Also,* $\mathrm{ind}(A) = \mathrm{ind}(\widehat{A})$, *where*

$$\widehat{A} := (D_A + D_A^*)\,|_{\Lambda^{\mathrm{even}}(\mathcal{H})} : \Lambda^{\mathrm{even}}(\mathcal{H}) \to \Lambda^{\mathrm{odd}}(\mathcal{H}),$$

i.e.,

$$\widehat{A} = \begin{pmatrix} D_0 & D_1^* & 0 & 0 \\ 0 & D_2 & D_3^* & 0 \\ 0 & 0 & D_4 & D_5^* & \cdots \end{pmatrix} \in M_{2^{n-1}}(\mathcal{L}(\mathcal{H})).$$

Proof: Observe that

$$\widehat{A}\widehat{A}^* \oplus \widehat{A}^*\widehat{A} = \square_A,$$

so that $A \in \mathcal{F}(\mathcal{H})$ if and only if $\widehat{A}$ is Fredholm. ∎

Remarks 6.7.

a) Although we have defined Fredholm-ness and index for commuting n-tuples, one can consider *almost* commuting n-tuples (commuting modulo the compacts). Some technical difficulties arise (due to the fact that $R(D_A)$ is no longer contained in $N(D_A)$), but much of the theory can be extended to that situation (see [36] and [82]). One can also deal with arbitrary Banach spaces instead of Hilbert spaces, and with arbitrary chain complexes instead of Koszul complexes (cf. [6], [99], [113]).

b) $\widehat{A}$ also determines invertibility, as the reader has already probably noticed (the proof above works almost verbatim in that case). One should not be misled, however, and think that $\sigma_T(A, \mathcal{H})$ can be described in terms of $\sigma(\widehat{A})$. It is true, however, that $\{(\lambda, 0, \ldots, 0) : \lambda \in \mathbb{R} \text{ and } \lambda \in \sigma(\widehat{A})\} \subseteq \sigma_T(A, \mathcal{H})$.

c) A notion of semi-Fredholm-ness can be given in terms of $\widehat{A}$, and index can be extended to that class ([36], [114]). However, semi-Fredholm n-tuples (or the semi-Fredholm domain of an n-tuple) are not as useful as in the $n = 1$ case (this is perhaps so because semi-Fredholm-ness should be defined differently). An exception is the case $n = 2$, where semi-Fredholm-ness already gives considerable information about the different stages of the Koszul complex (see, for instance, Theorem 6.8 below).

We shall now pause to list some basic facts about Fredholm n-tuples.

(i) If $A \in \mathcal{F}(\mathcal{H})$ and B is an arbitrary k-tuple such that $(A,B) \subset \mathcal{L}(\mathcal{H})$, then $(A,B) \in \mathcal{F}(\mathcal{H})$ and $\mathrm{ind}(A,B) = 0$. This can be seen as follows. $P_A \sigma_{Te}((A,B), \mathcal{H}) = \sigma_{Te}(A, \mathcal{H})$ and therefore $(0,0) \notin \sigma_{Te}((A,B), \mathcal{H})$. Moreover,

$$\gamma : [0,1] \longrightarrow \mathcal{F}(\mathcal{H}),$$

given by $\gamma(t) = (A,(1-t)B + tI)$, is a continuous path of Fredholm n-tuples joining (A,B) to (A,I). Therefore, $\mathrm{ind}(A,B) = \mathrm{ind}(A,I)$. (A,I) is, however, invertible (since $P_I \sigma_T((A,I), \mathcal{H}) = \sigma_T(I, \mathcal{H}) = \{1\}$), so that $\mathrm{ind}(A,I) = \mathrm{ind}((A,I)\hat{\ }) = 0$.

(ii) $\mathrm{ind}(U_+ \otimes I, I \otimes U_+) = 1$, as a straightforward calculation shows. Indeed, if K is the Koszul complex of $(U_+ \otimes I, I \otimes U_+)$, then K is exact at stages 0 and 1 and $(\ell_2 \otimes \ell^2)/R(D^1)$ is 1-dimensional. It is important to observe that if K_1 is the Koszul complex for U_+ on ℓ^2

$$(K_1 : 0 \to \ell^2 \overset{U_+}{\to} \ell^2 \to 0),$$

then $K = K_1 \otimes K_1$, i.e.,

$$K^0 = K_1^0 \otimes K_1^0$$
$$K^1 = (K_1^1 \otimes K_1^0) \oplus (K_1^0 \otimes K_1^1)$$

and

$$K^2 = K_1^1 \otimes K_1^1,$$

with boundary maps given by

$$D^0 = \begin{pmatrix} D_1^0 \otimes I \\ I \otimes D_1^0 \end{pmatrix}$$

and

$$D^1(x \otimes y, u \otimes v) = -(x \otimes D_1^0 y) + (D_1^0 u \otimes v).$$

This is the approach used by C. Grosu and F.-H. Vasilescu [61] to prove the Künneth formula for Hilbert complexes.

(iii) If $A \in \mathcal{F}(\mathcal{H})$ then $A^* \in \mathcal{F}(\mathcal{H})$, and

$$\mathrm{ind}(A^*) = (-1)^n \, \mathrm{ind}(A)$$

(cf. [36, Proposition 9.1]. For, $R_{A^*} = D_{A^*} + D_{A^*}^* = (D_A^*)^t + D_A^t = R_A^t$, from which one can derive that $A \in \mathcal{F}(\mathcal{H})$ if and only if $A^* \in \mathcal{F}(\mathcal{H})$. The index equality requires a careful description of $N(\square_{A^*}^k)$. A straightforward calculation reveals that

$$N(\square_{A^*}^k) = R((D_A^k)^t)^\perp \cap N((D_A^k)^t)$$

$$\simeq N(\square_A^{n-k}),$$

so that

$$\text{ind}(A^*) = \sum_{k=0}^n (-1)^k \dim N(\square_{A^*}^k)$$

$$= \sum_{k=0}^n (-1)^k \dim N(\square_A^{n-k}) = \sum_{k=0}^n (-1)^{n-k} \dim N(\square_A^k)$$

$$= (-1)^n \text{ind}(A).$$

(iv) Let $A \in \mathcal{F}(\mathcal{H})$ and let $V \in \mathcal{L}(\mathcal{H})$ be an invertible operator. Let $VAV^{-1} := (VA_1V^{-1}, \ldots, VA_nV^{-1})$. Then $VAV^{-1} \in \mathcal{F}(\mathcal{H})$ and $\text{ind}(VAV^{-1}) = \text{ind}(A)$ (cf. [36, Proposition 5.2]). For the first assertion, observe that $N(D_{VAV^{-1}}) = (V \otimes I)N(D_A)$ and $R(D_{VAV^{-1}}) = (V \otimes I)R(D_A)$; since $V \otimes I$ is invertible on $\Lambda(\mathcal{H})$, this also gives that $\text{ind}(VAV^{-1}) = \text{ind}(A)$. Another way to see the equality of indices is as follows: since the group of invertible operators on $\mathcal{H}$ is arcwise connected, we can join V to I by a path $\{V_t\}_{t\in[0,1]}$ of invertible operators. Then VAV^{-1} is joined to A by a path of Fredholm n-tuples. By the continuity of the index, we see that $\text{ind}(VAV^{-1}) = \text{ind}(A)$.

(v) If $A \subset \mathcal{L}(\mathcal{H})$ is invertible (resp. Fredholm), so are

$$A_1^* A_1 + A_2^* A_2 + \cdots + A_n^* A_n$$

$$A_1 A_1^* + A_2^* A_2 + \cdots + A_n^* A_n$$

$$A_1^* A_1 + A_2 A_2^* + \cdots + A_n^* A_n$$

$$\cdots$$

$$A_1 A_1^* + A_2 A_2^* + \cdots + A_n A_n^*$$

(cf. [39]). To see the validity of the assertion, let us consider for simplicity the case $n = 2$ (although the argument is quite general). Of course we know that $A_1^* A_1 + A_2^* A_2$ and $A_1 A_1^* + A_2 A_2^*$ are both invertible (resp. Fredholm), since they equal $D_0^* D_0$ and $D_1 D_1^*$, respectively (a fact first noted in [110], [111]). We shall see that $A_1 A_1^* + A_2^* A_2$ is inevrtible, the proofs for $A_1^* A_1 + A_2 A_2^*$ and for the Fredholm case being totally similar. Consider

$$\hat{A} = \begin{bmatrix} A_1 & -A_2^* \\ A_2 & A_1^* \end{bmatrix}.$$

We know that $\widehat{A}$ is invertible. Let $z \in \mathcal{H}$. Since $\widehat{A}$ is onto, there exist $x, y \in \mathcal{H}$ such that

$$\begin{cases} A_1 x - A_2^* y = z \\ A_2 x + A_1^* = 0 \end{cases}.$$

In particular, z is in the range of $(A_1 - A_2^*)$, so that $(A_1 - A_2^*) : \mathcal{H} \oplus \mathcal{H} \to \mathcal{H}$ is onto. It follows that

$$(A_1 \ - \ A_2^*) \begin{bmatrix} A_1^* \\ -A_2 \end{bmatrix}$$

is invertible, i.e., $A_1 A_1^* + A_2^* A_2$ is invertible.

The invertibility of the four operators $A_1^* A_1 + A_2^* A_2$, $A_1^* A_1 + A_2 A_2^*$, $A_1 A_1^* + A_2^* A_2$, $A_1 A_1^* + A_2 A_2^*$ does not guarantee, however, the invertibility of the pair (A_1, A_2), as we shall see later (Theorem 6.12).

We shall now give Chō and Takaguchi's Theorem on the topological boundary of the Taylor spectrum.

Theorem 6.8. ([24]). *Let* $A \in \mathcal{L}(\mathcal{H})_{\mathrm{com}}^{(2)}$. *Then*

$$\partial \sigma_T(A, \mathcal{H}) \subseteq \partial \sigma_H(A, \mathcal{H}),$$

so that $\sigma_T(A, \mathcal{H})$ *is the union of* $\sigma_H(A, \mathcal{H})$ *and some of the bounded components of the complement of* $\sigma_H(A, \mathcal{H})$ *in* $\mathbb{C}^2$.

Proof: We shall present the proof, based on the Fredholm index, given in [39]. It is considerably shorter than Chō and Takaguchi's original proof and it shows more clearly why the result is true.

Since $\sigma_H \subseteq \sigma_T$, it suffices to show that $\partial \sigma_T \subseteq \sigma_H$. Let $\lambda \in \partial \sigma_T(A, \mathcal{H})$ and let $\{\lambda_k\}_{k=1}^\infty$ be a sequence outside $\sigma_T(A, \mathcal{H})$ such that $\lambda_k \to \lambda (k \to \infty)$. If $\lambda \notin \sigma_H(A, \mathcal{H})$ then $A - \lambda$ is semi-Fredholm (i.e., $(A-\lambda)\widehat{\ }$ is semi-Fredholm) and $\mathrm{ind}(A-\lambda) = \lim_k \mathrm{ind}(A - \lambda_k) = 0$. Therefore, the Koszul complex $K(A - \lambda, \mathcal{H})$ is exact on the left and right stages and it has zero index, so $K(A - \lambda, \mathcal{H})$ must be exact at the middle stage as well, and therefore $\lambda \notin \sigma_T(A, \mathcal{H})$, a contradiction. Thus, $\lambda \in \sigma_H(A, \mathcal{H})$. ∎

Corollary 6.9. ([39, Corollary 3.3]) *Let* $A \in \mathcal{L}(\mathcal{H})_{\mathrm{com}}^{(2)}$. *Then*

$$\partial \sigma_{Te}(A, \mathcal{H}) \subseteq \partial \sigma_{He}(A, \mathcal{H})$$

so that $\sigma_{Te}(A, \mathcal{H})$ is the union of $\sigma_{He}(A, \mathcal{H})$ and some of the bounded components of the complement of $\sigma_{He}(A, \mathcal{H})$ in $\mathbb{C}^2$.

Proof: Let $\rho : \mathcal{Q}(\mathcal{H}) \to \mathcal{L}(\mathcal{K})$ be a faithful *-representation of the Calkin algebra on a Hilbert space $\mathcal{K}$. By the spectral permanence of σ_T and σ_H ([37]), we know that

$$\sigma_{Te}(A, \mathcal{H}) = \sigma_T(\rho(\pi(A)), \mathcal{K})$$

and

$$\sigma_{He}(A, \mathcal{H}) = \sigma_H(\rho(\pi(A)), \mathcal{K}),$$

from which the result follows by an application of Theorem 6.8. ■

The following corollary to the proof of Theorem 6.8 gives a criterion for a point to be isolated in the Taylor spectrum.

Corollary 6.10. ([39, Corollary 3.6]) *Let $A \in \mathcal{L}(\mathcal{H})^{(2)}_{com}$ and let λ be an isolated point of $\sigma_H(A, \mathcal{H})$. Assume that λ is in the semi-Fredholm domain of A (i.e., $(A - \lambda)$ is semi-Fredholm). Then λ is an isolated point of $\sigma_T(A, \mathcal{H})$ if and only if $\mathrm{ind}(A - \lambda) = 0$.*

Proof: The necessity is clear. Assume then that $\mathrm{ind}(A - \lambda) = 0$. For $\mu \neq \lambda, \mu$ near $\lambda, \mathrm{ind}(A - \mu) = 0$, and since $A - \mu$ is Harte invertible, we must have $\mu \notin \sigma_T(A, \mathcal{H})$. Therefore, λ is isolated in $\sigma_T(A, \mathcal{H})$. ■

Remarks 6.11.

(i) E. Albrecht has pointed out to us that Theorem 6.8 is still true when A is a pair of operators on a Banach space. The proof is similar to the one given above, but it uses the stability theorems for complexes of Banach spaces given in [113].

(ii) Theorem 6.8 fails when $n \leq 3$, as the following example taken from [39] shows. Let $A \in \mathcal{L}(\mathcal{H})^{(2)}_{com}$ be such that $\sigma_H(A, \mathcal{H}) \neq \sigma_T(A, \mathcal{H})(A_1 = U_+^* \otimes I, A_2 = I \otimes U_+$ acting on $\ell^2 \otimes \ell^2$ will do). Consider $\widetilde{A} = (A_1, A_2, A_2)$. By the spectral mapping theorem,

$$\sigma_T(\widetilde{A}, \mathcal{H}) = \{(\lambda_1, \lambda_2, \lambda_2) : (\lambda_1, \lambda_2) \in \sigma_T(A, \mathcal{H})\}$$

and

$$\sigma_H(\widetilde{A}, \mathcal{H}) = \{(\lambda_1, \lambda_2, \lambda_2) : (\lambda_1, \lambda_2) \in \sigma_H(A, \mathcal{H})\}.$$

Therefore,

$$\partial \sigma_T(\tilde{A}, \mathcal{H}) = \sigma_T(\tilde{A}, \mathcal{H})$$

so that $\partial \sigma_T(\tilde{A}, \mathcal{H})$ is not contained in $\sigma_H(\tilde{A}, \mathcal{H})$.

We have said before that the invertibility of the four operators $A_1^* A_1 + A_2^* A_2, \ldots,$ $A_1 A_1^* + A_2 A_2^*$ does not give, in general, the invertibility of A. Indeed, the example in Application 5.30 is such that if $(\lambda_1, \lambda_2) \in \mathbf{C}^2$ satisfies $0 < |\lambda_1|, |\lambda_2| < 1$, then $(A_1 - \lambda_1, A_2 - \lambda_2)$ is not invertible, but all four of the corresponding self-adjoint operators are. This fact is shown in [39], and it is used to answer a question of A. Dash and C. Davis on joint spectra ([48], [49, problem 1], or [81, C. Davis, Problem 2]). One can then ask: What else, in addition to the invertibility of those four operators, is needed for the invertibility of (A_1, A_2)? The following result gives a precise answer; it completely describes σ_T in terms of σ_H when $n = 2$.

Theorem 6.12. ([39, Theorem 3.10]) Let $A \in L(\mathcal{H})_{\mathrm{com}}^{(2)}$ be such that

(i) A is left invertible (let $H := (A_1^* A_1 + A_2^* A_2)^{-1}$);

(ii) A is right invertible (let $K := (A_1 A_1^* + A_2 A_2^*)^{-1}$);

(iii) $N(A_1) \cap N(A_2^*) = 0$;

(iv) $N(A_1^*) \cap N(A_2) = 0$. Then A is Taylor invertible if and only if $A_1 H A_2^* = A_2^* K A_1$.

Proof: $\Rightarrow$) If A is invertible, so is $\widehat{A}$, so that $(\widehat{A}^* \widehat{A})^{-1} \widehat{A}^*$, which is a left inverse for $\widehat{A}$, is also a right inverse for $\widehat{A}$, i.e., $\widehat{A}(\widehat{A}^* \widehat{A})^{-1} \widehat{A}^* = I_{\Lambda_2(\mathcal{H})}$. A straightforward matrix multiplication then gives

$$(*) \begin{cases} A_1 H A_1^* + A_2^* K A_2 = I \\ A_1 H A_2^* - A_2^* K A_1 = 0 \\ A_2 H A_2^* + A_1^* K A_1 = I, \end{cases}$$

from which it follows that

$$A_1 H A_2^* = A_2^* K A_1.$$

$\Leftarrow$) It is easy to see that if A_1 and A_2 satisfy the system of operator equations $(*)$ then (A_1, A_2) is invertible. (Observe that we need (i) and (ii) to define H and K.) We shall then show that (iii) and (iv) combined with the condition $A_1 H A_2^* = A_2^* K A_1$ imply the validity of the remaining two equations in $(*)$. Let $B := A_1 H A_1^* + A_2^* K A_2 - I$. Then

$$A_1^* B = A_1^* A_1 H A_1^* + A_1^* A_2^* K A_2 - A_1^*$$

$$= A_1^* A_1 H A_1^* + A_2^* (A_1^* K A_2) - A_1^*$$

$$= A_1^* A_1 H A_1^* + A_2^* (A_2 H A_1^*) - A_1^*$$

$$= (A_1^* A_1 + A_2^* A_2) H A_1^* - A_1^*$$

$$= 0,$$

so that $R(B) \subseteq N(A_1^*)$. Similarly, $A_2 B = 0$, or $R(B) \subseteq N(A_2)$. By (iv), $B = 0$. In a completely analogous manner one shows that (iii) implies that $C := A_2 H A_2^* + A_1^* K A_1 - I = 0$. ∎

Example 6.13. ([39, Example 4.1]). For A as in Application 5.30,

$$\sigma_H(A) = \{(0,0)\} \cup (\overline{D} \times T) \cup (T \times \overline{D})$$

and

$$\sigma_H(A_1^*, A_2) = (\overline{D} \times \{0\}) \cup (\overline{D} \times T) \cup (T \times \overline{D}),$$

so that the union

$$\sigma_H(A) \cup \{(\overline{\lambda}_1, \lambda_2) : (\lambda_1, \lambda_2) \in \sigma_H(A_1^*, A_2)\},$$

which represents the four above-mentioned self-adjoint operators, is much smaller than $\sigma_T(A)(= \overline{D} \times \overline{D})$.

The reader must have by now a pretty good idea of when a pair $(A_1, A_2) \subset \mathcal{L}(\mathcal{H})$ is invertible. Just in case this is not so, we shall present yet another way to describe the exactness of the Koszul complex associated with (A_1, A_2).

Let $\tilde{A}_1 := A_1|_{N(A_2)}$ and let

$$\tilde{\tilde{A}}_1 : \mathcal{H}/R(A_2) \to \mathcal{H}/R(A_2)$$

be given by $\tilde{\tilde{A}}(x + R(A_2)) = A_1 x + R(A_2)$. Although $\tilde{A}_1$ is a bounded linear operator on $N(A_2)$, $\tilde{\tilde{A}}_1$ is just a linear transformation on the vector space $\mathcal{H}/R(A_2)$. We claim that (A_1, A_2) is *invertible* on $\mathcal{H}$ if and only if both $\tilde{A}_1$ and $\tilde{\tilde{A}}_1$ are *linear isomorphisms*. For, suppose (A_1, A_2) is invertible. If $\tilde{A}_1 x = 0, x \in N(A_2)$, then $x \in N(A_1) \cap N(A_2) = \{0\}$; if $y \in N(A_2)$ then $-A_2 y + A_1 \cdot 0 = 0$, so that by the exactness of $K((A_1, A_2), \mathcal{H})$ at the

71

middle stage, there exists $x \in \mathcal{H}$ such that $0 = A_2 x$ and $y = A_1 x$, and therefore $y \in \tilde{A}_1(N(A_2))$. If $\tilde{\tilde{A}}_1(x + R(A_2)) = R(A_2)$, then $A_1 x \in R(A_2)$, say $A_1 x = A_2 y$, $y \in \mathcal{H}$; then $x = A_2 z$ and $y = A_1 z$ for some $z \in \mathcal{H}$, and in particular $x \in R(A_2)$, so that $\tilde{\tilde{A}}_1$ is one-to-one. Now, given $y + R(A_2)$, we find first x and z in $\mathcal{H}$ such that $-A_2 x + A_1 z = y$ (by the exactness of $K((A_1, A_2), \mathcal{H})$ at the right stage); then $\tilde{\tilde{A}}_1(z + R(A_2)) = A_1 z + R(A_2) = y + R(A_2)$, showing that $\tilde{\tilde{A}}_1$ is onto. Therefore, both $\tilde{A}_1$ and $\tilde{\tilde{A}}_1$ are isomorphisms.

For the converse, one proceeds in basically the same way. The only step that poses some difficulty is the proof of exactness at the middle stage, which we now explain: if $A_1 x = A_2 y$ then $\tilde{\tilde{A}}_1(x + R(A_2)) = R(A_2)$, so that $x = A_2 z$ for some $z \in \mathcal{H}$ (by the injectivity of $\tilde{\tilde{A}}_1$). Then $A_1 A_2 z = A_2 y$, or $y - A_1 z \in N(A_2)$. Since $\tilde{A}_1$ is onto, there exists $u \in N(A_2)$ such that $y - A_1 z = A_1 u$, or $y = A_1(z + u)$. Moreover, $A_2(z + u) = A_2 z = x$, so that $x = A_2(z + u)$ and $y = A_1(z + u)$, as desired.

Of course, the result extends to n-tuples (see [104, Lemma 1.3]): $(A_1, \ldots, A_n) \subseteq \mathcal{L}(\mathcal{H})$ is invertible if and only if A_1 acting on the homology spaces of $K((A_2, \ldots, A_n), \mathcal{H})$ gives rise to linear isomorphisms. One can use this to define invertibility by a recursive process. Here is a nice application: if $A, B, C \in \mathcal{L}(\mathcal{H}), (A, C), (B, C), (AB, C) \subset \mathcal{L}(\mathcal{H})$, and (A, C) and (B, C) are invertible, then so is (AB, C). For other results on σ_T vs. σ_H, see [66].

7. Subnormal and Bergman n-tuples

In this section we examine the notions of joint subnormality and of Bergman n-tuple. The results that we present are of a very general nature and are mainly intended as a sample of a variety of topics related to multiparameter spectral theory.

Definition 7.1. Let $N \subset \mathcal{L}(\mathcal{H})$. We say that N is *normal* if each N_i is a normal operator on $\mathcal{H}$.

Proposition 7.2. ([25], [37], [49]) Let $N \subset \mathcal{L}(\mathcal{H})$ be normal. Then
$$\sigma_\ell(N, \mathcal{H}) = \sigma_r(N, \mathcal{H}) = \sigma_H(N, \mathcal{H}) = \sigma_T(N, \mathcal{H})$$
$$= \sigma'(N) = \sigma''(N) = \sigma_{C^*(N)}(N).$$

Proof: It suffices to show that if N is left invertible then N is invertible with respect to the commutative C^*-algebra $C^*(N)$. Since N is left invertible, $N_1^* N_1 + \cdots + N_n^* N_n$ is invertible, and if $H := (N_1^* N_1 + \cdots + N_n^* N_n)^{-1}$, then

$$(HN_1^*)N_1 + \cdots + (HN_n^*)N_n = I.$$

Now, $HN_1^*, \ldots, HN_n^* \in C^*(N)$, so that N is invertible with respect to $C^*(N)$.

Remarks 7.3.

(i) It is interesting to see how $\sigma_{C^*(N)}(N)$ is related to $\sigma_{C^*(N)}(N, N^*)$. $C^*(N)$ is generated, as a Banach algebra, by N and N^*, and therefore $\sigma_{C^*(N)}(N, N^*)$ is polynomially convex by Theorem 1.8. Moreover, the map

$$\Phi : \sigma_{C^*(N)}(N) \to \sigma_{C^*(N)}(N, N^*)$$

$$\lambda \mapsto (\lambda, \bar\lambda)$$

is a homeomorphism (recall that if $\varphi \in M_{C^*(N)}$ then $\varphi(N^*) = \overline{\varphi(N)}$). Thus, although $\sigma_{C^*(N)}(N)$ is just an arbitrary compact nonempty set (for, given a compact nonempty set $K \subset \mathbb{C}^n, K = \sigma_T(M_z, L^2(K, m))$, where m is the volumetric Lebesgue measure and M_z is the operator of multiplication by the coordinate function z), $\sigma_{C^*(N)}(N, N^*)$ is always polynomially convex. We have, therefore, an operator theoretic proof of the following fact:

Given a compact nonempty set $K \subset \mathbb{C}^n$, the set $\{(\lambda, \bar\lambda) : \lambda \in K\}$ is polynomially convex in $\mathbb{C}^{2n}$.

In a completely analogous way, one also shows that if $K \subseteq \mathbb{R}^n + i0 \subseteq \mathbb{C}^n$, then K is polynomially convex.

(ii) Just as in the $n = 1$ case, $*$-cyclic normal n-tuples are modelled by M_z acting on $L^2(\mu)$, where μ is a positive Borel measure in $\mathbb{C}^n$. The support of μ is precisely $\sigma_T(M_z, L^2(\mu))$ (see the proof of Theorem 7.7 below).

Definition 7.4. Let $S \subset \mathcal{L}(\mathcal{H})$. We say that S is subnormal if there exist a Hilbert space $\mathcal{K} \supseteq \mathcal{H}$ and $N \subset \mathcal{L}(\mathcal{K})$ such that N is normal and $N_{|\mathcal{K}} = S$.

Several versions of the spectral inclusion property have been obtained for subnormal n-tuples ([67], [76]). As we mentioned in [39], they can all be derived from the spectral inclusion for the Taylor spectrum, obtained for doubly commuting n-tuples in [38], and generalized by M. Putinar to arbitrary subnormal n-tuples in [86].

Theorem 7.5. ([38, Theorems 1 and 2], [86]) *Let $S \subset \mathcal{L}(\mathcal{H})$ be subnormal and let N be the minimal normal extension of S, in symbols $N = \text{m.n.e.}(S)$. Then*

(i) $\sigma_T(N, \mathcal{H}) \subseteq \sigma_r(S, \mathcal{H}) \subseteq \sigma_T(S, \mathcal{H})$.

(ii) $\sigma_T(S, \mathcal{H}) \subseteq \hat{\sigma}(N)$.

(iii) $\sigma_p(N, \mathcal{K}) \subseteq \overline{\sigma_p(S^*, \mathcal{H})}$.

Proof: We refer to [86] for a proof of (i).

(ii) Let $\lambda \in \sigma_T(S, \mathcal{H})$ and let $p \in \mathbb{C}[z]$. Then $p(\lambda) \in \sigma_T(p(S), \mathcal{H}) = \sigma(p(S)) \subseteq \hat{\sigma}(p(N))$, so that

$$|p(\lambda)| \leq \|p(N)\| = r(p(N)) = \|p\|_{\sigma_T(N, \mathcal{H})},$$

and therefore $\lambda \in (\sigma_T(N, \mathcal{H}))^{\wedge} = \hat{\sigma}(N)$.

(iii) ([71, Lemma 1]) Let $\lambda \in \sigma_p(N, \mathcal{H})$ and let $y \in \mathcal{K}$ be such that $y \neq 0$ and $(N_i - \lambda_i)y = 0$ (all i). Since N is normal, we also have $(N_i^* - \bar{\lambda}_i)y = 0$, so that the subspace generated by y reduces N. Consider $x := Py$, where $P = \mathcal{K} \rightarrow \mathcal{H}$ is the orthogonal projection. If $x = 0$, then $y \in \mathcal{H}^{\perp}$, contradicting the minimality of N. Therefore, $x \neq 0$. Now

$$(N_i^* - \bar{\lambda}_i)Py = -(N_i^* - \bar{\lambda}_i)(I - P)y,$$

so that

$$(S_i^* - \bar{\lambda}_i)x = P(N_i^* - \bar{\lambda}_i)Py = -P(N_i^* - \bar{\lambda}_i)(I - P)y = 0,$$

and therefore $\bar{\lambda} \in \sigma_p(S^*, \mathcal{H})$. ∎

Definition 7.6. Let K be a compact nonempty subset of $\mathbb{C}^n$ and let μ be a positive Borel measure with support contained in K. We let $A^2(K, \mu)$ denote the $L^2(\mu)$-closure of $A(K)$, the algebra of germs of functions analytic in a neighborhood of K.

The following theorem is an extension to several variables of a well-known result about subnormal operators (cf. [30, Chapter III.5.2]).

Theorem 7.7. ([47, Theorem 2.1]) *Let $S \subset \mathcal{L}(\mathcal{H})$ be subnormal and let $N = $ m.n.e.(S) $\subset \mathcal{L}(\mathcal{K})$. Assume that S is analytically cyclic, i.e., there exists a vector $x_0 \in \mathcal{H}$ such that $\|x_0\| = 1$ and the closed linear span of $\{f(S)x_0 : f \in A(\sigma_T(S, \mathcal{H}))\}$ is all of $\mathcal{H}$. Then there exist a probability measure μ whose support is $\sigma_T(N, \mathcal{K})$, and an isometric isomorphism $U : \mathcal{H} \rightarrow A^2(\sigma_T(S, \mathcal{H})), \mu)$ such that $Ux_0 \equiv 1$ and $USU^* = M_z$.*

Proof: If $f \in A(\sigma_T(S, \mathcal{H}))$ then $f \in A(\sigma_T(N, \mathcal{K}))$, by Theorem 7.5(i). Also, $f(S) = f(N)|_{\mathcal{H}}$, by Application 5.25. Thus, x_0 is *-cyclic for N. Let $\varphi : C(\sigma_T(N, \mathcal{K})) \rightarrow \mathbb{C}$

be given by $\varphi(f) = (f(N)x_0, x_0)$. Then there exists a probability measure μ whose support is $\sigma_T(N, \mathcal{K})$ and such that

$$\varphi(f) = \int_{\varphi_T(N,\mathcal{K})} f \, d\mu.$$

Moreover,

$$(*) \qquad \|f\|_2^2 = \int_{\sigma_T(N,\mathcal{K})} |f|^2 \, d\mu = \varphi(|f|^2) = \|f(N)x_0\|^2 = \|f(S)x_0\|^2,$$

for every $f \in A^2(\sigma_T(S, \mathcal{H}), \mu)$. Therefore, we can define $U : \mathcal{H} \to A^2(\sigma_T(S, \mathcal{H}), \mu)$ by $U(f(S)x_0) = f, f \in A^2(\sigma_T(S, \mathcal{H}), \mu)$. Clearly, $Ux_0 \equiv 1$ and $USU^* = M_z$. Observe that if $f(S)x_0 = 0$, then f vanishes in $\sigma_T(S, \mathcal{H})$ and not only in $\sigma_T(N, \mathcal{K})$ as $(*)$ above indicates: For, if $f(S)x_0 = 0$ then $f(S)g(S)x_0 = 0$ for every $g \in A(\sigma_T(S, \mathcal{H}))$, and then $f(S) = 0$. Then $f(\sigma_T(S, \mathcal{H})) = \sigma(f(S)) = \{0\}$, so that $f|_{\sigma_T(S,\mathcal{H})} \equiv 0$. ∎

We shall now restrict attention to Bergman n-tuples. For Ω Reinhardt or strongly pseudoconvex, for instance, the Bergman n-tuple for Ω is analytically cyclic and agrees with the model given by Theorem 7.7.

Definition 7.8. Let Ω be a bounded domain in $\mathbf{C}^n$. We shall let $A^2(\Omega)$ denote the *Bergman space* of Ω, i.e., $A^2(\Omega)$ is the $L^2(\Omega, m)$-closure of $A(\Omega)$, where m is volumetric Lebesgue measure on Ω. The n-tuple $M_z = (M_{z_1}, \cdots, M_{z_n})$ of multiplications by the coordinate functions is called *the Bergman n-tuple for Ω.*

When $n = 1$, the spectral theory of M_z on $A^2(\Omega)$ has been described in detail by S. Axler, J. Conway, and G. McDonald in [12]. There, $\sigma(M_z) = \bar{\Omega}$ and $\sigma_e(M_z) = \partial_{2-e}\Omega$, the essential boundary of Ω, obtained from $\partial\Omega$ by deleting the 2-removable points.

Definition 7.9. Let $\Theta : \Omega \to \mathcal{M}_\Omega$ be the canonical embedding of Ω into its envelope of holomorphy $\mathcal{M}_\Omega$, and let $p : \mathcal{M}_\Omega \to \mathbf{C}^n$ be the holomorphic projection associated with $\mathcal{M}_\Omega$ (see [62], [69], or [72] for these and related facts on envelopes of holomorphy). We shall let $\widetilde{\Omega} := p(\mathcal{M}_\Omega)$. $\Omega \subseteq \widetilde{\Omega}$; moreover, $\rho : A(\mathcal{M}_\Omega) \to A(\Omega)$, given by $\rho(F)(z) = F(\Theta(z)), z \in \Omega, F \in A(\mathcal{M}_\Omega)$, is a bijection.

Lemma 7.10. ([95, Lemma 4.6]) *Let Ω be a bounded domain in $\mathbf{C}^n$ and let $\omega \in \mathcal{M}_\Omega$. Then $f \to \rho^{-1}(f)(\omega)$ defines a bounded linear functional on $A^2(\Omega)$.*

Corollary 7.11. ([47, Corollary 3.2]) *Let Ω be a bounded domain in $\mathbf{C}^n$ and let $\lambda \in \widetilde{\Omega}$. Then $N(M_z^* - \bar{\lambda}) \neq \{0\}$. In particular, $\widetilde{\Omega} \subseteq \sigma_T(M_z, A^2(\Omega))$.*

Proof: Let $\omega \in \mathcal{M}_\Omega$ be such that $\lambda = p(\omega)$. By Lemma 7.10 there exists a nonzero function $g_\lambda \in A^2(\Omega)$ such that $\rho^{-1}(f)(\omega) = (f, g_\lambda)$, for all $f \in A^2(\Omega)$. Then

$$(M^*_{z_i} g_\lambda, f) = (g_\lambda, M_{z_i} f)$$
$$= \overline{\rho^{-1}(z_i)(\omega)} \cdot \overline{\rho^{-1}(f)(\omega)}$$
$$= \overline{p(\omega)_i}(g_\lambda, f) = (\bar{\lambda}_i g_\lambda, f)$$

for all $f \in A^2(\Omega), i = 1, \ldots, n$. Thus, $g_\lambda \in N(M^*_{z_1} - \bar{\lambda}_1) \cap \cdots \cap N(M^*_{z_n} - \bar{\lambda}_n)$, so that $\bar{\lambda} \in \sigma_\ell(M^*_z, A^2(\Omega))$, or $\lambda \in \sigma_r(M_z, A^2(\Omega)) \subseteq \sigma_T(M_z, A^2(\Omega))$.

Theorem 7.12. ([47]) *Let Ω be a bounded domain in $\mathbf{C}^n$ and assume that $\Omega \subseteq \mathcal{M}_\Omega \subseteq \mathbf{C}^n$. The following statements hold.*

(i) $\sigma_\ell(M_z, A^2(\Omega)) \subseteq \partial\Omega \cap \partial\mathcal{M}_\Omega$.

(ii) *If $\lambda \in \mathcal{M}_\Omega$ and $n = 2$, we have $N(D^1_{M_z - \lambda}) = R(D^0_{M_z - \lambda})$.*

(iii) *If $\lambda \in \bar{\Omega}$ then $N(D^1_{M_z - \lambda}) = R(D^0_{M_z - \lambda})$.*

Proof: We shall only give an outline of (ii), which illustrates the basic technique needed for proofs of exactness of D_{M_z}. For Ω a polydisc, the result follows from [23] and [61] (see [47, Lemma 3.3]), since $A^2(\Omega)$ is in that case isomorphic to the tensor product of two copies of the 1-variable Bergman space $A^2(\mathbf{D})$ (or it can be directly derived by a power series argument). We shall use this fact and the rudiments of sheaf theory to prove (ii). Let $f, g \in A^2(\Omega)$ be such that

$$-M_{z_2 - \lambda_2} f + M_{z_1 - \lambda_1} g = 0.$$

Let P be a small polydisc of multiradius $\delta > 0$ centered at λ and such that $P \subseteq \mathcal{M}_\Omega$. Then $\tilde{f}|_P, \tilde{g}|_P \in A^2(P)$ and

$$-M_{z_2 - \lambda_2}(\tilde{f}|_P) + M_{z_1 - \lambda_1}(\tilde{g}|_P) = 0,$$

where $\tilde{f}$ and $\tilde{g}$ denote $\rho^{-1}(f)$ and $\rho^{-1}(g)$, respectively. Then there exists a unique $h_P \in A^2(P)$ such that

$$\begin{cases} \tilde{f}|_P = M_{z_1 - \lambda_1} h_P \\ \tilde{g}|_P = M_{z_2 - \lambda_2} h_P. \end{cases}$$

Let

$$\Omega_1 := \left\{ z \in \Omega \ : \ |z_1 - \lambda_1| \geq \frac{\delta}{2} \right\}$$

$$\Omega_2 := \left\{ z \in \Omega \ : \ |z_1 - \lambda_1| < \frac{\delta}{2} \text{ and } |z_2 - \lambda_2| \geq \frac{\delta}{2} \right\},$$

and define h on Ω by

$$
h(z) := \begin{cases} \dfrac{f(z)}{z_1 - \lambda_1} & z \in \Omega_1 \\[2mm] \dfrac{g(z)}{z_2 - \lambda_2} & z \in \Omega_2 \\[2mm] h_P(z) & z \in P \cap \Omega. \end{cases}
$$

h is well-defined, analytic in Ω, and

$$
\int_\Omega |h(z)|^2 \, dm(z) \leq \int_{\Omega_1} \frac{|f(z)|^2}{|z_1 - \lambda_1|^2} \, dm(z) + \int_{\Omega_2} \frac{|g(z)|^2}{|z_2 - \lambda_2|^2} \, dm(z) + \int_P |h_P(z)|^2 \, dm(z)
$$
$$
\leq \frac{4}{\delta^2} \left(\|f\|^2 + \|g\|^2 \right) + \int_P |h_P(z)|^2 \, dm(z).
$$

Thus, $h \in A^2(\Omega)$ and

$$
\begin{cases} M_{z_1 - \lambda_1} h = f \\ M_{z_2 - \lambda_2} h = g, \end{cases}
$$

as desired.

Remark 7.13. The results in Theorem 7.12 make no assumptions on the domain Ω. If one restricts attention to special classes of domains, those results can be improved. For instance, if Ω is *strongly pseudoconvex* then one can use Theorem 7.12 above together with the fact that M_z is essentially normal (indeed, $C^*(M_z)/\mathcal{K}(A^2(\Omega)) \simeq C(\partial\Omega)$ (see for instance [70], [92], [117], [121])) and the Index Theorem in [17] to conclude that

$$
\sigma_T(M_z, A^2(\Omega)) = \bar{\Omega} \text{ and } \sigma_{Te}(M_z, A^2(\Omega)) = \partial\Omega
$$

(see [47, Theorem 6.1]). Recently, M. Putinar has used his sheaf model theory for commuting n-tuples ([85], [88], and [54]) to prove that the same formulas for σ_T and σ_{Te} hold for M_z acting on Bergman spaces associated with plurisubharmonic functions in bounded pseudoconvex domains ([88, Proposition 4.1]). His methods, however, do not give information on $C^*(M_z)$; in case Ω is a *pseudo-regular* domain, N. Salinas has shown that M_z gives rise to an element of $\text{Ext}(\partial\Omega)$:

$$
0 \to \mathcal{K}(A^2(\Omega)) \to C^*(M_z) \to C(\partial\Omega) \to 0,
$$

and again, $\sigma_T(M_z) = \bar{\Omega}$ and $\sigma_{Te}(M_z) = \partial\Omega$ (see [96, Theorem 2.3], [97]).

8. M_z on Reinhardt Domains

If instead of boundary regularity (see Remark 7.13), one insists on the rotational symmetry of Ω, Theorem 7.12 can again be improved. At the same time, M_z becomes a multivariable weighted shift.

Definition 8.1. Let Ω be a domain in $\mathbf{C}^n$. Ω is said to be *Reinhardt* if it is invariant under the action of the n-torus $\mathbf{T}^n$, i.e., if whenever $z \in \Omega$ and $e^{i\theta} \in \mathbf{T}^n$, it follows that $e^{i\theta}z := (e^{i\theta_1}z_1, \ldots, e^{i\theta_n}z_n) \in \Omega$. Reinhardt domains are important in several complex variables because of their intrinsic relationship to power series. For instance, pseudoconvex Reinhardt domains containing the origin are the natural domains of convergence of power series. Moreover, if Ω is a complete Reinhardt domain containing the origin, then int. $\widehat{\Omega} = \mathcal{M}_\Omega = \widetilde{\Omega}$, where $\widetilde{\Omega}$ is the pseudoconvex hull of Ω. (For more on Reinhardt domains, see [69] or [74].)

In what follows, we shall assume that $0 \in \Omega$.

Observation 8.2. For $\alpha \in \mathbf{Z}_+^n$ let $z^\alpha = z_1^{\alpha_1} \ldots z_n^{\alpha_n}$; let us also denote by z^α the function whose value at z is z^α. If $c_\alpha = \|z^\alpha\|_{A^2(\Omega)}$ then

$$\left\{ \frac{z^\alpha}{c_\alpha} \right\}_{\alpha \in \mathbf{Z}_+^n}$$

is an orthonormal basis for $A^2(\Omega)$. Moreover,

$$M_{z_i}\left(\frac{z^\alpha}{c_\alpha} \right) = \frac{c_{\alpha+\epsilon_i}}{c_\alpha} \left(\frac{z^{\alpha+\epsilon_i}}{c_{\alpha+\epsilon_i}} \right) \quad (\text{all } \alpha \in \mathbf{Z}_+^n),$$

where $\epsilon_i = (0, \ldots, \overset{i}{1}, \ldots, 0)$, so that M_{z_i} is a weighted shift with weights

$$w_i(\alpha) = \frac{c_{\alpha+\epsilon_i}}{c_\alpha}, \alpha \in \mathbf{Z}_+^n, i = 1, \ldots, n.$$

Example 8.3. Let $0 < p, q < \infty$ and let

$$\Omega_{p,q} = \left\{ z \in \mathbf{C}^2 : |z_1|^p + |z_2|^q < 1 \right\}.$$

$\Omega_{p,q}$ is pseudoconvex (because $\log\|\Omega_{p,q}\| := \{(\log|z_1|, \log|z_2|) : (z_1, z_2) \in \Omega_{p,q}$ and $z_1 \neq 0, z_2 \neq 0\}$ is a convex set); when $p, q \geq 2, = \Omega_{p,q}$ is *Levi* pseudoconvex, and $\Omega_{p,q}$ is strongly pseudoconvex if and only if $p = q = 2$. For $\Omega_{p,q}$ the coefficients c_α can be calculated explicitly:

$$c_\alpha = \frac{2\pi^2}{p(\alpha_2 + 1)} B\left(\frac{2\alpha_1 + 2}{p}, \frac{2\alpha_2 + 2}{q} + 1 \right),$$

where B is the Beta function.

The study of multivariable weighted shifts was initiated by N. Jewell and A. Lubin in [72], extending in various ways the results in [100]; additional properties were derived in [47], especially in the case of M_z on Reinhardt domains. The following theorem summarizes most of the known results on Bergman n-tuples for Reinhardt domains.

Theorem 8.4. ([47, Theorem 4.9]) *Let $\Omega \subseteq \mathbb{C}^2$ be a Reinhardt domain, $0 \in \Omega$. Then*

(i) $\sigma_\ell(M_z) = \sigma_{\ell_e}(M_z) \subseteq \partial\Omega \cap \partial\widehat{\Omega}$.

(ii) $\sigma_r(M_z) = \widehat{\Omega}$.

(iii) $(\widehat{\Omega})^\circ \subseteq \sigma_p(M_z^*) = \left\{ \lambda \in \mathbb{C}^2 : \sum_\alpha \frac{|\lambda^\alpha|^2}{c_\alpha^2} < \infty \right\}$ *(σ_p is the point spectrum (i.e.,*
   the set of joint eigenvalues) of M_z^*).

(iv) $N(D^1_{M_z - \lambda}) = R(D^\circ(M_z - \lambda)(\lambda \in \text{int}.\widehat{\Omega})$.

(v) $\partial\widehat{\Omega} \subseteq \sigma_{re}(M_z)$.

(vi) *There exists $w > 0$ such that*

$$\left\{ \lambda \in \text{int}.\widehat{\Omega} : \|\lambda\| < w \right\} \cap \sigma_{Te}(M_z) = \phi.$$

(vii) $\text{ind}(M_z) = 1$.

(viii) *If Ω is complete (i.e., $(z \in \Omega, t_1, t_2 \in [0,1]) \Rightarrow (t_1 z_1, t_2 z_2) \in \Omega$) and not pseudoconvex, then M_z is not essentially normal.*

We shall conclude this section with an example of a domain that has turned out to be quite an interesting one.

Example 8.5. Let $0 < \delta, \epsilon < 1$ and let

$$\Omega_{\delta,\epsilon} := \{ z \in \mathbb{C}^2 : (|z_1| < \delta \text{ and } |z_2| < 1) \text{ or } (|z_1| < 1 \text{ and } |z_2| < \epsilon) \}$$

$\Omega_{\delta,\epsilon}$ is an L-shaped domain, and

$$\widehat{\Omega}_{\delta,\epsilon} = \left\{ z \in \mathbb{C}^2 : |z_1|, |z_2| \leq 1 \text{ and } |z_1|^{\log\epsilon} |z_2|^{\log\delta} \leq \delta^{\log\epsilon} \right\}.$$

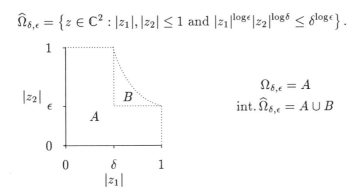

79

M_z on $A^2(\Omega_\delta, \epsilon)$ is a subnormal pair such that $\sigma_T(M_z) = \widehat{\Omega}_{\delta,\epsilon}$ and $\sigma_T(\text{m.m.e.}(M_z)) = \overline{\Omega}_{\delta,\epsilon}$, so that although $\Omega_{\delta,\epsilon}$ is topologically trivial, the Bergman pair for $\Omega_{\delta,\epsilon}$ has spectrum strictly larger than $\overline{\Omega}_{\delta,\epsilon}$. Moreover, $\sigma_{Te}(M_z) \supseteq \partial\widehat{\Omega}_{\delta,\epsilon}$ and M_z is Fredholm of index 1 ([47, Example 5.1]). It is unknown whether the previous containment is actually an equality. The weights for M_z are associated with the coefficients

$$c_\alpha = \pi \left(\frac{\delta^{2\alpha_1+2} + \epsilon^{2\alpha_2+2} - \delta^{2\alpha_1+2}\epsilon^{2\alpha_2+2}}{(\alpha_1 + 1)(\alpha_2 + 1)} \right)^{1/2}.$$

In [45] it was shown that the theory of groupoids ([77], [93]) can be used to study the ideal structure of the C^*-algebra generated by multivariable weighted shifts. As an application, the following result was obtained.

Theorem 8.6. ([45, Theorem 4.2]) *Let $\Omega_{\delta,\epsilon}$ be as above. Then $C^*(M_z)$ is type I if and only if $\frac{\log \delta}{\log \epsilon}$ is a rational number.*

Remark 8.7. $\Omega_{\delta,\epsilon}$ is the first known example of a domain for which the Bergman C^*-algebra is not type I. Previous works indicated that $C^*(M_z)$ might always be type I ([14], [15], [28], [52], [77], [108], [109]).

9. Concluding Remarks and Some Open Problems

Throughout these notes we have tried and emphasized the role of σ_T in the study of commuting n-tuples of Hilbert spaces operators. Although Taylor invertibility of an n-tuple $A \subset \mathcal{L}(\mathcal{H})$ is determined by a single operator ($R_A, \square_A$ or $\widehat{A}$), it is still difficult to calculate σ_T for specific examples. In [39] a first step was taken in making precise the connection between σ_T and σ_H, the latter usually easier to calculate. In case $n = 2$, that connection is completely described by Theorem 6.12 above, but in the general case it remains an open problem.

Problem 9.1. How does σ_H sit inside σ_T? Or, in the language of Section 3, how does M_{σ_H} sit inside M_{σ_T}, where M_σ is calculated with respect to a commutative Banach algebra containing the n-tuple A (for instance, (A))?

σ_T is certainly a preferred joint spectrum, since on one hand it is big enough to carry an analytic functional calculus, and, on the other hand, it is small enough to allow that functional calculus to be rich. From the axiomatic viewpoint, σ_T seems to have both minimal and maximal properties.

Problem 9.2. Let $\tilde{\sigma}$ be a spectral system on $\mathcal{H}$. Assume that $\tilde{\sigma}$ satisfies the projection property and $\tilde{\sigma} \subseteq \hat{\sigma}$ (i.e., $\tilde{\sigma}(A) \subseteq \hat{\sigma}(A)$ for all $A \subseteq \mathcal{L}(\mathcal{H})$). Must $\tilde{\sigma} \subseteq \sigma_T$? In other words, is σ_T maximal among spectral systems contained in $\tilde{\sigma}$ with the projection property? (Problem 9.2 is basically due to W. Żelazko.)

Problem 9.3. Let $\tilde{\sigma}$ be a spectral system on $\mathcal{H}$. Assume that $\tilde{\sigma}$ carries an analytic functional calculus, i.e., for $A \subset \mathcal{L}(\mathcal{H})$ there exists a continuous homomorphism $\Phi_A :$ $A(\tilde{\sigma}(A)) \to \mathcal{L}(\mathcal{H})$ such that $\Phi_A(1) = I, \Phi_A(z_i) = A_i (i = 1, \ldots, n)$ and $\Phi_A(f) \in (A)''$ for every $f \in A(\tilde{\sigma}(A))$. Must $\sigma_T \subseteq \tilde{\sigma}$?

Let $\mathcal{I}_n(\mathcal{H}) := \{A \in \mathcal{L}(\mathcal{H})^{(n)}_{\text{com}} : 0 \notin \sigma_T(A, \mathcal{H})\}$, the collection of invertible n-tuples on $\mathcal{H}$. It is, of course, well-known that $\mathcal{I}_1(\mathcal{H})$ is contractible (Kuiper's Theorem), but even the path-connectedness of $\mathcal{I}_n(\mathcal{H}) (n \geq 2)$ remains an open question. (For partial results, see [35].)

Problem 9.4. [80, D. Voiculescu (p. 600)] Is $\mathcal{I}_n(\mathcal{H})$ path-connected?

It is interesting to observe that $\mathcal{I}_n(\mathcal{H}) (n \geq 2)$ is not obviously locally path-connected; for, although we know that $\mathcal{I}_n(\mathcal{H})$ is open in the norm topology, the line segment joining two *close* n-tuples may contain noncommuting n-tuples! Therefore, one needs more than connectedness in Problem 9.4.

For Fredholm n-tuples on $\mathcal{H}$, one can ask similar questions.

Problem 9.5. Let $A, B \in \mathcal{F}(\mathcal{H}) \cap \mathcal{L}(\mathcal{H})^{(n)}_{\text{com}}$ and assume that $\text{ind}(A) = \text{ind}(B)$. Can A and B be joined by a path in $\mathcal{F}(H)$?

The answer to this question is probably no. In [36] we established that if two Fredholm *doubly* commuting pairs of a certain kind have equal index, then they *can* be path-connected in $\mathcal{F}(\mathcal{H})$. The result actually showed the need for more invariants in the general case; M. Putinar described in [82] some new invariants, which seems to indicate that index alone does not suffice (see also [91]).

Extension theory is relevant to problem 9.5. Let us mention briefly an important step in the proof of Theorem 9 in [36], that is, the fact that on $H^2(\mathsf{T}^2)$, T_z and T_z^* can be joined by a path of Fredholm pairs. (Here $T_z = (U_+ \otimes I, I \otimes U_+)$.) First, one considers M_z on $H^2(S^3)$, the Hardy space for the sphere, and writes down the separate polar decompositions $M_{z_i} = V_i P_i (i = 1, 2)$. One then joins M_z to $V := (V_1, V_2)$ and M_z^* to V^* by showing that $[V_i, P_j] \in K(H^2(S^3))$, and then using the line segment from VP to V (the more general theory of almost commuting Fredholm n-tuples ([36], [82])

81

is needed. Now V is unitarily equivalent to T_z, so that to join T_z to T_z^*, we need to complete the following diagram

$$T_z \longleftrightarrow M_z$$
$$\vdots$$
$$T_z^* \longleftrightarrow M_z^* \,.$$

Here is where the BDF-theory ([18]) enters: M_z and M_z^* give rise to equivalent extensions of S^3. Indeed, ind : $\mathrm{Ext}(S^3) \to \mathbb{Z}$ is a group isomorphism, and $\mathrm{ind}(M_z) = \mathrm{ind}(T_z) = 1 = \mathrm{ind}(T_z^*) = \mathrm{ind}(M_z^*)$. Therefore, there exists a unitary operator U such that $U M_z U^* - M_z^*$ is compact on $H^2(S^3)$. Now join U to I to complete the argument.

Concerning M_z on $A^2(\Omega)$ (or more generally, multiplication operators on the Bergman space), a number of results have recently been obtained for various kinds of domains Ω. There is still, however, much to learn about M_z on general Ω's.

Problem 9.6. Find a complete description of the Koszul complex $K(M_z, A^2(\Omega))$.

The above-mentioned problems represent a small sample of open questions in multiparameter spectral theory. We have stated them here because of their relevance to the topics discussed in these lecture notes. Since we were committed to an elementary level of discussion, we have not had occasion to touch on many important developments that have taken place in the last few years. We shall briefly mention some of them:

(i) Building on S. Frunza's work ([58]), J. Eschmeier has developed a local analytic spectral theory ([53]), later refined by M. Putinar and F.-H. Vasilescu ([90]).

(ii) A theory of decomposable n-tuples exists, due to the work of many authors, most notably E. Albrecht, S. Frunza and F.-H. Vasilescu (see [1] and [114]; see also [40], where a summary of the main results is given).

(iii) M. Putinar has found a sheaf theoretic model for commuting n-tuples, which seems to provide a better generalization of decomposability, and is relevant to the study of Bergman n-tuples ([54], [85], [88]).

(iv) R. Carey and J. D. Pincus have discovered an index theory for n-tuples which extends their work on trace class commutators and makes contact with geometric measure theory ([20], [21]); alternate descriptions have been independently proposed by M. Putinar and by R. Levi ([75], [89]).

(v) The (Taylor) spectral picture of left and right multiplications on Hilbert space has been described in detail in [42], building on previous work of various authors ([22], [36], [55], [56], [61]). (See also [124] and [125] for extensions to

Banach spaces.)

(vi) M. Cowen and R. G. Douglas have extended to several variables their work on the classes $B_n(\Omega)$ ([31], [32], [33]); a functional analytic approach to these classes, which includes several variables as well as infinite dimensional joint kernels has been given in [46].

(vii) The structure of the Toeplitz C^*-algebra $\mathcal{T}$ associated with bounded symmetric domains has been described in terms of Jordan triples by H. Upmeier ([108], [109]); in particular, Upmeier shows that $\mathcal{T}$ is type I.

(viii) A theory of closures of joint similarity orbits, in the spirit of the single variable theory (due to C. Apostol, D. Herrero, and D. Voiculescu ([9]))has been initiated in [44].

(ix) Non-analytic functional calculi have been considered by various authors ([4], [5], [6]; see also [80, F.-H. Vasilescu's problem, p. 601]).

Finally, we would like to say a word about non-commuting n-tuples. Their study is of a different nature, and it involves techniques different from those explained here. For the interested reader, we mention the works in [3], [63], [64], [65], and [107].

Acknowledgements: I am thankful to John B. Conway for his invitation to deliver these talks and for his efforts in organizing the conference and publishing the proceedings. I am also grateful to D. Herrero, Q. Lin, V. Paulsen, and M. Putinar for useful comments and discussions on the topics of these notes.

This work was partially supported by a grant from NSF.

References

1. Albrecht, E., Generalized spectral operators, Proceedings Paderborn Conference on Functional Analysis, North Holland Math. Studies **27** (1977), 259–277.

2. _____, On joint spectra, Studia Math. **64** (1979), 263–271.

3. _____, Several variables spectral theory in the non-commutative case, Banach Center Publications, vol. **8** (Spectral Theory) (1982), 9–30.

4. _____ and S. Frunza, Non-analytic functional calculi in several variables, Manuscripta Math. **18** (1976), 327–336.

5. _____ and M. Neumann, On the continuity of non-analytic functional calculi, J. Operator Theory **5** (1981), 109–117.

6. _____ and F.-H. Vasilescu, Non-analytic local spectral properties in several variables, Czech. Math. J. **24** (1974), 430–443.

7. ———, Semi-Fredholm complexes, Op. Theory: Adv. Appl. **11** (1983), 15–39.

8. Apostol, C., Inner derivations with closed range, Revue Roum. Math. Pures Appl. **21** (1976), 249–265.

9. ——— and L. Fialkow, D. Herrero, and D. Voiculescu, *Approximation of Hilbert Space Operators, II*, Research Notes in Math., vol. **102**, Pitman Books Ltd., Boston–London–Melbourne, 1984.

10. Arens, R., The analytic functional calculus in commutative topological algebras, Pacific J. Math. **11** (1961), 405–429.

11. ——— and A. P. Calderón, Analytic functions of several Banach algebra elements, Annals of Math. **62** (1955), 204–216.

12. Axler, S., J. Conway, and G. McDonald, Toeplitz operators on Bergman spaces, Canad. J. Math. **34** (1982), 466–483.

13. Ball, J., Rota's theorem for general functional Hilbert spaces, Proc. Amer. Math. Soc. **64** (1977), 55-61.

14. Berger, C. and L. Coburn, Wiener-Hopf operators on U_2, Integral Eq. Op. Theory **2** (1979), 139–173.

15. ———, ———, and A. Koranyi, Opérateurs de Wiener-Hopf sur les spheres de Lie, C. R. Acad. Sci. Paris **290** (1980),Série A, 989–991.

16. Bingen, F., J. Tits, and L. Waelbroeck, Séminaire sur les Algèbres de Banach, Université Libre de Bruxelles, Inst. de Mathematique (1962–1963).

17. Boutet de Monvel, L., On the index of Toeplitz operators of several complex variables, Inventiones Math. **50** (1979), 249–272.

18. Brown, L. G., R. G. Douglas, and P. A. Fillmore, Extensions of C*-algebras and K-homology, Annals of Math. **105** (1977), 265–324.

19. Bunce, J., The joint spectrum of commuting non-normal operators, Proc. Amer. Math. Soc. **29** (1971), 499–505.

20. Carey, R. and J. Pincus, Operator theory and boundaries of complex curves, preprint.

21. ———, Principal currents, Integral Eq. Op. Theory **8** (1985), 614–640.

22. Carrillo, A. and C. Hernández, Spectra of constructs of a system of operators, Proc. Amer. Math. Soc. **91** (1984), 426–432.

23. Ceausescu, Z. and F.-H. Vasilescu, Tensor products and the joint spectrum in Hilbert spaces, Proc. Amer. Math. Soc. **72** (1978), 505–508.

24. Chō, M. and M. Takaguchi, Boundary of Taylor's joint spectrum for two commuting operators, Sc. Rep. Hirosaki Univ. **28** (1981), no. 1, 1–4.

25. _____, Identity of Taylor's joint spectrum and Dash's joint spectrum, Studia Math. **70** (1982), 225–229.

26. Choi, M.-D. and C. Davis, The spectral mapping theorem for joint approximate spectrum, Bull. Amer. Math. Soc. **80** (1974), 317–321.

27. Clark, D. N., On commuting contractions, J. Math. Anal. Appl. **32** (1970), 590–596.

28. Coburn, L. A., Singular integral operators and Toeplitz operators on odd spheres, Indiana Univ. Math. J. **23** (1973), 433–439.

29. _____ and M. Schechter, Joint spectra and interpolation of operators, J. Funct. Anal. **2** (1968), 226–237.

30. Conway, J. *Subnormal Operators*, Research Notes in Math., vol. **51**, Pitman Books Ltd., Boston–London–Melbourne, 1981.

31. Cowen, M. and R. G. Douglas, Complex geometry and operator theory, Acta Math. **141** (1978), 187–261.

32. _____, Operators possessing an open set of eigenvalues, Proc. Féjer-Riesz Conference, Budapest, 1980.

33. _____, Equivalence of connections, preprint.

34. Curto, R. E., Fredholm and invertible tuples of bounded linear operators, Ph.D. Dissertation, S.U.N.Y. at Stony Brook, New York, 1978.

35. _____, On the connectedness of invertible n-tuples, Indiana Univ. Math. J. **29** (1980), 393–406.

36. _____, Fredholm and invertible n-tuples of operators. The deformation problem, Trans. Amer. Mth. Soc. **266** (1981), 129–159.

37. _____, Spectral permanence for joint spectra, Trans. Amer. Math. Soc. **270** (1982), 659–665.

38. _____, Spectral inclusion for doubly commuting subnormal n-tuples, Proc. Amer. Math. Soc. **83** (1981), 730–734.

39. _____, Connections between Harte and Taylor spectra, Revue Roum. Math. Pures Appl. **31** (1986), 203–215.

40. _____, Book Review: Analytic Functional Calculus and Spectral Decompositions, by. F.-H. Vasilescu (reference [116] below), Bull. Amer. Math. Soc. **14** (1986), 136–145.

41. _____ and A. T. Dash, Browder spectral systems, Proc. Amer. Math. Soc., to appear.

42. _____ and L. Fialkow, The spectral picture of (L_A, R_B), J. Funct. Anal. **71** (1987), 371–392.

43. _____ and L. Fialkow, Elementary operators with H^∞-symbols, Int. Eq. Op. Theory **10** (1987), 707–720.

44. _____ and D. A. Herrero, On closures of joint similarity orbits, Integral Eq. Op. Theory **8** (1985), 489–556.

45. _____ and P. S. Muhly, C^*-algebras of multiplication operators on Bergman spaces, J. Funct. Anal. **64** (1985), 315–329.

46. _____ and N. Salinas, Generalized Bergman kernels and the Cowen-Douglas theory, Amer. J. Math. **106** (1984), 447–488.

47. _____ and N. Salinas, Spectral properties of cyclic subnormal m-tuples, Amer. J. Math. **107** (1985), 113–138.

48. Dash, A. T., On a conjecture concerning joint spectra, J. Funct. Anal. **6** (1970), 165–171.

49. _____, Joint spectra, Studia Math. **45** (1973), 225–237.

50. _____, Joint essential spectra, Pacific J. Math. **64** (1976), 119–128.

51. Douglas, R. G., On majorization, factorization and range inclusion of operators on Hilbert space, Proc. Amer. Math. Soc. **17** (1966), 413–415.

52. _____ and R. Howe, On the C^*-algebra of Toeplitz operators on the quarter plane, Trans. Amer. Math. Soc. **158** (1971), 203–217.

53. Eschmeier, J., Local properties of Taylor's analytic functional calculus, Invent. Math. **68** (1982), 103–116.

54. _____ and M. Putinar, Spectral theory and sheaf theory III, J. reine angew. Math. **354** (1984), 150–163.

55. Fialkow, L. A., Spectral properties of elementary operators II, Trans. Amer. Math. Soc. **290** (1985), 415–429.

56. _____, The index of an elementary operator, Indiana Univ. Math. J. **35** (1986), 73–102.

57. Foias, C. and W. Mlak, The extended spectrum of completely non-unitary contractions and the spectral mapping theorem, Studia Math. **26** (1966), 239–245.

58. Frunza, S., On the localization of the spectrum for systems of operators, Proc. Amer. Math. Soc. **69** (1978), 233–239.

59. Gamelin, T., *Uniform Algebras*, Prentice-Hall, New Jersey, 1969.

60. Grauert, H. and K. Fritzsche, *Several Complex Variables*, Springer-Verlag, Berlin, 1976.

61. Grosu, C. and F.-H. Vasilescu, The Künneth formula for Hilbert complexes, Integral Eq. Op. Theory **5** (1982), 1-17.

62. Gunning, R. and H. Rossi, *Analytic Functions of Several Complex Variables*, Prentice-Hall, Englewood Cliffs, N.J., 1965.

63. Harte, R. E., Spectral mapping theorems, Proc. Royal Irish Acad. **72A** (1972), 89–107.

64. _____, Tensor products, multiplication operators and the spectral mapping theorem, Proc. Royal Irish Acad. **73A** (1973), 285–302.

65. _____, The spectral mapping theorem for quasicommuting systems, Proc. Royal Irish Acad. **73A** (1973), 7–18.

66. _____, Invertibility, singularity, and Joseph L. Taylor, Proc. Royal Irish Acad. **81A** (1981), 71–79.

67. Hastings, W., Commuting subnormal operators simultaneously quasisimilar to unilateral shifts, Illinois J. Math. **22** (1978), 506–619.

68. Hernández Garciadiego, C., A note on the spectral mapping theorem, Studia Math. **83** (1985).

69. Hörmander, L., *An Introduction to Complex Analysis in Several Variables*, North-Holland, Amsterdam, 1973.

70. Janas, J., Toeplitz operators related to certain domains in $\mathbb{C}^n$, Studia Math. **54** (1975), 73–79.

71. _____, Spectral inclusion theorems for commuting subnormal pair, Annales Soc. Math. Pol.. Series I: Comment. Math. **23** (1983), 219–229.

72. Jewell, N. and A. Lubin, Commuting weighted shifts and analytic function theory in several variables, J. Operator Th., 207–223.

73. Kaup, L. and B. Kaup, Holomorphic Functions of Several Variables, de Gruyter Studies in Mathematics, **3**, Berlin, 1983.

74. Krantz, S., *Function Theory of Several Complex Variables*, J. Wiley & Sons, New York, 1982.

75. Levi, R., Cohomological invariants for essentially commuting systems of operators (Russian), Funct. Anal. Appl. **17** (1983), 79–80; English translation by Plenum Press, New York, Funct. Anal. Appl. **17** (1984), 229–230.

76. Lubin, A., Spectral inclusion and c.n.e., Canadian J. Math. **34** (1982), 883–887.

77. Muhly, P. S. and J. Renault, C^*-algebras of multivariable Wiener-Hopf operators, Trans. Amer. Math. Soc. **274** (1982), 1–44.

78. Paulsen, V., Completely bounded homorphisms of operator algebras, Proc. Amer. Math. Soc. **92** (1984), 225–228.

79. ———, private communication.

80. Problems, Banach Center Publ., vol. **8** (Spectral Theory) (1982), 597–603.

81. Problems, Sesquicentennial Seminar in Operator Theory, Indiana University, November 1969.

82. Putinar, M., Some invariants for semi-Fredholm systems of essentially commuting operators, J. Operator Theory **8** (1982), 65–90.

83. ———, The superposition property for Taylor's functional calculus, J. Operator Theory **7** (1982), 149–155.

84. ———, Uniquenesss of Taylor's functional calculus, Proc. Amer. Math. Soc. **89** (1983), 647–650.

85. ———, Spectral theory and sheaf theory, I, Operator Theory: Adv. Appl. **11** (1983), 283–297.

86. ———, Spectral inclusion for subnormal n-tuples, Proc. Amer. Math. Soc. **90** (1984), 405–406.

87. ———, Functional calculus and the Gelfand transformation, Studia Math. **79** (1984), 83–86.

88. ———, Spectral theory and sheaf theory, II, Math. Z. **192** (1986), 473–490.

89. ———, Base change and the Fredholm index, Integral Eq. Op. Theory **8** (1985), 674–692.

90. ———, Spectral theory of Stein algebra representations (in Romanian): I, St. Cerc. Mat. **36** (1984), 193–220; II, *ibid.* **36** (1984), 293–310; III, *ibid.* **36** (1984), 387–408.

91. ——— and F.-H. Vasilescu, Continuous and analytic invariants for deformations of Fredholm complexes, J. Operator Theory **9** (1983), 3–26.

92. Raeburn, I., On Toeplitz operators associated with strongly pseudoconvex domains, Studia Math. **63** (1978), 253–258.

93. Renault, J., A groupoid approach to C^*-algebras, in Lecture Notes in Math., vol. **793**, Springer-Verlag, New York, 1980.

94. Rudol, K., On spectral mapping theorems, J. Math. Anal. and Appl. **97** (1983), 131–139.

95. Salinas, N., Quasitriangular extensions of C^*-algebras and problems on joint quasitriangularity of operators, J. Operator Theory **10** (1983), 167–205.

96. _____, The $\bar{\partial}$-formalism and the C^*-algebra of the Bergman n-tuple, preprint.

97. _____, Toeplitz operators on bounded pseudoconvex domains, preprint.

98. Sebastião e Silva, J., Sur le calcul symbolique á une ou plusieurs variables pour une algèbre localement convexe, Rendiconti Acad. Naz. Lincei **8.30** (1961), 167–172.

99. Segal, G., Fredholm complexes, Quart J. Math. Oxford Ser. 2 **21** (1970), 385–402.

100. Shields, A., Weighted shift operators and analytic function theory, Math. Surveys **13** (1974), 49–128.

101. Shilov, G., On the decomposition of a normed ring into a direct sum of ideals, Amer. Math. Soc. Transl. **1** (1955), 37–48.

102. Slodkowski, Z., An infinite family of joint spectra, Studia Math. **61** (1977), 239–255.

103. _____ and W. Żelazko, On joint spectra of commuting families of operators, Studia Math. **50** (1974), 127–148.

104. Taylor, J. L., A joint spectrum for several commuting operators, J. Funct. Anal. **6** (1970), 172–191.

105. _____, The analytic functional calculus for several commuting operators, Acta Math. **125** (1970), 1–38.

106. _____, A general framework for a multi-operator functional calculus, Adv. in Math. **9** (1972), 183–252.

107. _____, Functions of several noncommuting variables, Bull. Amer. Math. Soc. **79** (1973), 1–34.

108. Upmeier, H., Toeplitz operators on bounded symmetric domains, Trans. Amer. Math. Soc. **280** (1983), 221–237.

109. _____, Toeplitz C^*-algebras on bounded symmetric domains, Annals of Math. **119** (1984), 549–576.

110. Vasilescu, F.-H., A characterization of the joint spectrum in Hilbert spaces, Rev. Roum. Math. Pures Appl. **22** (1977), 1003–1009.

111. _____, On pairs of commuting operators, Studia Math. **62** (1978), 203–207.

112. _____, A Martinelli type formula for the analytic functional calculus, Rev. Roum. Math. Pures Appl. **23** (1978), 1587–1605.

113. ———, Stability of the index of a complex of Banach spaces, J. Op. Theory **2** (1979), 247–275.

114. ———, The stability of the Euler characteristic for Hilbert complexes, Math. Ann. **248** (1980), 109–116.

115. ———, A multidimensional spectral theory in C^*-algebras, Banach Center Publ., vol. **8** (Spectral Theory) (1982), 471–491.

116. ———, *Analytic Functional Calculus and Spectral Decompositions*, Math. and Its Appl., East European Series, vol. **1**, D. Reidel Publishing Co., Dordrecht, Holland, 1982.

117. Venugopalkrisna, U., Fredholm operators associated with strongly pseudoconvex domains in C^n, J. Funct. Anal. **9** (1972), 349–373.

118. Voiculescu, D., A noncommutative Weyl-von Neumann theorem, Rev. Roum. Math. Pures Appl. **21** (1976), 97–113.

119. Waelbroeck, L., Le calcul symbolique dans les algèbres commutatives, J. Math. Pures et Appl. (9) **33** (1954), 147–186.

120. ———, The holomorphic functional calculus as an operational calculus, Banach Center Publications, vol. **8** (Spectral Theory) (1982), 513–552.

121. Yabuta, K., A remark to a paper of Janas, Toeplitz operators related to certain domains in C^n, Studia Math. **62** (1978), 73–74.

122. Zame, W., Existence and uniqueness of functional calculus homomorphisms, Bull. Amer. Math. Soc. **82** (1976), 123–125.

123. Żelazko, W., An axiomatic approach to joint spectra, I, Studia Math. **64** (1979), 250–261.

124. Eschmeier, J., Tensor products and elementary operators, preprint.

125. Fainstein, A. S., Fredholmness and index of a function of left and right multiplication operators, Dokl. Akad. Nauk Azerbaidzhan SSR, Dokl. **40** (11) (1984), 3–7 (Russian).

Raúl E. Curto
Department of Mathematics
University of Iowa
Iowa City, Iowa 52242
U.S.A.

Another Look at Real-Valued Index Theory

by

Ronald G. Douglas

0. Introduction

As the sixties ended, index theory was being stretched in all directions by researchers around the world. The index theorem of Atiyah and Singer [3], obtained near the start of the decade, had convinced many that interesting topological information could be obtained from the index of natural operators on manifolds. This resulted in a dramatic increase in the level of topological sophistication in index theory (cf. [34]). About the same time another look was taken at Toeplitz operators on the unit circle using methods from index theory [24]. Here the topology was elementary but the level of analytical and functional analytical sophistication was high. One consequence of the juxtaposition of these developments for index theory was the collaboration of Brown-Douglas-Fillmore on extensions of operator algebras [9], [10] which combined with Atiyah's generalized elliptic operators [1] and Kasparov's work on the Novikov conjecture using K-homology [32] to lead to the non-commutative topology and geometry program (cf. [18]) presently underway in operator algebras.

However, we have already discussed these developments in [25]. In this note we want to look at some other work done in the early seventies on real-valued index theory [15] and relate it to current research. We will slightly recast this work in the context of the study of Toeplitz operators on the torus where the topology and the geometry are relatively straightforward and can be easily visualized. Most of what we will describe can be generalized but that is not our purpose here. We want to concentrate on the interaction of index theory with the topology and geometry in this very simple situation.

My thoughts on these matters have been strongly influenced by my many collaborators on these topics but much of my understanding of this view toward real-valued index theory was gained in work currently underway with S. Hurder and J. Kaminker [27], an aspect of which we will describe at the end. In particular, most of the proofs omitted here will appear in [28], as will a more careful treatment of all of the technicalities.

We begin our exposition by recalling the salient features of the familar case of Toeplitz operators on the unit circle.

1. Toeplitz Operators on the Unit Circle

Let T denote the unit circle in the complex plane $\mathbb{C}$ and let $L^2(T)$ denote the Lebesgue space defined relative to normalized arc-length measure on T. A natural orthonormal basis for $L^2(T)$ is given by the functions $\{\exp in\theta\}_{n\epsilon Z}$. If $l^2(Z)$ denotes the Hilbert space of complex sequences indexed by the integers Z and l_n denotes the sequence in $l^2(Z)$ with a 1 in the n-th entry and a 0 elsewhere, then the correspondence

$$\exp in\theta \longleftrightarrow l_n$$

sets up an isomorphism between the Hilbert spaces $L^2(T)$ and $l^2(Z)$.

If we let $H^2(T)$ denote the subspace spanned by $\{\exp in\theta : n \in Z^+\}$, then $H^2(T)$ is the Hardy space for the unit disk and the orthogonal projection P of $L^2(T)$ onto $H^2(T)$ is the Szëgo projector. For ϕ a function in the algebra $C(T)$ of continuous complex-valued functions defined on T, we let M_ϕ denote the operator on $L^2(T)$ defined by $M_\phi f = \phi f$. Then the Toeplitz operator T_ϕ is defined by $T_\phi f = P(M_\phi f)$ and T_ϕ is in the algebra $\mathcal{L}(H^2(T))$ of bounded linear operators defined on $H^2(T)$.

Although there are many interesting problems concerning Toeplitz operators that one can consider, we concentrate here on that of the invertibility or the *near-invertibility* of such operators.

It can be readily shown that a necessary condition for the operator T_ϕ to be invertible is for its symbol ϕ to be invertible. If ϕ is an invertible continuous function, then there exist an integer k and a continuous function ψ on T such that $\phi(e^{i\theta}) = \exp ik\theta \exp \psi(e^{i\theta})$ for $e^{i\theta}$ in T. The integer k is the winding number of ϕ with respect to the origin and is the obstruction to defining a continuous logarithm for ϕ. If ψ is the trigonometric polynomial

$$\psi(e^{i\theta}) = \sum_{n\leq N} a_n \exp in\theta,$$

and we write

$$\psi_+(e^{i\theta}) = \sum_{n\geq 0} a_n \exp in\theta$$

and

$$\psi_-(e^{i\theta}) = \sum_{n<0} a_n \exp in\theta,$$

then

$$T_{\exp \psi} = T_{\exp \psi_-} T_{\exp \psi_+}$$

and it is straightforward to show that

$$T_{\exp -\psi_+} T_{\exp -\psi_-} = T_{\exp \psi}^{-1}.$$

An easy approximation argument extends this to show that $T_{\exp \psi}$ is always invertible. Further, since $T_{\exp(ik\theta)}1 = P(\exp ik\theta) = 0$ for $k < 0$, and $T_{\exp(ik\theta)}{}^* = T_{\exp(-ik\theta)}$, we see that $T_{\exp \psi}$ is invertible precisely when $k < 0$. Finally, since

$$T_\phi = \begin{cases} T_{\exp \psi} T_{\exp \ ik\theta} & \text{for } k \geq 0; \\ T_{\exp \ ik\theta} T_{\exp \psi} & \text{for } k < 0 \end{cases}$$

we have proved

Theorem 1. *If ϕ is an invertible continuous complex-valued function on T, then T_ϕ is invertible if and only if ϕ has a continuous logarithm.*

This statement says nothing about the nature of T_ϕ for the case of an invertible ϕ with non-zero winding number. Of course, we know that such an operator is *nearly-invertible* in the sense that it is Fredholm, but before we can discuss this we need to recall some standard ideas from index theory (cf. [22]).

For $\mathcal{H}$ a complex separable Hilbert space, we let $\mathcal{L}(\mathcal{H})$ denote the algebra of bounded linear operators defined on $\mathcal{H}$. An operator is said to be of finite rank if its range is contained in a finite dimensional subspace. The collection of finite rank operators forms a two-sided ideal in $\mathcal{L}(\mathcal{H})$ and the norm closure of this ideal is the unique *closed* two-sided ideal $\mathcal{K}(\mathcal{H})$ of compact operators on $\mathcal{H}$. The quotient algebra $\mathcal{Q}(\mathcal{H})$ is called the Calkin algebra and the relationship of these algebras can be expressed in terms of the short exact sequence

$$0 \to \mathcal{K}(\mathcal{H}) \to \mathcal{L}(\mathcal{H}) \xrightarrow{\pi} \mathcal{Q}(\mathcal{H}) \to 0,$$

where π denotes the natural quotient map.

Definition 1. An operator T in $\mathcal{L}(\mathcal{H})$ is said to be a Fredholm operator if $\pi(T)$ is invertible in $\mathcal{Q}(\mathcal{H})$.

Let $\mathcal{T}$ denote the C^*-span of $\{T_\phi : \phi \in C(T)\}$, that is, $\mathcal{T}$ is the smallest norm closed, self-adjoint subalgebra of $\mathcal{L}(H^2(T))$ containing $\{T_\phi : \phi \in C(T)\}$. The C^*-algebra $\mathcal{T}$ can be shown to contain $\mathcal{K}(H^2(T))$ and the correspondence $\sigma(T_\phi) = \phi$ to extend to a *-homomorphism, yielding the following result due to Gohberg [29] and Coburn [12].

Theorem 2. *The sequence*

$$0 \to \mathcal{K}(H^2(T)) \to \mathcal{T} \overset{\sigma}{\to} C(T) \to 0$$

is short exact.

If we consider the diagram

$$
\begin{array}{ccccccccc}
0 & \longrightarrow & \mathcal{K}(H^2(T)) & \longrightarrow & \mathcal{T} & \overset{\sigma}{\longrightarrow} & C(T) & \longrightarrow & 0 \\
& & \| & & \cap\,| & & \tau\,\vdots\downarrow & & \\
0 & \longrightarrow & \mathcal{K}(H^2(T)) & \longrightarrow & \mathcal{L}(H^2(T)) & \overset{\pi}{\longrightarrow} & (H^2(T)) & \longrightarrow & 0
\end{array}
$$

then an easy argument enables one to define the *-monomorphism τ. Finally, for T in $\mathcal{T}, \pi(T)$ is invertible in $\mathcal{Q}(H^2(T))$ if and only if it is invertible in the range of τ, since the latter is a C^*-subalgebra of $\mathcal{Q}(H^2(T))$. Therefore, we have proved that

Corollary 1. *The operator T_ϕ is Fredholm if and only if ϕ is invertible in $C(T)$.*

Thus the Toeplitz operator T_ϕ is *nearly-invertible* for an invertible function ϕ in the sense that it is invertible modulo the ideal of compact operators.

In addition to deciding when a Toeplitz operator is Fredholm, we can calculate its index. Recall that by a theorem of Atkinson (cf. [22]), we know that an operator T on $\mathcal{H}$ is Fredholm if and only if it has closed range and the dimensions of both the null spaces of T and T^* are finite. (Actually this was the original definition of a Fredholm operator.)

Definition 2. For T a Fredholm operator on $\mathcal{H}$, we define

$$\text{index } T = \dim \text{ null } T - \dim \text{ null } T^*.$$

It can be shown that the index function is continuous and determines the connected components of the space of Fredholm operators in the relative norm topology (cf [22]).

Since the null space of $T_{\exp(ik\theta)}$ can be seen to be the span of $\{\exp im\theta : 0 \leq m < -k\}$ for $k < 0$ and $\{0\}$ otherwise, we have that

$$\text{index } T_{\exp(ik\theta)} = -k.$$

This enables us to prove using the simple homotopy argument of deforming ϕ to $\exp ik\theta$, the following result of Gohberg [29].

Theorem 2. *If ϕ is an invertible complex-valued continuous function on T, then T_ϕ is Fredholm and*

$$\text{index } T_\phi = - \text{ winding number of } \phi.$$

94

2. Toeplitz Operators on the Torus

In this exposition we want to consider Toeplitz operators on the two torus T^2. There are several possibilities for this with the most obvious being obtained by considering the Toeplitz operators defined on the Hardy space of the bidisk D^2 viewed as a subset of $\mathbf{C}^2$. Since D^2 has T^2 as its distinguished boundary, the symbol of such operators are functions on T^2. This is the so-called *quarter-plane* case considered by Howe and myself [26]. That is not our interest here. We shall concentrate instead on the family of *half-plane* cases.

The Lebesgue space $L^2(T^2)$ again has a natural orthonormal basis

$$\{\exp i(m\theta_1 + n\theta_2)\}_{(m,n)\in Z^2}$$

as does the Hilbert space $l^2(Z^2)$ of square summable complex sequences indexed by Z^2. If $l_{(m,n)}$ denotes the sequence having a 1 as its (m,n)-entry and 0 elsewhere, then the correspondence

$$\exp i(m\theta_1 + n\theta_2) \longleftrightarrow l_{(m,n)}$$

establishes an isomorphism between the two Hibert spaces $L^2(T^2)$ and $l^2(Z^2)$. For α a real number or $\pm\infty$, let H_α^2 be the span of

$$\{\exp i(m\theta_1 + n\theta_2) : -\alpha m + n \geq 0\}.$$

(The cases $\alpha = \pm\infty$ correspond to (m,n) being in the right and left half planes, respectively.) One can see that under the above isomorphism H_α^2 corresponds to the sequences in $l^2(Z^2)$ which are supported in the half-plane $\{(m,n) \in Z^2 : -\alpha m + n \geq 0\}$, and hence the terminology *half-plane case*. We let P_α denote the orthogonal projection of $L^2(T^2)$ onto H_α^2 and for ϕ in $C(T^2)$ we define the Toeplitz operator T_ϕ^α on H_α^2 such that

$$T_\phi^\alpha f = P_\alpha M_\phi f \text{ for } f \text{ in } H_\alpha^2.$$

We want to investigate the invertibility of T_ϕ^α. Again, it is not difficult to show the necessity of the invertibility of φ. If $\phi = \exp \psi$, and hence has a continuous logarithm, then we can proceed as in the case of the circle. Assume that ψ is a trigonometric polynomial and write $\psi = \psi_+ + \psi_-$, where ψ_+ and ψ_- are polynomials supported in the two half planes. Again, we see that

$$T_\phi^\alpha = T_{\exp \psi_-}^\alpha T_{\exp\psi_+}^\alpha$$

95

and we obtain the inverse

$$T_{\exp\psi_+}^{\alpha} T_{\exp\psi_-}^{\alpha} = T_{\psi}^{\alpha-1}.$$

Finally, an easy approximation argument shows

Theorem 3. *If $\phi = \exp\psi$ for ψ in $C(T^2)$, then T_{ϕ}^{α} is invertible.*

But what about the converse and what happens when ϕ has no continuous logarithm? Let us consider the case $\alpha = 0$ or that of the upper half plane $\{(m,n) : n \geq 0\}$. In this case we have

$$H_0^2 = L^2(T_{\theta_1}) \otimes H^2(D_{\theta_2}) = L^2(T_{\theta_1}, H^2(D_{\theta_2}))$$

and where the subscript thetas determine the two disks and circles

$$T_{\phi}^0 = T_{\phi}(e^{i\theta_1}, \cdot) \text{ is in } C(T_{\theta_1}, \mathcal{L}(H^2(D_{\theta_2})))$$

for ϕ in $C(T^2)$. But it is clear that such an operator-valued function is invertible if and only if it is invertible pointwise and hence we have

Theorem 4. *For an invertible function ϕ in $C(T^2)$, the operator T_{ϕ} is invertible if and only if the winding number of the function defined by $e^{i\theta_2} \to \phi(e^{i\theta_1}, e^{i\theta_2})$ is zero for some $e^{i\theta_1}$ in T or, equivalently, for all $e^{i\theta_1}$ in T.*

To avoid the circumlocution necessary to describe such winding numbers, let us introduce a definition and some notation. A non-vanishing continuous function ϕ in T^2 determines an element $[\phi]$ of the first Čech cohomology group $H^1(T^2, Z)$. The latter is isomorphic to Z^2 and under this isomorphism $[\phi] = m \oplus n$, where m is the winding number of the function defined by

$$e^{i\theta_1} \longrightarrow \phi(e^{i\theta_1}, e^{i\theta_2}) \text{ for } e^{i\theta_1} \text{ in } T$$

and n is the winding number of the function defined by

$$e^{i\theta_2} \longrightarrow \phi(e^{i\theta_1}, e^{i\theta_2}) \text{ for } e^{i\theta_2} \text{ in } T.$$

To obtain the appropriate *winding number* for the half-plane case determined by α, we introduce the homomorphism

$$i_{\alpha} : H^1(T^2, Z) \longrightarrow R$$

defined by $i_{\alpha}(m \oplus n) = \alpha m - n$ for α in R and

$$i_{\pm\infty}(m \oplus n) = \pm m \text{ for } \alpha = \pm\infty.$$

If α is in $Q^* = Q \cup \{\pm\infty\}$, then the same analysis as before, where T^2 is fibered into circles of slope α, yields the following analogous result.

96

Theorem 5. *For α in Q^* and ϕ a non-vanishing continuous function on T^2, the operator T_ϕ^α is invertible if and only if $i_\alpha[\phi] = 0$.*

Although this solves the invertibility problem for α in Q^*, it says nothing about any possible *near-invertibility* of T_ϕ^α in case $i_\alpha[\phi] \neq 0$. However, before considering this problem, we want to consider the case of irrational α.

For a non-vanishing continuous function ϕ on T^2, it is always possible to find integers a and b and a continuous function ψ on T^2 such that

$$\phi(e^{i\theta_1}, e^{i\theta_2}) = e^{i(a\theta_1 + b\theta_2)} \, \exp \, \psi(e^{i\theta_1}, e^{i\theta_2})$$

for $(e^{i\theta_1}, e^{i\theta_2})$ in T^2. In fact, this holds for $a \oplus b = [\phi]$. If $a - \alpha b \geq 0$, then

$$T_\phi^\alpha = T_{\exp\psi}^\alpha T_{\exp i(a\theta_1 + b\theta_2)}^\alpha$$

or

$$T_\phi^\alpha = T_{\exp i(a\theta_1 + b\theta_2)}^\alpha T_{\exp\psi}^\alpha$$

for $a - \alpha b < 0$. If $a - \alpha b < 0$, then for (c, d) in Z^2 such that $a - \alpha b < c - \alpha d < 0$, we can show that the function $\exp i(c\theta_1 + d\theta_2)$ is in the null space of $T_{\exp i(a\theta_1 + b\theta_2)}^\alpha$. Therefore, $T_{\exp i(a\theta_1 + b\theta_2)}^\alpha$ is not invertible for $a - \alpha b > 0$. A similar argument after taking adjoints completes the proof that $T_{\exp i(a\theta_i + b\theta_2)}^\alpha$ is not invertible for $a - \alpha b > 0$. Since $a - \alpha b = 0$, is equivalent to $a = b = 0$, we can combine the preceding for all α to state

Theorem 6. *For ϕ a non-vanishing continuous function on T^2, the operator T_ϕ^α is invertible if and only if $i_\alpha[\phi] = 0$.*

Again, this does not yield any understanding of the operator T_ϕ^α when $i_\alpha[\phi] \neq 0$. We continue to analyze this situation as in the case of the circle, by letting $\mathcal{T}_\alpha$ denote the C^*-span of $\{T_\phi^\alpha : \phi \in C(T^2)\}$. As before, we want to obtain a *-homomorphism $\sigma_\alpha : \mathcal{T}_\alpha \to C(T^2)$ satisfying $\sigma_\alpha(T_\phi^\alpha) = \phi$. That this is possible was established in [14].

Theorem 7. *If $\mathcal{C}_\alpha$ denotes the commutator ideal in $\mathcal{T}_\alpha$, then σ_α extends to a *-homomorphism such that the sequence*

$$0 \longrightarrow \mathcal{C}_\alpha \longrightarrow \mathcal{T}_\alpha \xrightarrow{\sigma_\alpha} C(T^2) \longrightarrow 0$$

is short exact. Moreover, we have $\mathcal{C}_\alpha \cap \mathcal{K}(H^2) = (0)$.

Since an operator T is Fredholm if and only if it is invertible modulo the ideal of compact operators, Theorem 7 implies that

97

Corollary 2. *An operator T in $\mathcal{T}_\alpha$ is Fredholm if and only if it is invertible.*

The point of this corollary is to observe that ordinary Fredholm theory can not help in understanding any *near-invertibility* of T_ϕ^α for $i_\alpha[\psi] \neq 0$. In addition to this there is a more specific problem that remains. For an invertible operator T in $\mathcal{T}_\alpha$, it follows that $\sigma_\alpha(T)$ is invertible in $C(T^2)$ but it is not obvious that we must have $i_\alpha[\sigma_\alpha(T)] = 0$. In the case of the circle, this is proved using index theory. The solution to this problem and to the *near-invertibility* problem was given in [15] and depended on the generalized index theory of Breuer which we now recall.

3. Index Theory in a von Neumann Algebra

If $\mathcal{W}$ is a von Neumann algebra, then an equivalence relation can be defined on the projections in $\mathcal{W}$ which reflects the internal structure of $\mathcal{W}$ (cf. [31]). Projections E_1 and E_2 in $\mathcal{W}$ which satisfy $E_1 = V^*V$ and $E_2 = VV^*$ for some V in $\mathcal{W}$ are viewed as being equivalent or as having the same dimension relative to $\mathcal{W}$. If $\mathcal{W}$ is a factor, that is, if $\mathcal{W}$ has trival center C, then the possibilities for the range of this dimension function are very limited. If the identity projection is equivalent to a proper projection, that is, if $\mathcal{W}$ is infinite, then the range of this dimension function can be identified with $\{0, 1, \ldots, \infty\}, [0, \infty]$, or $\{0, \infty\}$. In the first case $\mathcal{W}$ is said to be a type I_∞ factor and is isomorphic to $\mathcal{L}(\mathcal{H})$. The second and third cases are the type II_∞ and the type III factors, respectively, and are highly nonunique (cf. [31]).

In [8] Breuer proposed considering the closed two-sided ideal $\mathcal{I}$ generated by the projections having finite dimension relative to $\mathcal{W}$ and then studying the *Fredholm operators* which are defined to be the operators in $\mathcal{W}$ which are invertible modulo $\mathcal{I}$. We restrict attention to the case in which $\mathcal{W}$ is a II_∞ factor. If $\mathcal{Q}_{II}$ denotes the quotient C^*-algebra, then we again have the short exact sequence

$$0 \longrightarrow \mathcal{I}_{II} \longrightarrow \mathcal{W}_{II} \xrightarrow{\pi_{II}} \mathcal{Q}_{II} \longrightarrow 0.$$

Although a Fredholm operator T relative to $\mathcal{W}_{II}$ need not have closed range, it is the case that the projections onto null T and null T^* are finite dimensional relative to $\mathcal{W}_{II}$ and we define

$$\text{index}_{II} \, T = \dim_{II} \text{ null } T - \dim_{II} \text{ null } T^*.$$

Breuer showed [8] that this index determines the connected components in the space of Fredholm operators in $\mathcal{W}_{II}$.

98

Since for our Toeplitz operators T_ϕ^α we have the real-valued topological index $i_\alpha[\phi]$, it is reasonable to seek a real-valued analytical index for T_ϕ^α in the sense of Breuer. However, the algebra $\mathcal{T}_\alpha$ can be shown to act irreducibly on H_α^2 implying that the obvious von Neumann algebra generated by it is $\mathcal{L}(H_\alpha^2)$ which certainly does not give the desired result. The secret is to seek a different representation of $\mathcal{T}_\alpha$. We do this here using the generalization obtained in [23] of the characterization by Coburn [12] of the C^*-algebra generated by an isometry.

Let Γ be a subgroup of R and set $\Gamma^+ = \Gamma \cap R^+$. A one-parameter semigroup of isometries is a homomorphism $\gamma \to V_\gamma$ such that $V_0 = I$, each V_γ is an isometry on $\mathcal{H}$, and $V_{\gamma_1 + \gamma_2} = V_{\gamma_1} V_{\gamma_2}$ for γ_1, γ_2 in Γ^+.

Theorem 8. *If V_γ and $\tilde{V}_\gamma$ are non-unitary one-parameter semigroups of isometries for Γ^+, then the correspondence $V_\gamma \longleftrightarrow \tilde{V}_\gamma$ extends to a $*$-isomorphism between the corresponding C^*-algebras. If $\mathcal{T}_\Gamma$ denotes C^*-span $\{V_\gamma : \gamma \in \Gamma^+\}$ and $\mathcal{C}_\Gamma$ is the commutator ideal in $\mathcal{T}_\Gamma$, then there is a short exact sequence*

$$0 \longrightarrow \mathcal{C}_\Gamma \longrightarrow \mathcal{T}_\Gamma \to C(\widehat{\Gamma}) \longrightarrow 0,$$

where $\widehat{\Gamma}$ denotes the compact dual of the discrete group Γ.

If Γ is a discrete subgroup of R, that is, if Γ is not dense in R, then Γ is isomorphic to Z and we recover Coburn's result for the shift algebra [12].

If $\mathcal{J}$ is an ideal in $\mathcal{C}_\Gamma$, then $\mathcal{J}$ is an ideal in $\mathcal{T}_\Gamma$ and a faithful representation ρ of $\mathcal{T}_\Gamma / \mathcal{J}$ and the correspondence $\gamma \to \rho(V_\gamma + \mathcal{J})$ yield a one-parameter semigroup of isometries. But $V_\gamma \to \rho(V_\gamma + \mathcal{J})$ must extend to an isomorphism of C^*-algebras. This implies that $\mathcal{J} = 0$ and hence we have generalized the fact that $\mathcal{K}(\mathcal{H}) = \mathcal{C}_Z$ is simple.

Corollary 3. *The C^*-algebra $\mathcal{C}_\Gamma$ is simple.*

We shall discuss the C^*-algebra $\mathcal{C}_\Gamma$ more a little later and, in particular, its dependence on Γ. Although it is not difficult to see that $\mathcal{T}_\Gamma$ determines Γ as an abstract group, with a little more reflection one can see that it actually determines Γ as an ordered group. The same is true for $\mathcal{C}_\Gamma$.

If for α in *Irr* we set $\Gamma_\alpha = \{-\alpha m + n : (m, n) \in Z^2\}$, then $\Gamma_\alpha \subseteq R$ and $\mathcal{T}_\alpha = \mathcal{T}_{\Gamma_\alpha}$. Therefore, identifying a one-parameter semigroup of isometries for Γ_α^+ yields a representation of $\mathcal{T}_\alpha$. For example, translation on $L^2(R^+)$ or on $l^2(\Gamma_\alpha^+)$ both yield representations of $\mathcal{T}_\alpha$. The latter is equivalent to that on H_α^2 in view of the isomorphism between H_α^2 and $l^2(\Gamma_\alpha^+)$ as subspaces of $L^2(T^2)$ and $l^2(Z^2)$, respectively.

In [15] a variant of the Murray-von Neumann group-measure space construction was used to obtain the desired representation of $\mathcal{T}_\alpha$. The representation was on the Hilbert space $L^2(R^+) \otimes l^2(\Gamma_\alpha)$. To obtain this we set

$$(V_\gamma f)(x, \lambda) = f(x + \gamma, \lambda + \gamma) \text{ for } \gamma \text{ in } \Gamma_\alpha^+.$$

Clearly, $\gamma \to V_\gamma$ is a one-parameter semigroup of isometries on $L^2(R^+) \otimes l^2(\Gamma_\alpha)$ and we let $\mathcal{W}_\alpha$ denote the W^*-span of $\{V_\gamma : \gamma \in \Gamma_\alpha^+\}$. The von Neumann algebra $\mathcal{W}_\alpha$ can be shown to be a II_∞ factor. If we set $(M_\phi f)(x, \lambda) = \phi(x) f(x, \lambda)$ for ϕ in $L^\infty(R^+)$, then the ideal $\mathcal{I}_\alpha$ generated by the finite operators in $\mathcal{W}_\alpha$ contains the C^*-span of $\{V_\gamma M_\phi : \gamma \in \Gamma_\alpha^+, \phi \in L_C^\infty(R^+)\}$, where $L_C^\infty(R^+)$ denotes the functions in $L^\infty(R^+)$ having compact support. Calculating, we see that $[V_\gamma, V_\gamma^*] = -1_{[0,\gamma)}$ and hence $\mathcal{C}_\alpha$ is contained in $\mathcal{I}_\alpha$. Moreover, the trace Tr_α defined on $\mathcal{W}_\alpha$ acts on these commutators such that

$$Tr_\alpha(V_\gamma M_\phi) = \begin{cases} 0, & \gamma \neq 0 \\ \int_{R^+} \phi & \gamma = 0 \end{cases}$$

and hence $Tr_\alpha([V_\gamma, V_\gamma^*]) = -\gamma$. As a consequence of the first statement we have the existence of the *-homomorphism $\tau_\alpha : C(T^2) \to \mathcal{Q}_\alpha$ which makes the following diagram commute:

$$
\begin{array}{ccccccccc}
0 & \longrightarrow & \mathcal{C}_\alpha & \longrightarrow & \mathcal{T}_\alpha & \overset{\sigma_\alpha}{\longrightarrow} & C(T^2) & \longrightarrow & 0 \\
 & & \| & & \cap| & & \downarrow \tau_\alpha & & \\
0 & \longrightarrow & \mathcal{I}_\alpha & \longrightarrow & \mathcal{W}_\alpha & \longrightarrow & \mathcal{Q}_\alpha & \longrightarrow & 0.
\end{array}
$$

This shows that T in $\mathcal{T}_\alpha$ is Fredholm relative to $\mathcal{W}_\alpha$ if $\sigma_\alpha(T)$ doesn't vanish. To see the converse or equivalently, that $\mathcal{T}_\alpha$ is a monomorphism, note that R acts as automorphisms on $\mathcal{T}_\alpha$ such that it induces the action on $C(T^2)$ of translation along the lines of slope $-1/\alpha$ on T^2 and that this action can be unitarily implemented on $L^2(R^+) \otimes \ell^2(\Gamma_\alpha)$. But this implies that the maximal ideal space of $\mathcal{T}_\alpha/\mathcal{T}_\alpha \cap \mathcal{I}_\alpha$ must be a closed subset of T^2 which is invariant under such translations. Since α is irrational, this subset must be all of T^2. Therefore, $\mathcal{T}_\alpha \cap \mathcal{I}_\alpha = \mathcal{C}_\alpha$ and hence τ_α is a *-monomorphism. Finally, since $T_{\exp i(m\theta_1 + n\theta_2)}$ is an isometry if $am - n > 0$, it follows that

$$
\begin{aligned}
\text{Index}_\alpha T_{\exp i(m\theta_1 + n\theta_2)}^\alpha &= \\
&= Tr_\alpha\left[T_{\exp i(m\theta_1 + n\theta_2)}^\alpha, T_{\exp i(m\theta_1 + n\theta_2)}^{\alpha}{}^* \right] \\
&= -\int_{R^+} 1_{[0,\alpha m - n)} dt = -i_\alpha\left(\left[e^{i(m\theta_1 + n\theta_2)} \right] \right) \\
&= -i_\alpha\left(\left[\sigma_\alpha\left(T_{\exp i(m\theta_1 + n\theta_2)}^\alpha \right) \right] \right).
\end{aligned}
$$

Theorem 9. *An operator T in $\mathcal{T}_\alpha$ is Fredholm relative to $\mathcal{W}_\alpha$ if and only if $\sigma_\alpha(T)$ is non-vanishing in which case*

$$\mathrm{Index}_\alpha\, T = -i_\alpha([\sigma_\alpha(T)]).$$

This completely resolves the problems stated earlier. First, we have exhibited a natural sense in which the operator T_ϕ^α is nearly-invertible for non-vanishing ϕ. Further, if T is an invertible operator in $\mathcal{I}_\alpha$, then

$$0 = \mathrm{Index}_\alpha(T) = -i_\alpha([\sigma_\alpha(T)])$$

which resolves the other problem in exactly the same way as was done for the case of the circle, by the use of index theory.

Before going on, we want to point out that analogous results can be obtained for rational α, where the relevant von Neumann algebra is no longer a factor. Let us consider the case $\alpha = 0$.

In this case, we have $\mathcal{T}_0 = C(T, \mathcal{I})$ and $\mathcal{C}_0 = C(T, \mathcal{K})$. Therefore, the relevant von Neumann algebra is $\mathcal{W}_0 = L^\infty(T, \mathcal{L})$ with $\mathcal{I}_0 = L^\infty(T, \mathcal{K})$. Then T in $\mathcal{T}_0$ is Fredholm relative to $\mathcal{W}_0$ if and only if $\sigma_0(T)$ is non-vanishing. Moreover,

$$\mathrm{Index}_0(T) = -i_0([\sigma_0(T)]),$$

where we define Index_0 to be the center-valued pointwise index on $L^\infty(T, \mathcal{K})$ followed by integration over T. We omit the details. Similar results can be obtained for all α in Q^* and the statement of Theorem 9 then applies to all α with this interpretation.

These are the results that were obtained for these Toeplitz operators in the early seventies. As we indicated earlier, one can generalize this framework to higher dimensional tori or correspondingly, to arbitrary one-parameter semigroups $\Gamma^+ \subseteq R^+$ and ultimately to the Bohr compactification of the reals, and hence to the semigroup R^+ itself. However, despite considerable effort at the time, it was not clear how to generalize these results to a larger class of examples. (The notion of pseudo-differential operator with almost periodic coefficients was developed in this context [16], however.) Such a generalization can now be realized, based on the work of Connes and Skandalis [20]. To that end we must reexamine two important aspects of the preceding discussion. First we must ask how does one arrive, intrinsically, at an analytical index from a short exact sequence. We did it in the preceding discussion using the index theory of Breuer.

However, it should be clear that in more complicated examples, finding a representation of the algebra generated by the Toeplitz operators into an appropriate von Neumann algebra may not be so easy. Secondly, and even more fundamental, is the question of obtaining such a short exact sequence in the first place. These questions will occupy the remainder of this exposition.

4. K-Theory for Operator Algebras

The first question is answered using the K-theory for operator algebras [36], a topic yet to be developed in the early seventies since much of the impetus for its development came from the study of K-homology in [11] and [32]. We shall sketch enough K-theory to make clear our intention but refer the reader to [7] for a comprehensive treatment of this topic and many other related matters.

Let $\mathcal{A}$ be a separable C^*-algebra with unit and $\text{Proj}(\mathcal{A})$ denote the collection of (orthogonal) projections in $\mathcal{A}$. If $\mathcal{A} \otimes M_n(\mathbf{C})$ denotes the C^*-algebra consisting of the $n \times n$ matrices with entries from $\mathcal{A}$, then there is a natural inclusion

$$\text{Proj}(\mathcal{A} \otimes M_n(\mathbf{C})) \hookrightarrow \text{Proj}(\mathcal{A} \otimes M_{n+1}(\mathbf{C})),$$

where

$$\begin{pmatrix} e_{11} & \cdots & e_{1n} \\ \vdots & & \vdots \\ e_{n1} & \cdots & e_{nn} \end{pmatrix} \longrightarrow \begin{pmatrix} e_{11} & \cdots & e_{1n} & 0 \\ \vdots & & \vdots & \vdots \\ e_{n1} & \cdots & e_{nn} & 0 \\ 0 & \cdots & 0 & 0 \end{pmatrix}.$$

If we let $\text{Proj}_\infty(\mathcal{A})$ denote the direct union of the $\text{Proj}(\mathcal{A} \otimes M_n(\mathbf{C}))$, then we can define an equivalence relation on $\text{Proj}_\infty(\mathcal{A})$ such that for e in $\text{Proj}(\mathcal{A} \otimes M_k(\mathbf{C}))$ and f in $\text{Proj}(\mathcal{A} \otimes M_j(\mathbf{C}))$, we have $e \sim f$ if there exists a unitary element u in $\mathcal{A} \otimes M_{k+j+m}(\mathbf{C})$ for some m satisfying

$$u^*(e \oplus 0 \oplus 1)u = 0 \oplus f \oplus 1.$$

We define an addition on $\text{Proj}_\infty(\mathcal{A})$ such that for e in $\text{Proj}(\mathcal{A} \otimes M_k(\mathbf{C}))$ and f in $\text{Proj}(\mathcal{A} \otimes M_m(\mathbf{C}))$, we set

$$e + f = e \oplus f \text{ in } \text{Proj}(\mathcal{A} \otimes M_{k+m}(\mathbf{C})).$$

This addition is actually defined on the equivalence classes $\text{Proj}_\infty(\mathcal{A})/\sim$ making the latter into a commutative semigroup with identity. We let $K_0(\mathcal{A})$ denote the abelian group completion of this commutative semigroup; that is, $K_0(\mathcal{A})$ is the abelian group consisting of formal differences of elements in $\mathrm{Proj}_\infty(\mathcal{A})/\sim$.

There is a second group defined using the unitary elements. For $\mathcal{A}$ a C^*-algebra with unit, let $\mathrm{Unit}(\mathcal{A})$ denote the collection of unitary elements in $\mathcal{A}$. Again, there is a natural embedding

$$\mathrm{Unit}(\mathcal{A} \otimes M_n(\mathbb{C})) \hookrightarrow \mathrm{Unit}(\mathcal{A} \otimes M_{n+1}(\mathbb{C}))$$

where

$$\begin{pmatrix} u_{11} & \cdots & u_{1n} \\ \vdots & & \vdots \\ u_{n1} & \cdots & u_{nn} \end{pmatrix} \longrightarrow \begin{pmatrix} u_{11} & \cdots & u_{1n} & 0 \\ \vdots & & \vdots & \vdots \\ u_{n1} & \cdots & u_{nn} & 0 \\ 0 & \cdots & 0 & 1 \end{pmatrix}.$$

If we let $\mathrm{Unit}_\infty(\mathcal{A})$ denote the directed union, then it naturally forms a topological group. We let $\mathrm{Unit}_\infty^0(\mathcal{A})$ denote the connected component containing the identity and define

$$K_1(\mathcal{A}) = \mathrm{Unit}_\infty(\mathcal{A})/\mathrm{Unit}_\infty^0(\mathcal{A}).$$

The correspondences $\mathcal{A} \longrightarrow K_0(\mathcal{A})$ and $\mathcal{A} \longrightarrow K_1(\mathcal{A})$ are covariant functors which have many important properties. In particular, if $\psi : \mathcal{A} \longrightarrow \mathcal{B}$ is a unital homomorphism between the unital C^*-algebras $\mathcal{A}$ and $\mathcal{B}$, then it is straightforward to define

$$\psi_* : K_0(\mathcal{A}) \longrightarrow K_0(\mathcal{B}) \text{ and } \psi_* : K_1(\mathcal{A}) \longrightarrow K_1(\mathcal{B}).$$

Such maps enable us to extend the definition of K-theory to nonunital C^*-algebras which is necessary for our application.

For $\mathcal{A}$ a C^*-algebra, let $\widehat{\mathcal{A}} = \mathcal{A} \oplus \mathbb{C}$ denote the C^*-algebra to which a unit has been adjoined. Then we have the short exact sequence

$$0 \longrightarrow \mathcal{A} \longrightarrow \widehat{\mathcal{A}} \overset{j}{\longrightarrow} \mathbb{C} \longrightarrow 0$$

and we define

$$K_0(\mathcal{A}) = \ker j^* \subset K_0(\widehat{\mathcal{A}})$$
$$K_1(\mathcal{A}) = \ker j^* = K_1(\widehat{\mathcal{A}}).$$

Although there is no difference btween $K_1(\mathcal{A})$ and $K_1(\widehat{\mathcal{A}})$ since $K_1(\mathbb{C}) = 0, K_0 = (\mathcal{A})$ and $K_0(\widehat{\mathcal{A}})$ differ by the non-canonical direct summand $K_0(\mathbb{C}) = Z$.

Let us state the results of calculating the K-theory for the C^*-algebras we will need. If X is a locally compact metrizable space, and $C_0(X)$ denotes the C^*-algebra of complex-valued continuous functions on X which vanish at infinity, then $K_0(C_0(X)) = K^0(X)$, and $K_1(C_0(X)) = K^1(X)$, where $K^*(X)$ denotes the topological K-theory of X with compact supports. In particular,

$$K_0(C(T)) = Z \text{ and } K_1(C(T)) = Z,$$

where a positive integer n in the first isomorphism corresponds to the trivial bundle of rank n or to the identity matrix in $C(T) \otimes M_n(\mathbb{C})$, and an integer in the second isomorphism represents the winding number of a function in

$$M_n(C(T)) = C(T) \otimes M_n(\mathbb{C}).$$

Further, we have

$$K_0(C(T^2)) = Z \oplus Z \text{ and } K_1(C(T^2)) = Z \oplus Z,$$

while

$$K_0(C(S^2)) = Z \oplus Z \text{ and } K_1(C(S^2)) = 0.$$

In the groups $K_0(C(T^2))$ and $K_0(C(S^2))$, the first integer again corresponds to the identity matrix but the second is related to the Bott bundle on S^2. The calculation $K_1(C(T^2)) = Z \oplus Z$ is the same as that of $H^1(T^2, Z)$ with the two integers being the two winding numbers. For the C^*-algebra $\mathcal{K}$ of compact operators, one can prove (cf. [7]) that $K_0(\mathcal{K}) = Z$, where a finite rank projection F in $\mathcal{K} \otimes M_n(\mathbb{C})$ corresponds to its rank, and that $K_1(\mathcal{K}) = 0$.

Another important property of the K-theory for C^*-algebras is Bott periodicity which yields a six-term periodic exact sequence in K-theory for each short exact sequence of C^*-algebras. More precisely, given a C^*-algebra $\mathcal{A}$ and a closed two-sided ideal $\mathcal{I}$ in $\mathcal{A}$, we have the short exact sequence

$$0 \longrightarrow \mathcal{I} \xrightarrow{i} \mathcal{A} \xrightarrow{p} \mathcal{A}/\mathcal{I} \longrightarrow 0,$$

where i is inclusion and p the quotient map. Then there exist natural connecting maps

$$\partial_0 : K_0(\mathcal{A}/\mathcal{I}) \longrightarrow K_1(\mathcal{I}) \text{ and } \partial_1 : K_1(\mathcal{A}/\mathcal{I}) \longrightarrow K_0(\mathcal{I})$$

which yield the six-term cyclic exact sequence

$$
\begin{array}{ccccc}
K_0(\mathcal{I}) & \xrightarrow{i_*} & K_0(\mathcal{A}) & \xrightarrow{p_*} & K_0(\mathcal{A}/\mathcal{I}) \\
\partial_1 \uparrow & & & & \downarrow \partial_0 \\
K_1(\mathcal{A}/\mathcal{I}) & \xleftarrow{p_*} & K_1(\mathcal{A}) & \xleftarrow{i_*} & K_1(\mathcal{I})
\end{array}
$$

This sequence is closely related to index theory, especially the connecting map ∂_1, which can be seen as follows.

Given u in $\text{Unit}((\mathcal{A}/\mathcal{I}) \otimes M_k(\mathbf{C}))$, the inverse of $[u]$ in $K_1(\mathcal{A}/\mathcal{I})$ is given by $[u^*]$ and therefore, $[u \oplus u^*]$ must represent the zero element in $K_1(\mathcal{A}/\mathcal{I})$. Actually, one can show that $u \oplus u^*$ lifts to a unitary w in $\mathcal{A} \otimes M_{2k}(\mathbf{C})$ and that

$$\partial_1[u] = [w^*(1_k \oplus 0_k)w] - [1_k \oplus 0_k] \text{ in } K_0(\mathcal{I}),$$

where 0_k and 1_k denote the zero and identity matrix in $\mathcal{A} \otimes M_k(\mathbf{C})$, respectively.

If we apply this to the Toeplitz sequence

$$0 \longrightarrow \mathcal{K} \xrightarrow{i} \mathcal{T} \xrightarrow{\sigma} C(T) \longrightarrow 0,$$

we obtain

$$
\begin{array}{ccccc}
K_0(\mathcal{K}) & \xrightarrow{i_*} & K_0(\mathcal{T}) & \xrightarrow{\sigma_*} & K_0(C(T)) \\
\partial_1 \uparrow & & & & \downarrow \partial_0 \\
K_1(C(T)) & \xleftarrow{\sigma_*} & K_1(\mathcal{T}) & \xleftarrow{i_*} & K_1(\mathcal{K})
\end{array}
$$

In this case the map ∂_0 is without interest being the zero map since $K_1(\mathcal{K}) = 0$, but we want to examine more carefully the map $\partial_1 : K_1(C(T)) \to K_0(\mathcal{K})$ which is a homomorphism from Z to Z.

For $\exp ik\theta$ in $C(T)$ we have that $[\exp ik\theta]$ in $K_1(C(T))$ corresponds to k in Z or equivalently, the unitary $\sigma(T_{\exp ik\theta})$ in $\mathcal{T}/\mathcal{K}$ corresponds to k. Moreover, if we decompose $M_{\exp ik\theta}$ on $L^2(T) = H^2(D) \oplus H^2(D)^{\perp}$, then

$$M_{\exp ik\theta} = \begin{pmatrix} T_{\exp ik\theta} & * \\ * & T_{\exp ik\theta}{}^* \end{pmatrix},$$

where the asterisks denote compact operators. Hence, $M_{\exp ik\theta}$ can be seen to be a lift of $\sigma(T_{\exp ik\theta} \oplus T_{\exp ik\theta}{}^*)$ to a unitary in $\mathcal{T} \otimes M_2(\mathbf{C})$. Thus

$$\partial_1([\exp ik\theta]) = \partial_1([\sigma(T_{\exp ik\theta})]) =$$
$$= [M_{\exp ik\theta}{}^* P \, M_{\exp ik\theta}] - [P] =$$
$$= [P_{\exp ik\theta \, H^2}] - [P_{H^2}] = -k$$

and hence ∂_1 corresponds to the index map.

Suppose we apply the same analysis to the short exact sequence for $\mathcal{T}_\alpha$. Then we obtain

$$
\begin{array}{ccccc}
K_0(\mathcal{C}_\alpha) & \xrightarrow{i^*} & K_0(\mathcal{T}_\alpha) & \xrightarrow{\sigma_{\alpha *}} & K_0(C(T^2)) \\
\partial_1 \uparrow & & & & \downarrow \partial_0 \\
K_1(C(T^2)) & \xleftarrow{\sigma_{\alpha *}} & K_1(\mathcal{T}_\alpha) & \xleftarrow{i^*} & K_1(\mathcal{C}_\alpha).
\end{array}
$$

105

and $\partial_1 : K_1(C(T^2)) \to K_0(\mathcal{C}_\alpha)$ with

$$K_1(C(T^2)) = K^1(T^2) = H^1(T^2, Z) = Z \oplus Z.$$

If T in $\mathcal{T}_\alpha$ is unitary, then $\sigma_\alpha(T)$ is unitary in $C(T^2)$ and $\partial_1([\sigma_\alpha(T)]) = \partial_1 \sigma_{\alpha^*}[T] = 0$ by exactness. Therefore, a necessary condition for T_φ^α to be invertible[1] in $\mathcal{T}_\alpha$ is that $\partial_1[\phi] = 0$ in $K_0(\mathcal{C}_\alpha)$. But what does this mean? In the case of the circle, our index map had range in $K_0(\mathcal{K}) = Z$, but as we have discussed, this K_0-group can be identified with the ordinary Hilbert space dimension and hence with the classical index. It turns out that one can always view $K_0(\mathcal{A})$ for a C^*-algebra $\mathcal{A}$ as an attempt at defining a range for a dimension function for the projections in $\mathcal{A}$ and hence the same interpretation can be applied to the index of T_φ^α in $K_0(\mathcal{C}_\alpha)$. Still, in a way, this is begging the question since we would like, if possible, a numerical index for T_ϕ^α, or at least something that we can calculate.

5. The Pairing of K_0 with a Trace

Although a C^*-algebra $\mathcal{A}$ may possess a trace defined on all of $\mathcal{A}$, it is more likely that the trace will be defined (and finite) on just a dense subalgebra. For example, this is the case for the usual trace on $\mathcal{K}(\mathcal{H})$ or for ordinary integration on $C_0(R)$; the trace is defined on the subalgebra of trace class operators in the first instance, and on the subspace $C_0(R) \cap L^1(R)$ in the second. Suppose τ is a trace which is defined and finite on a self-adjoint subalgebra $\mathcal{A}^\infty$ of $\mathcal{A}$. Then $\tau \otimes Tr_k$ is naturally defined on the subalgebra $\mathcal{A}^\infty \otimes M_k(\mathbb{C})$ of $\mathcal{A} \otimes M_k((\mathbb{C})$ for each positive integer k, where Tr_k is the trace on $M_k(\mathbb{C})$. It can be shown that the correspondence

$$[e] \longrightarrow (\tau \otimes Tr_k)(e),$$

defined for e in $\mathrm{Proj}(\mathcal{A}^\infty) \otimes M_k(\mathbb{C})$, extends to a homomorphism $K_0(\mathcal{A}^\infty) \xrightarrow{\otimes[\tau]} R$ and we set $< [e], [\tau] >= (\tau \otimes Tr_k)(e)$.[2] The inclusion map $\mathcal{A}^\infty \subseteq \mathcal{A}$ yields a homomorphism $K_0(\mathcal{A}^\infty) \longrightarrow K_0(\mathcal{A})$ which, in general, is not an isomorphism.

---

[1] A reasonably straightforward argument shows that the map which takes an invertible element of $\mathcal{A} \otimes M_k(\mathbb{C})$ to the unitary factor in its polar decomposition is well-defined for K-theory. Hence we need not restrict our discussion to unitary operators in $\mathcal{A}$.

[2] Although we defined K_0 for C^*-algebras, the same definition using idempotents instead of projections and similarity via invertible elements instead of unitary equivalence yields the same group (cf. [33]). Indeed, this is just the K_0 functor of algebraic K-theory.

106

When it is, the trace yields a homomorphism $K_0(\mathcal{A}) \xrightarrow{\otimes[\tau]} R$. The preceding analysis is due to Connes [18] which he then combined with an idea of Karoubi to yield the following result.

Theorem 10. *If $\mathcal{A}^\infty$ is a subalgebra of $\mathcal{A}$ such that $\mathcal{A}^\infty \otimes M_k(\mathbb{C})$ is closed in $\mathcal{A} \otimes M_k(\mathbb{C})$ under the holomorphic functional calculus, then the map $K_0(\mathcal{A}^\infty) \to K_0(\mathcal{A})$ is an isomorphism and hence, in this case, a trace τ on $\mathcal{A}^\infty$ induces a homomorphism*

$$K_0(\mathcal{A}) \xrightarrow{\otimes[\tau]} R.$$

Let us apply these ideas to the case $\alpha = 0$. If we set $\mathcal{C}_0^\infty = \mathcal{C}(T, \mathcal{L}^1(H^2(D)))$, where $\mathcal{L}^1(H^2(D))$ denotes the ideal of trace class operators on $H^2(D)$, then

$$\tau_0(f) = \frac{1}{2\pi} \int_0^2 Tr\ f(e^{i\theta}) d\theta$$

is a finite trace on $\mathcal{C}_0^\infty$. Moreover, $\mathcal{C}_0^\infty \otimes M_k(\mathbb{C})$ is closed under the holomorphic functional calculus in $\mathcal{C}_0 \otimes M_n(\mathbb{C})$, where $\mathcal{C}_0 = \mathcal{C}(T, \mathcal{K}(H^2(D)))$. The analogous results hold for all α in Q^* and we have

Theorem 11. *If T is an operator in $\mathcal{T}_\alpha \otimes M_k(\mathbb{C})$ such that $(\sigma_\alpha \otimes 1_k)(T)$ is invertible in $C(T^2) \otimes M_k(\mathbb{C})$, then*

$$\text{Index}_\alpha\ T = < \partial_1[(\sigma_\alpha \otimes 1_k)(T)], [\tau_\alpha] > .$$

Further, for α in Irr and τ_α the trace on $\mathcal{W}_\alpha$, we set $\mathcal{I}_\alpha^\infty = \{T \in \mathcal{I}_\alpha : \tau_\alpha(|T|) < \infty\}$ and $\mathcal{C}_\alpha^\infty = \mathcal{C}_\alpha \cap \mathcal{I}_\alpha^\infty$. Then $K_0(\mathcal{C}_\alpha^\infty) \to K_0(\mathcal{C}_\alpha)$ is an isomorphism and the above formula holds in this case also. Hence the important thing for obtaining this real-valued, analytical index is the original index which takes values in $K_0(\mathcal{C}_\alpha)$, the trace on $\mathcal{C}_\alpha^\infty$, and the homomorphism induced from $K_0(\mathcal{C}_\alpha)$ to R by the trace. These results can be summarized in the following commutative diagram

$$
\begin{array}{ccc}
K_1(C(T^2)) & \xrightarrow{\partial_1} & K_0(\mathcal{C}_\alpha) \\
\text{Ch} \downarrow & & \downarrow \otimes[\tau_\alpha] \\
H_1(T^2, R) & \xrightarrow{i_\alpha} & R
\end{array}
$$

where Ch denotes the Chern character which in this case is an isomorphism. Then the map from $K_1(C(T^2))$ to R via the upper right hand corner is the analytical index, while that around the bottom left hand corner is the topological index.

Now that we have shown how the index results can be made to depend directly on the short exact sequence

$$0 \longrightarrow \mathcal{C}_\alpha \longrightarrow \mathcal{T}_\alpha \xrightarrow{\sigma_\alpha} C(T^2) \longrightarrow 0$$

and the trace τ_α on $\mathcal{I}_\alpha$, we want to reconsider the source of such structure. We return first to the case of the circle.

6. Spectral Half Spaces for Differential Operators

Let D denote the first order differential operator

$$D = -i\frac{d}{d\theta}$$

defined on T. Then D defines a self-adjoint operator on $L^2(T)$ which has the complete set of eigenfunctions $\{\exp in\theta\}$ with

$$D \exp in\theta = n \exp in\theta.$$

Thus $H^2(D)$ is the span of the eigenfunctions for D which correspond to non-negative eigenvalues. Such a construction can be shown to yield a projection for which the necessary commutators are compact if the operator is elliptic. For the given operator D on T the symbol is

$$(D)(e^{i\theta}, \xi) = \xi$$

for $(e^{i\theta}, \xi)$ in the cotangent bundle $T^*T = T \times R$ for R, and D is elliptic since for $\xi \neq 0$ it follows that $\sigma(D)(e^{i\theta}, \xi)$ is invertible.

Let us now consider the case of a smooth compact manifold M without boundary and a smooth complex vector bundle E over M. If we fix a metric on M and on E, then we can form the Hilbert space $L^2(E)$ of measurable sections of E. Suppose D is an elliptic differential operator which defines a self-adjoint operator on $L^2(E)$. Then the projection P onto the span of the non-negative eigenfunctions can be shown [2] to be a pseudo-differential operator of order zero on $L^2(E)$. Therefore, since ϕ in $C^\infty(M)$ also defines the zero order pseudo-differential operator M_ϕ, the commutator $[M_\phi, P]$ is a pseudo-differential oeprator of order -1 and hence is compact. An easy approximation argument then shows that $[M_\phi, P]$ is compact for all ϕ in $C(M)$. If we set $T_\phi = PM_\phi P$

108

for ϕ in $C(M)$ and let $\mathcal{T}_D$ denote the C^*-span of $\{T_\phi + K : \phi \in C(M), K \in \mathcal{K}(\mathrm{ran}\ P)\}$, then we have the extension

$$0 \longrightarrow \mathcal{K}(\mathrm{ran}\ P) \longrightarrow \mathcal{T}_D \longrightarrow C(M) \longrightarrow 0.$$

Thus we have a situation generalizing that of Toeplitz operators on the unit circle. This is discussed in [5], [6], where it is shown that such an operator D determines an element of $K_1(M)$ with the indices of the Toeplitz operators defining the pairing of $K_1(M)$ and $K^1(M)$. We do not pursue this any further here but go on to the case we have been considering on T^2 which is different.

For each α in R there is again the differential operator [3]

$$D_\alpha = i \left(\alpha \frac{\partial}{\partial \theta_1} - \frac{\partial}{\partial \theta_2} \right)$$

which is self-adjoint on $L^2(T^2)$. However, D_α is not elliptic since its symbol is

$$\sigma(D_\alpha)(e^{i\theta_1}, e^{i\theta_2}; \xi_1, \xi_2) = -\alpha\xi_1 + \xi_2$$

for $(e^{i\theta_1}, e^{i\theta_2}; \xi_1, \xi_2)$ in the cotangent bundle $T^*T^2 = T^2 \times R^2$ for T^2 and this symbol vanishes in the direction $(\xi, \alpha\xi)$ at each point of T^2. Nevertheless, D_α still possesses the complete set of eigenfunctions

$$\{\exp\ i(m\theta_1 + n\theta_2) : (m, n) \in Z^2\}.$$

The eigenvalues $\{-\alpha m + n : (m, n) \in Z^2\}$ form a discrete subset of R for α in Q^* but the spectrum of D_α is dense in R for α in Irr. The spectral projection for D_α corresponding to the half-interval $[0, \infty)$ is P_α and the Toeplitz operator T_ϕ^α was defined in §2 to be $T_\phi^\alpha = P_\alpha M_\phi P_\alpha$ for ϕ in $C(T^2)$. However, in this case, the commutator $[P_\alpha, M_\phi]$ is not compact but lies in the algebra $\mathcal{C}_\alpha$. In any case, we obtain the short exact sequence

$$0 \longrightarrow \mathcal{C}_\alpha \longrightarrow \mathcal{T}_\alpha \xrightarrow{\sigma_\alpha} C(T^2) \longrightarrow 0.$$

We want to investigate why such a sequence exists.

---

[3] For $\alpha = \pm\infty$, we have $D_\alpha = i\frac{\partial}{\partial \theta_1}$ or $-i\frac{\partial}{\partial \theta_1}$ and the following discussion is valid with the obvious modifications.

7. Foliation Algebras

Many operators acting on an L^2-space of functions on some space X can be defined using a kernel on $X \times X$, especially if something like distributional kernels are allowed. In the following paragraph, which is intended to be only heuristic, we shall act as though all operators can be defined by kernels and we will completely ignore technicalities.

For an elliptic self-adjoint operator, the kernel for the commutator of the spectral projection onto the positive space with a smooth multiplier on the space has the property that it is *absolutely continuous* in all directions or equivalently, for all pairs of points in $X \times X$, and hence this kernel defines a compact operator between the appropriate L^2-spaces. For general pseudo-differential operators, such a commutator is only *absolutely continuous* in the directions for which the operator is elliptic and there is no restriction in the non-elliptic or charcteristic directions. Let us try to make this intuitive discussion a little more precise.

The operator D_α has symbol defined at $(e^{i\theta_1}, e^{i\theta_2}; \xi_1, \xi_2)$ to be

$$\sigma(D_\alpha)(e^{i\theta_1}, e^{i\theta_2}; \xi_1, \xi_2) = -\alpha\xi_1 + \xi_2.$$

Hence, it is characteristic precisely in the direction $(t, \alpha t)$ at each point of T^2. Moreover, locally D_α is just differentiation along the curve tangent to the direction $(-\alpha t, t)$ at each point. The latter curves fill T^2 and define what is called a foliation of T^2, where the leaves of this foliation are these curves. In general, a smooth foliation $\mathcal{F}$ is a special kind of decomposition of a smooth manifold V into smooth submanifolds called the leaves of $\mathcal{F}$. This decomposition has the property that at each point of V we can find an open set of V about that point which we can express as the product $R^p \times R^q$, where the R^p-variable parametrizes the intersection of the leaves with the open set. In the preceding example of the Kronecker foliation $\mathcal{F}_\alpha$, the open set can be taken to be the interior of a small rectangle with sides having slopes α and $-1/\alpha$, which we can express as $R \times R$. Observe that the leaves determined by different values of the R-parameter are not all distinct. Note that the leaves of $\mathcal{F}_\alpha$ are compact if and only if α is in Q^* in which case they are copies of T twisted around T^2. If α is in Irr, then the leaves are copies of R. These leaves inherit a metric from that of T^2 and this metric defines arc-length measure on the leaf.

As stated above, D_α is just differentiation along the leaves of $\mathcal{F}_\alpha$ and D_α is an elliptic operator restricted to each leaf. Thus the study of $P_\alpha M_\phi P_\alpha$ depends on the fact that the kernel defining the commutator for $[M_\phi, P_\alpha]$ is absolutely continuous along the leaves and hence should define an operator which is *compact-like* along the leaves of $\mathcal{F}_\alpha$ and *function-like* in the transverse direction. To go further we must define an algebra of such operators, which is the foliation C^*-algebra of Connes [19].

Before doing that let us digress and consider how to realize the compact operators on an L^2-space in terms of absolutely continuous kernels.

Let X be a locally compact metrizable space and let μ be a positive measure on X. We make the algebra $C_c(X \times X)$ of continuous complex-valued functions on $X \times X$ having compact support into an algebra with involution as follows. For k_1, k_2 in $C_c(X \times X)$ set

$$(k_1 \circ k_2)(x, z) = \int_X k_1(x, y)k_2(y, z)d\mu(y)$$

and

$$k_1^*(x, y) = \overline{k_1(y, x)}.$$

Then it is straightforward to demonstrate that $C_c(X \times X)$ is an algebra with involution which has the natural *-representation π on $L^2(\mu)$ defined by

$$((\pi k)f)(x) = \int_X k(x, y)f(y)d\mu(y).$$

If we complete $C_c(X \times X)$ in the norm induced by π, we obtain a C^*-algebra isomorphic to $\mathcal{K}$ and the image of this C^*-algebra under π is precisely $\mathcal{K}(L^2(\mu))$.

Now let us consider how one realizes the operators defined to be multiplication by a continuous function on $L^2(X)$ in terms of distribution kernels. For ϕ in $C_c(X)$ let δ_ϕ denote the measure supported on the diagonal $\{(x, x) : x \in X\}$ in $X \times X$ such that

$$\delta_\phi(f) = \int_X f(x, x)\phi(x)d\mu(x)$$

for f in $C_c(X \times X)$. Then the bounded operator defined on $L^2(X)$ by the kernel δ_ϕ is just M_ϕ.

Now what we want to do is to define an algebra of functions for a foliation $\mathcal{F}$ in such a way that the operator algebra product operation is convolution on each leaf of $\mathcal{F}$, but is *function-like* in the transverse direction. To understand how we should proceed, let us consider the very special case of the foliation $R^p \times R^q$, where the leaves are precisely the submanifolds $\lambda \times R^q$ for λ in R^p. A general kernel operator on $L^2(R^p \times R^q)$ would be defined by a distribution on the product $(R^p \times R^q) \times (R^p \times R^q)$. However, as we said above, the support of the kernel for a *function-like* operator should lie in the diagonal.

Therefore, since we are interested in operators that are *function-like* in the R^p-variable, we can restrict our attention to kernels defined on $R^p \times R^q \times R^q$, that is, to the product of the diagonal of the parameter space with the product of the leaf space. For k_1 and k_2 in $C_c(R^p \times R^q \times R^q)$ we define

$$(k_1 \circ k_2)(x, y, z) = \int_{R^q} k_1(x, y, w) k_2(x, w, z) dw$$

and

$$k_1^*(x, y, z) = \overline{k_1(x, z, y)}.$$

Now, by combining techniques from both of the cases described above, one can show that $C_c(R^p \times R^q \times R^q)$ is an algebra with involution. For each x in R^p there is a *-representation π_x on $L^2(R^q)$ defined by

$$((\pi_x k)f)(x, y) = \int_{R^q} k(x, y, z) f(x, z) dz.$$

In this case, these representations are all different. We use these representations to complete $C_c(R^p \times R^q \times R^q)$ in the norm defined by

$$\|k\| = \sup_{x \in R^p} \|\pi_x(k)\| \text{ for } k \text{ in } C_c(R^p \times R^q \times R^q)$$

to obtain a C^*-algebra of operators which can be identified with $C_0(R^p, \mathcal{K}(L^2(R^q)))$. These operators are *function-like* in the R^p-direction and *compact-like* in the R^q direction along the leaves. This is the foliation C^*-algebra of Connes [19] for the case of the foliation $R^p \times R^q$. The analogue of this C^*-algebra for the Kronecker foliation $\mathcal{F}_\alpha$ is what we will now define. With a little thought it is clear how to do this for α in Q^* but we want a uniform approach defined strictly in terms of the foliation.

To begin we must introduce the notion of the graph of a foliation. In the case of $\mathcal{F}_\alpha$ the graph $\mathcal{G}_\alpha$ is somewhat simplified since there is no holomomy (cf. [19]). We set

$$\mathcal{G}_\alpha = \{(\gamma_1, \gamma_2) : \gamma_1, \gamma_2 \text{ lie on the same leaf of } \mathcal{F}_\alpha\}$$

and give it the relative differential structure as a subset of the product of the space T^2 of $\mathcal{F}_\alpha$ with itself. For the case $\alpha = 0$ we have

$$\mathcal{G}_0 = \{((e^{it_1}, e^{it_2}), (e^{it_3}, e^{it_4})) : e^{it_1} = e^{it_3}\}$$
$$= T \times T \times T = T^3.$$

In general, we have

$$\mathcal{G}_\alpha = \begin{cases} T^3 & \alpha \in Q^*; \\ T^2 \times R & \alpha \in Irr. \end{cases}$$

The graph $\mathcal{G}_\alpha$ can be partitioned into the sets which consist of the pairs of points on a fixed leaf of $\mathcal{F}_\alpha$. Fixing the first point of this set determines the *leaf* $\ell_{(e^{i\theta_1}, e^{i\theta_2})}$ of $\mathcal{G}_\alpha$ which can be parametrized as follows:

$$\ell_{(e^{i\theta_1}, e^{i\theta_2})} = \{((e^{i\theta_1}, e^{i\theta_2}), (e^{i(-\alpha t + \theta_1)}, e^{i(t+\theta_1)})) :$$

$$0 \le t < p \text{ for } \alpha = p/q \text{ or } t \in R \text{ for } \alpha \in Irr\}.$$

We let $C_c^\infty(\mathcal{G}_\alpha)$ denote the algebra of smooth functions on $\mathcal{G}_\alpha$ having compact support with the following multiplication and involution:

$$(k_1 \circ k_2)(e^{i\theta_1}, e^{i\theta_2}, t) = \int_{\ell_{(e^{i\theta_1}, e^{i\theta_2})}} k_1(e^{i\theta_1}, e^{i\theta_2}, t - s) k_2(e^{i(-\theta_1 - \alpha(t-s))}, e^{i(\theta_2 + t - s)}, t) ds$$

and

$$k^*(e^{i\theta_1}, e^{i\theta_2}, t) = \overline{k(e^{i(\theta_1 - \alpha t)}, e^{i(\theta_2 + t)}, -t)}.$$

Then $C_c^\infty(\mathcal{G}_\alpha)$ is an algebra with involution and there is a natural *-representation $\pi_{\ell(e^{i\theta_1}, e^{i\theta_2})}$ on $L^2(\ell(e^{i\theta_1}, e^{i\theta_2}))$. For α in Irr all these representations are equivalent but for α in Q^*, they are not. Let $C^*(\mathcal{F}_\alpha)$ denote the completion of $C_c^\infty(\mathcal{G}_\alpha)$ in the norm induced by all of these representations. This is the foliation C^*-algebra of Connes [19] for $\mathcal{F}_\alpha$. We have used the metric on T^2 to induce the metric on the leaves of $\mathcal{F}_\alpha$ to define the measures on the leaves.

Lebesgue measure on T defines an invariant transverse measure for $\mathcal{F}_\alpha$ which allows us to define the trace on $C_c^\infty(\mathcal{G}_\alpha)$

$$Tr_\alpha[k] = \int_{T^2} k(e^{i\theta_1}, e^{i\theta_2}, 0) d\theta_1 d\theta_2.$$

With this trace we can define the foliation W^*-algebra $W^*(\mathcal{F}_\alpha)$ of Connes [19] which is closely related with the construction of $\mathcal{W}_\alpha$. In fact, $W^*(\mathcal{F}_\alpha)$ is defined on

$$L^2(R) \otimes l^2(\Gamma_\alpha)$$

and

$$\mathcal{W}_\alpha = 1_{[0,\infty)} W^*(\mathcal{F}_\alpha) 1_{[0,\infty)}.$$

Moreover, the restriction of the trace on $W^*(\mathcal{F}_\alpha)$ to $\mathcal{W}_\alpha$ agrees with τ_α on $\mathcal{W}_\alpha$.

Before moving on we should point out that there is also a natural *-representation of $C_c^\infty(\mathcal{G}_\alpha)$ on $L^2(T^2)$ via convolution. One can use this *-representation to define a norm which one can use to complete $C_c^\infty(\mathcal{G}_\alpha)$. In general, one obtains a C^*-algebra which is only a homomorphic image of $C^*(\mathcal{F}_\alpha)$. In the case of $\mathcal{F}_\alpha$, however, these two foliation C^*-algebras are isomorphic.

113

8. Smooth Approximate Projections

It is important for us to understand the relation between $C^*(\mathcal{F}_\alpha)$ and $\mathcal{C}_\alpha$. To that end let us introduce the C^*-algebra $\mathcal{S}_\alpha$ defined to be the C^*-span of $C(T^2)$ and P_α acting on $L^2(T^2)$. If $\tilde{\mathcal{C}}_\alpha$ denotes the commutator ideal in $\mathcal{S}_\alpha$, then we have the short exact sequence

$$0 \longrightarrow \tilde{\mathcal{C}}_\alpha \longrightarrow \mathcal{S}_\alpha \xrightarrow{\tilde{\sigma}_\alpha} C(T_+^2 \vee T_-^2) \longrightarrow 0$$

where $T_+^2 \vee T_-^2$ denotes the disjoint union of two copies of T^2, one for P_α and one for $I - P_\alpha$. Thus in this sequence $\tilde{\sigma}_\alpha(P_\alpha)$ is 1 on T_+^2 and 0 on T_-^2.

Now both $\mathcal{C}_\alpha \subseteq \tilde{\mathcal{C}}_\alpha$ and $C^*(\mathcal{F}_\alpha) \subseteq \tilde{\mathcal{C}}_\alpha$. It is perhaps easiest to compare them in the translation representation on $L^2(R)$. Recall that V_γ denotes translation by γ and we let B_{Γ_α} denote the complex functions on R vanishing at infinity generated by the characteristic functions of the half-open intervals $\{[\gamma_1, \gamma_2) : \gamma_1, \gamma_2 \text{ in } \Gamma_\alpha\}$. Then we have

$$\tilde{\mathcal{C}}_\alpha = \text{clos}\left\{ \sum_{i=1}^N \phi_i V_{\gamma_i} \psi_i : \gamma_i \in \Gamma_\alpha, \phi_i, \psi_i \in B_{\Gamma_\alpha} \right\}$$

$$\mathcal{C}_\alpha = \text{clos}\left\{ \sum_{i=1}^N \phi_i V_{\gamma_i} \psi_i : \gamma_i \in \Gamma_\alpha, \phi_i, \psi_i \in 1_{[0,\infty)} B_{\gamma_\alpha} \right\}$$

$$C^*(\mathcal{F}_\alpha) = \text{clos}\left\{ \sum_{i=1}^N \phi_i V_{\gamma_i} \psi_i : \gamma_i \in \Gamma_\alpha, \phi_i, \psi_i \in C_0(R) \right\}$$

$$\mathcal{C}_\alpha \cap C^*(\mathcal{F}_\alpha) = \text{clos}\left\{ \sum_{i=1}^N \phi_i V_{\gamma_i} \psi_i : \gamma_i \in \Gamma_\alpha, \phi_i, \psi_i \in C_0(R^+) \right\}.$$

Moreover, the inclusion map of $\mathcal{C}_\alpha \cap C^*(\mathcal{F}_\alpha)$ into $C^*(\mathcal{F}_\alpha)$ induces an isomorphism on K_0 and K_1. The inclusion of $\mathcal{C}_\alpha$ in $\tilde{\mathcal{C}}_\alpha$ also induces an isomorphism on both K_0 and K_1 but while both $\mathcal{C}_\alpha \cap C^*(\mathcal{F}_\alpha)$ in $\mathcal{C}_\alpha$ and $C^*(\mathcal{F}_\alpha)$ in $\tilde{\mathcal{C}}_\alpha$ induce isomorphisms on K_0, they do not on K_1.

All of the algebras $\mathcal{C}_\alpha, \tilde{\mathcal{C}}_\alpha, C^*(\mathcal{F}_\alpha) \cap \mathcal{C}_\alpha$ are simple C^*-algebras and each has K_0 group equal to Γ_α. It is obvious that the natural order on $K_0(\mathcal{C}_\alpha)$ and $K_0(\tilde{\mathcal{C}}_\alpha)$ induced by the semigroup of elements represented by projections, agrees with that of $\Gamma_\alpha \subseteq R$ but it requires a construction analogous to that of Rieffel [35] to see that the same is true for $K_0(C^*(\mathcal{F}_\alpha))$ and $K_0(C^*(\mathcal{F}_\alpha) \cap \mathcal{C}_\alpha)$. This argument shows that $\mathcal{C}_\alpha$ determines Γ_α as an ordered group, something which eluded us in [23]. For more general subgroups Γ of R, the situation is more complicated as has been recently shown by R. Ji and J. Xia, but the algebra still determines Γ as an ordered subgroup of R.

Now the point of all of this is that the algebra $C^*(\mathcal{F}_\alpha)$, which has been defined in terms of the foliation $\mathcal{F}_\alpha$, consists of operators which are *compact-like* along the leaves and *function-like* in the transverse direction. Moreover, $C^*(\mathcal{F}_\alpha)$ has a trace which is defined using the metric on T^2. Finally, since D_α is a differential operator which acts elliptically along the leaves of $\mathcal{F}_\alpha$, we might expect the commutator of the projection P_α onto the spectral half-space for D_α with a smooth multiplier to lie in $C^*(\mathcal{F}_\alpha)$. Then we would be able to define the Toeplitz operators T_ϕ^α using P_α and study them using $C^*(\mathcal{F}_\alpha)$ and its K-theory as above. Unfortunately, this is not always possible since the commutators of the smooth multipliers with the *sharp* spectral projection P_α do not lie in $C^*(\mathcal{F}_\alpha)$ unless there is a gap at zero in the spectrum of D_α acting on $L^2(T^2)$. Since the spectrum of D_α is the closure of $\Gamma_\alpha \subseteq R$, that will be the case exactly when α is in Q^*. To use $C^*(\mathcal{F}_\alpha)$ for the case of α in *Irr* requires that we replace P_α by an *approximate spectral projection* onto R^+.

We proceed as follows. Let p be a function in $C^\infty(R)$ satisfying

$$p(t) = \begin{cases} 1 & t \geq \varepsilon; \\ 0 & t \leq -\varepsilon \end{cases}$$

for some $\varepsilon > 0$. Then $p(D_\alpha)$, which can be defined using the functional calculus for self-adjoint operators, is a smooth approximation to P_α. Indeed, for α in Q^*, it is equal to P_α if ε is chosen to be smaller than the gap in the spectrum of D_α at zero. In general, we have that the operators

$$p(D_\alpha) - P_\alpha,$$
$$p(D_\alpha)^2 - p(D_\alpha), \text{ and}$$
$$P_\alpha M_\phi P_\alpha - p(D_\alpha)M_\phi p(D_\alpha) \text{ for } \phi \text{ in } C(T^2)$$

all lie in the finite ideal $\tilde{\mathcal{I}}_\alpha$ in $W^*(\mathcal{F}_\alpha)$. Therefore, $p(D_\alpha)M_\phi p(D_\alpha)$ is a kind of *approximation* to the Toeplitz operator T_ϕ. But in what sense and on what space is it defined?

The projection P_α determines the space H_α^2 on which the Toeplitz operator is defined but clearly $p(D_\alpha)$ can not do this nor does it determine some substitute space. Rather, we define our *substitute* Toeplitz operator[4] on $L^2(T^2)$ such that

$$\tilde{T}_\phi^\alpha = p(D_\alpha)M_\phi p(D_\alpha) + \phi(1,1)(I - p(D_\alpha)).$$

---

[4] The factor $\phi(1,1)$ is added to make the map $\phi \longrightarrow \tilde{T}_\phi^\alpha$ multiplicative modulo $C^*(\mathcal{F}_\alpha)$.

This operator has many of the same properties as T_ϕ^α; in particular, both are Fredholm at the same time in which case they have the same index. This follows since

$$\Pi_{II}(\widetilde{T}_\phi^\alpha) = \Pi_{II}(T_\phi^\alpha) \oplus \phi(1,1)\Pi_{II}(I - p(D_\alpha))$$

in $\mathcal{W}^*(\mathcal{F}_\alpha)/\widetilde{\mathcal{I}}_\alpha$.

Let $\widetilde{S}_\alpha$ be the C^*-span of $C(T^2), C^*(\mathcal{F}_\alpha)$, and $p(D_\alpha)$ in $\mathcal{L}(L^2(T^2))$. One can show that $[p(D_\alpha), M_\phi]$ for ϕ in $C(T^2)$ and $p(D_\alpha)^2 - p(D_\alpha)$ lie in $C^*(\mathcal{F}_\alpha)$. Therefore, we have the short exact sequence

$$0 \longrightarrow C^*(\mathcal{F}_\alpha) \longrightarrow \widetilde{S}_\alpha \longrightarrow C(T_+^2 \vee T_-^2) \longrightarrow 0$$

which is completely analogous to that for $\mathcal{T}_\alpha$ except this time the kernel ideal is the foliation C^*-algebra $C^*(\mathcal{F}_\alpha)$. Moreover, the following diagram expresses the fact that the analytical index defined in this manner agrees with that defined using the Breuer index or the K-theory of $\mathcal{C}_\alpha$:

$$
\begin{array}{ccccc}
K_0(C^*(\mathcal{F}_\alpha)) & \longrightarrow & K_0(\widetilde{\mathcal{C}}_\alpha) & \longleftarrow & K_0(\mathcal{C}_\alpha) \\
\otimes[Tr_\alpha] & \searrow & \otimes[\widetilde{\tau}_\alpha]\downarrow & \nearrow \otimes[\tau_\alpha] & \\
& & R & &
\end{array}
$$

after one notes that the maps on the top line are isomorphisms.

9. Longitudinal and Transverse Index Theorems

What we have done is to develop a very special case of the odd analogue of the longitudinal index theorem of Connes-Skandalis [20]. More generally, if $\mathcal{F}$ is a foliation with a metric on its tangent bundle and an invariant transverse measure and D is a self-adjoint differential operator which acts elliptically along the leaves of $\mathcal{F}$, then we can define Toeplitz operators for continuous multipliers defined on the space V of $\mathcal{F}$ (T^2 in the case of $\mathcal{F}_\alpha$) using an *approximate spectral half-space projection* as above. A short exact sequence exists with first term $C^*(\mathcal{F})$ and hence the index of such Toeplitz operators will lie in $K_0(C^*(\mathcal{F}))$. The trace on $C^*(\mathcal{F})$, which is defined using the invariant transverse measure, gives a real-valued index which agrees with the odd analogue of the formula given by Connes-Skandalis in [20]. Moreover, the Breuer index defined for these Toeplitz operators relative to the foliation W^*-algebra will also coincide with the real-valued index obtained from the K-theory of $C^*(\mathcal{F})$ and the trace. More detail will be given in [28] where much of this point of view originated.

Before concluding, we want to proceed a little further to show how this development of real-valued index theory can shed light on other more recent matters. We saw above that the kernel of the commutator of a function multiplier with an *approximate spectral half-space projection* for a longitudinally elliptic operator is *absolutely continuous* in the longitudinal direction and hence such a commutator will lie in the foliation C^*-algebra of Connes.

There is another possiblity for obtaining *smooth commutators* in this situation. If we start with a multiplier defined by a kernel which is already *absolutely continuous* in the transverse direction to the foliation, then the commutator with an *approximate spectral half-space projection* would be *absolutely continuous* in all directions, and hence it would define a compact operator. Then ordinary Fredholm index theory would apply to the study of such Toeplitz operators. That is indeed the case and this is the heuristic framework for the transverse index theory of Connes [17]. We provide a description in the case of the Kronecker foliation on the torus omitting the considerable technical details.

If α and β are distinct real numbers, then D_β is transversely elliptic to $\mathcal{F}_\alpha$, that is, it is elliptic in the transverse direction to the leaves of $\mathcal{F}_\alpha$. If we let $\mathcal{R}_{\alpha,\beta}$ denote the C^*-span of $C^*(\mathcal{F}_\alpha), I, p(D_\beta)$ and $\mathcal{K}(L^2(T^2))$, then we have the short exact sequence

$$0 \longrightarrow \mathcal{K}(L^2(T^2)) \longrightarrow \mathcal{R}_{\alpha,\beta} \longrightarrow C^*(\widehat{\mathcal{F}_\alpha}) \oplus C^*(\widehat{\mathcal{F}_\alpha}) \longrightarrow 0.$$

Hence the Toeplitz operators obtained by compressing elements k in $C^*(\widehat{\mathcal{F}_\alpha})$ to obtain the operators $p(D_\beta)k \ p(D_\beta)$ can be studied using ordinary Fredholm index theory, where $C^*(\widehat{\mathcal{F}_\alpha})$ is the C^*-algebra obtained from $C^*(\mathcal{F}_\alpha)$ by adjoining a unit to it. A formula for the index of such operators can be obtained, at least in this case, from the transverse index theory of Connes [17].

However, we have lost something in passing from the index formula for the Toeplitz operators defined by continuous multipliers in $C(T^2)$ to that defined for multipliers in $C^*(\widehat{\mathcal{F}_\alpha})$. In particular, the index of the Toeplitz operators for multipliers in $C^*(\widehat{\mathcal{F}_\alpha})$ lies in $K_0(\mathcal{K}) \cong Z$. This is not the case for the Toeplitz operator T_ϕ^β for ϕ in $C(T^2)$ since its index lies in $K_0(\mathcal{C}_\beta) \cong Z \oplus Z$ or in $K_0(C^*(\mathcal{F}_\beta)) \cong Z \oplus Z$. Restricting the multiplier defining the Toeplitz operator to lie in $C^*(\widehat{\mathcal{F}_\alpha})$ doesn't allow the possibility of incommensurate indices. It is natural to ask if it is somehow possible to recover all the *index information* for the operator D_β from the transverse index problem. The answer is "yes" and provides one source of the motivation for [27].

If ϕ is in $C(T^2)$, then $(\exp -tD_\alpha^2)T_\phi^\beta$ will lie in $C^*(\mathcal{F}_\alpha)$ due to the ellipticity of D_α along the leaves of $\mathcal{F}_\alpha$. Hence, we can try to approach the index of the Toeplitz operator T_ϕ^β by way of that for $k_t = (\exp -tD_\alpha^2)T_\phi^\beta$ and then take the limit as t approaches 0. However, this won't work since such a limit doesn't make sense in K-theory. However, if one uses the cyclic cohomology of Connes [17] and, in particular, the odd cocycle which corresponds to the short exact sequence for $\mathcal{R}_{\alpha,\beta}$, then one can take such a limit and obtain an expression for a limit cyclic cocycle. In [27] it is shown that this limit yields the same cyclic cocycle as that for the longitudinal index theorem and hence one can recover all the index information as before.

The reader is referred to [27] for more details and the relation of these latter remarks to the subject of secondary invariants for differential operators. It should not be surprising that such a connection exists, since what we have just described can be viewed as a regularizing procedure involving the eigenvalues of a differential operator, which is, of course, what the η-function provides also.

References

1. Atiyah, M. F., Global theory of elliptic operators. In *Proc. Int. Cong. on Functional analysis and related topics, Tokyo, 1969*. Univ. of Tokyo Press, 1970, 21-29.

2. ———, V. K. Patodi, and I. M. Singer, Spectral asymmetry and Riemannian geometry I, II, III. Math. Proc. Camb. Phil. Soc. **77** (1975), 43-69, **78** (1975), 405-432, **79** (1976), 71-99.

3. ——— and I. M. Singer, The index of elliptic operators on compact manifolds, Bull. Amer. Math. Soc. **69** (1963), 484-433.

4. ———, The index of elliptic operators I, Ann. Math. **87** (1968), 484-530.

5. Baum, P. and R. G. Douglas, Index theory, Bordism, and K-homology, *Contemportary Mathematics 10*. Amer. Math. Soc., Providence, 1982, 1-31.

6. ———, Toeplitz operators and Poincaré duality. In *Proc. Toeplitz Memorial Conference, Tel Aviv, 1981* (ed. I. C. Gohberg) Basel, Birkhauser, 1982, 137-166.

7. Blackadar, B., *K-Theory for Operator Algebras*, Math. Sci. Res. Inst. Publ. No. 5, Springer Verlag, New York, 1986.

8. Breuer, M., Fredholm theories in von Neumann algebras I, Math. Ann. **178** (1968), 243-254; II, Math. Ann. **180** (1969), 313-325.

9. Brown, L. G., R. G. Douglas, and P. A. Fillmore, Extensions of C^*-algebras, operators with compact self-commutators, and K-homology, Bull. Amer. Math. Soc. **79** (1973), 973-978.

10. _____, Unitary equivalence modulo the compact operators and extensions of C^*-algebras. In *Lect. Notes Math.* 345, Springer Verlag 1973, 58-128.

11. _____, Extensions of C^*-algebra and K-homology, Ann. Math. (2) **105** (1977), 265-324.

12. Coburn, L. A., The C^*-algebra generated by an isometry I, Bull. Amer. Math. Soc. **73** (1967), 722-726; II, Trans. Amer. Math. Soc., **137** (1969), 211-217.

13. _____ and R. G. Douglas, Translation operators on the half-line, Proc. Nat. Acad. Sci. U.S.A. **62** (1969), 1010-1013.

14. _____ and R. G. Douglas, On C^*-algebras of operators on a half-space I, Inst. Hautes Etudes Sci. Publ. Math. **40** (1971), 59-67.

15. _____, D. G. Schaeffer, and I. M. Singer, On C^*-algebras of operators on a half-space II. Index Theory, Inst. Hautes Etudes Sci. Pub. Math. **40** (1971), 69-79.

16. _____, R. D. Moyer, and I. M. Singer, C^*-algebras of almost-periodic pseudo-differential operators, Acta Math. **130** (1973), 279-307.

17. Connes, A., Cyclic cohomology and the transverse fundamental class of a foliation, Preprint IHES/M/84/7.

18. _____, Non-commutative differential geometry, *Inst. Hautes Etu. des Sci. Publ. Math.* **62** (1985), 257-360.

19. _____, A survey of foliations and operator algebras, In *Operator Algebras and Applications* (ed. R. V. Kadison), Proc. Symp. Pure Math. **38**, Amer. Math. Soc., Providence, 1981.

20. _____ and G. Skandalis, The longitudinal index theorem for foliations, Publ. Res. Inst. Math. Sci. Kyoto Univ. **20** (1984), no. 6, 1139-1183.

21. Douglas, R. G., Toeplitz and Wiener-Hopf operators in $H^\infty + C$, Bull. Amer. Math. Soc. **74** (1968), 895-899.

22. _____, *Banach algebra techniques in operator theory*, Academic Press, New York, 1972.

23. _____, On the C^*-algebra of a one-parameter semi-group of isometries, Acta Math. **128** (1972), 143-151.

24. ———, *Banach Algebra Techniques in the Theory of Toeplitz Operators*, Conf. Bd. Math. Sci., No. 15, Amer. Math. Soc., Providence, 1973.

25. ———, *C*-algebra Extensions and K-homology*, Ann. Math. Studies 95, Princeton, 1973.

26. ——— and R. Howe, On the C*-algebra of Toeplitz operators on the quarter-plane, Trans. Amer. Math. Soc. **158** (1971), 203-217.

27. ———, S. Hurder, and J. Kaminker, The η-invariant, foliation algebras and cyclic cocycles, Math. Sci. Res. Inst., Preprint 14711-85.

28. ———, in preparation.

29. Gohberg, I. C., On an application of the theory of normed rings to singular integral equations, Uspehi Mat. Nauk **7** (1952), 149-156 (Russian).

30. ——— and L. S. Goldenstein, On a multi-dimensional integral equation on half-space whose kernel is a function of the difference of the arguments and on a discrete analogue of this equation, Dokl. Akad. Nauk SSSR, **131** (1960), 9-12 (Russian); Soviet Math. Dokl. **1** (1960), 173-176.

31. Kadison, R. V. and J. R. Ringrose, *Fundamentals of the Theory of Operator Algebras II*, Academic Press, New York, 1986.

32. Kasparov, G. G., Topological invariants of elliptic operators, I: K-homology, Math. USSR-Izv. **9** (1975), 751-792.

33. Milnor, J., *Introduction to Algebraic K-theory*, Ann. Math. Studies 72, Princeton, 1971.

34. Palais, R., Seminar on the Atiyah-Singer Index Theorem, Ann. Math. Studies 57, Princeton, 1965.

35. Rieffel, M., *C*-Algebras associated with irrational rotations*, Pacific J. Math. **93** (1981), 415-429.

36. Taylor, J. L., Banach algebras and topology, In *Algebras in Analysis*, (ed. J. H. Williamson), Academic Press, London, 1975, 118-186.

Ronald G. Douglas

Department of Mathematics

State University of New York at Stony Brook

Stony Brook, New York, 11794 U.S.A.

The author was partially supported by a grant from the National Science Foundation.

Triangularity in Operator Algebras

by

David R. Larson

Introduction

This series of talks will be concerned with the subject of operator algebras, acting on a complex Hilbert space, which are not necessarily assumed to be selfadjoint. They are intended to be primarily expository, with some emphasis on certain developments that have taken place in the past few years.

In the course of preparation, it soon became apparent that the subject had grown much too large in recent years to enable one to cover, in five lectures, suitable exposition of more than a few selected topics. We have decided to devote primary attention to two aspects of the theory: the "roots" of the subject, and the similarity theory of nest algebras. A primary purpose of these notes is to make these aspects of the subject accessible to beginning research students interested in working in this area. We necessarily omit, regretably, more than causal references to many topics of importance, including some which are central to our own research program. To supplement somewhat, we have expanded our reference list to include many articles we would have liked to speak about, had time permitted. Even here, this list is by no means exhaustive.

Our goal will be to give a description of many of the basic ideas involved, with some account of much of what is known to date, some of the directions that have been pursued in the past, and with a view toward some exposition of open problems and reasons behind certain current directions of interest. The order of presentation will not always be chronological.

In these talks, all Hilbert spaces will be complex and separable, all operators will be bounded, all subspaces will be closed, and all projections will be self-adjoint. We write $B(H)$ for the collection of all bounded linear operators from a Hilbert space H into itself, and $\mathcal{K}(H)$, or simply $\mathcal{K}$, for the ideal of compact operators in $B(H)$.

We will introduce most of the relevant concepts and definitions in the context of the talks. However, the notation, and terminology, of reflexivity of operator algebras and of subspace lattices is so frequently used that we find it desirable to introduce the relevant ideas first, without elaboration.

Let $\mathcal{L}$ be a collection of subspaces containing $\{0\}$ and H which form a lattice under the operations $\vee$ and $\wedge$, where $M \wedge N$ is the closed linear span of M and N while $M \wedge N$ is the intersection $M \cap N$. $\mathcal{L}$ is *commutative* if the projections on the subspaces commute pairwise. $\mathcal{L}$ is a *nest* (usually denoted by $\mathcal{N}$) if the lattice is linearly ordered by inclusion; thus a nest is commutative, in particular.

For convenience we shall disregard the distinction between a subspace of H and the orthogonal projection onto it. Thus a lattice will consist of either subspaces or projections depending on the context in which it is used. As usual we write Lat $\mathcal{S}$ for the lattice of all projections left invariant under every operator in a subset $\mathcal{S}$ of $B(H)$, and dually Alg $\mathcal{Q}$ denotes the algebra of all operators leaving each projection in a subset $\mathcal{Q}$ of $\mathcal{P}(H)$ invariant. Here, $\mathcal{P}(H)$ denotes the set of all selfadjoint projections in $B(H)$. The term *subspace lattice* will denote a strongly closed lattice of projections containing O and I. Unless otherwise stated, all lattices will be strongly closed. An algebra $\mathcal{A}$ is reflexive if $\mathcal{A} = Alg \operatorname{Lat} \mathcal{A}$, and dually a lattice $\mathcal{L}$ is reflexive if $\mathcal{L} = \operatorname{Lat} Alg \mathcal{L}$. Subspace lattices need not be reflexive; however, *commutative* subspace lattices are reflexive. Strongly closed lattices are complete as lattices, and the converse is true for *commutative* lattices; arbitrary complete lattices need not be strongly closed, however. A *nest algebra* is a reflexive algebra whose lattice of invariant subspaces is a nest.

We wish to thank the Mathematics Department of Indiana University for hosting the Special Year in Operator Theory, and this conference. Special thanks are given to John B. Conway for his work in organization. Appreciation is given to the National Science Foundation for funding this event.

0. Some Basic Properties

We will begin an account of some familiar properties of the most elementry triangular operator algebra, which will be useful for perspective in the exposition of the topics we will consider. These topics will be expanded upon in the context of the lectures, filling in with appropriate exposition as desirable.

Let H be a complex *finite* dimensional Hilbert space, let

$$\{e_1, \ldots, e_n\}$$

be an orthonormal basis for H, and let $\mathcal{A}$ be the algebra of all operators in $B(H)$ whose matrices with respect to this ordered basis are upper triangular. So if, say, $n = 3$, we would descriptively have

$$A = \begin{pmatrix} * & * & * \\ 0 & * & * \\ 0 & 0 & * \end{pmatrix}.$$

To say that an operator T is in $\mathcal{A}$, so has upper triangular form with respect to the ordered basis $\{e_1, \ldots, e_n\}$, simply means that for each i the vector Te_i, which is the ith column vector, is a linear combination of

$$\{e_j : j \le i\}.$$

Equivalently, for each i, the subspace

$$N_i = \text{span}\{e_1, \ldots, e_i\}$$

is invariant under T.

The algebra $\mathcal{A}$ has a number of properties which relate to basic concepts we will be concerned with in an infinite dimensional setting.

One of the most fundamental of these is that every operator T contained in $B(H)$ is unitarily equivalent to an operator contained in $\mathcal{A}$. (Just obtain a basis for H with respect to which T has triangular matrix.) So $\mathcal{A}$ is a *model* set for elements of $B(H)$. In an infinite dimensional setting, one can hope for *approximate* models for Hilbert space operators, and *generalized* triangular forms.

A second property is that

$$\mathcal{A} \cap \mathcal{A}^*$$

is a maximal abelian selfadjoint subalgebra of $B(H)$. (We will use the standard abbreviation m.a.s.a. for this.) This is relevant to the class of *triangular* operators of R. V. Kadison and I. M. Singer.

A third property is that the lattice of invariant subspaces of the algebra $\mathcal{A}$ is totally ordered by inclusion. That is, $\text{Lat}\,(\mathcal{A})$ is a *nest*, or *chain* of subspaces. We have

$$\text{Lat}\,(\mathcal{A}) = \{0, N_1, \ldots, N_n\}$$

where

$$N_i = \text{span}\,\{0, e_1, \ldots, e_i\} \text{ for } 1 \le i \le n.$$

Thus, this is the elementary prototype of the class of *nest algebras* of John Ringrose.

A fourth property, which is not difficult to establish here, is that if $\mathcal{B}$ is any subalgebra of $B(H)$ which contains $\mathcal{A}$, then $\mathcal{B}$ is the *full* algebra of block-triangular matrices for some subnest of $\text{Lat}(\mathcal{A})$. (In general, it is true that any algebra which contains a nest algebra *is* a nest algebra.) If the inclusion is proper, then at least one *gap* in the subnest must have dimension greater than one, so

$$\text{Diag}(\mathcal{B}) = \mathcal{B} \cap \mathcal{B}^*$$

is not abelian. Thus $\mathcal{A}$ is a subalgebra of $B(H)$ which is maximal with respect to the property that its diagonal $\mathcal{A} \cap \mathcal{A}^*$ is abelian.

It is elementary to see in this setting that $\mathcal{A}$ and $\mathcal{B}$ are *reflexive* in the sense that each contains any operator which leaves invariant the lattice of invariant subspace for the algebra.

A fifth property is that

$$\mathcal{A} + \mathcal{A}^* = B(H).$$

This is relevant to the class of *subdiagonal* operator algebras of W. B. Arveson.

A sixth property is that $\mathcal{A}$ admits a Wedderburn decomposition

$$\mathcal{A} = \mathcal{D} + \mathcal{R}, \ \mathcal{D} \cap \mathcal{R} = (0),$$

where $\mathcal{D}$ is a semisimple subalgebra of $\mathcal{A}$ (here, $\mathcal{D} = \mathcal{A} \cap \mathcal{A}^*$), and $\mathcal{R}$ is the Jacobson radical of $\mathcal{A}$ (here, $\mathcal{R}$ is the 2-sided ideal of all operators in $\mathcal{A}$ with zero diagonal). So the quotient $\mathcal{A}/\mathcal{R}$ is abelian and is isomorphic to $\mathcal{D}$.

Recall, if $\mathcal{A}$ is a unital Banach algebra then $\mathcal{A}$ contains a unique (necessarily norm closed) two-sided ideal maximal with respect to the property that all of its elements are quasinilpotent, and this coincides with the Jacobson radical of $\mathcal{A}$, which is abstractly defined to be the intersection of the kernels of all irreducible representations of $\mathcal{A}$.

A seventh property is that if T is an invertible operator in $B(H)$, then the algebra

$$T\mathcal{A}T^{-1} = \{T A T^{-1} : A \in \mathcal{A}\},$$

(the similarity transform of $\mathcal{A}$ under T), is unitarily equivalent to $\mathcal{A}$. There exists a, perhaps different, orthonormal basis

$$\{\widetilde{e}_1, \widetilde{e}_2, \ldots, \widetilde{e}_n\}$$

for H such that with respect to this new basis the algebra $\mathcal{B} = T\mathcal{A}T^{-1}$ has the triangular form of $\mathcal{A}$.

The proof of this is simple, but revealing: Let

$$\mathcal{N} = \{0, N_1, \ldots, N_n\}$$

be the nest of $\mathcal{A}$, and let $\mathcal{M}$ be the *image* nest under T. That is,

$$\mathcal{M} = \{0, TN_1, \ldots, TN_n\},$$

where for each i, TN_i is the subspace of H defined by $TN_i = \{Tx : x \in N_i\}$. Then $\mathcal{M}$ is also a nest of subspaces. Each member of $\mathcal{M}$ also has codimension one in its immediate successor, because for each i the subspace TN_i is the span of TN_{i-1} together with the vector Te_i. (Here, we denote $N_0 = (0)$.)

Let $M_i = TN_i, 0 \leq i \leq n$. The Gram-Schmidt process applied to $\{Te_1, \ldots, Te_n\}$ yields an orthonormal basis $\{\tilde{e}_1, \ldots, \tilde{e}_n\}$. We have

$$M_i = \operatorname{span}\{\tilde{e}_1, \ldots, \tilde{e}_i\}, 1 \leq i \leq n.$$

If $A \in \mathcal{A}$, then A leaves each N_i invariant, so

$$(TAT^{-1})(TN_i) = TAN_i \subseteq TN_i = M_i,$$

and so TAT^{-1} leaves each member of $\mathcal{M}$ invariant. Reversing this argument shows that the converse is also true: if TAT^{-1} leaves each member of $\mathcal{M}$ invariant then $A \in \operatorname{Alg}(\mathcal{N}) = \mathcal{A}$, so $TAT^{-1} \in T\mathcal{A}T^{-1}$. Thus $\mathcal{B}$ is the full upper triangular algebra with respect to the ordered orthonormal basis $\{\tilde{e}_1, \ldots, \tilde{e}_n\}$. So $\mathcal{B}$ is unitarily equivalent to $\mathcal{A}$.

If we now let U be the unitary operator for which $U\tilde{e}_i = e_i, 1 \leq i \leq n$, and let $A = UT$, then for each i,

$$TN_i = \operatorname{span}\{\tilde{e}_1, \ldots, \tilde{e}_i\},$$
$$\text{so } AN_i = UTN_i = \operatorname{span}\{e_1, \ldots, e_i\} = N_i.$$

Since $AN_i = N_i, 1 \leq i \leq n$, we have $A \in \mathcal{A}$. Also,

$$A^{-1}N_i = A^{-1}(AN_i) = N_i.$$

So $A^{-1} \in \mathcal{A}$ also.

Writing $T = U^*A$, we have an *eighth* property: every invertible operator $T \in B(H)$ admits a factorization

$$T = VA$$

with V unitary and $A \in \mathcal{A} \cap \mathcal{A}^{-1}$.

Equivalently, every positive invertible $S \in B(H)$ admits a factorization

$$S = A^*A, \text{ with } A \in \mathcal{A} \cap \mathcal{A}^{-1}.$$

To see that these are equivalent, given S let $T = S^{1/2}$ be the positive square root and factor $T = VA$ as above. Then $T = T^*$, so

$$S = T^*T = A^*V^*VA = A^*A,$$

as required.

Conversely, if one assumes the factorization property, then if T is a given invertible operator let

$$S = T^*T$$

and factor $S = A^*A$ for $A \in \mathcal{A} \cap \mathcal{A}^{-1}$.

Then

$$T^*T = A^*A$$

so $\quad TA^{-1} = (A^*)^{-1}T^* = (TA^{-1})^*,$

and thus TA^{-1} is a unitary operator. Call this V. Then $T = VA$ is the desired factorization.

A *ninth* property is that if $P \in \mathcal{A}$ is an idempotent (that is, $P^2 = P$), then there is an element $A \in \mathcal{A} \cap \mathcal{A}^{-1}$, (the group of invertible elements of $\mathcal{A}$), with the property that

$$APA^{-1} \text{ is selfadjoint}$$

(and hence contained in $\mathcal{D} = \mathcal{A} \cap \mathcal{A}^*$). For this, simply let T be *any* invertible operator in $B(H)$ for which TPT^{-1} is selfadjoint. For instance, we may let

$$T = [P^*P + (I - P^*)(I - P)]^{1/2}.$$

Now factor $T = UA$ with U unitary and $A \in \mathcal{A} \cap \mathcal{A}^{-1}$. Then
$$APA^{-1} = (U^*T)P(U^*T)^{-1}$$
$$= U^*(TPT^{-1})U,$$

so APA^{-1} is selfadjoint, as desired.

Remark 1. At one point in the above argument we knew that A was an element of $\mathcal{A}$ and that A was invertible in $B(H)$. Since $\dim(H) < \infty$ here, the algebra $\mathcal{A}$ of upper triangular operators with respect to a fixed basis is of course inverse-closed, so the argument given that $A^{-1} \in \mathcal{A}$ also was unnecessary. If H is infinite dimensional, however, a nest algebra contained in $B(H)$ will be inverse-closed if and only if its nest is either finite, order isomorphic to the extended positive integers, or order isomorphic to the extended negative integers. This is useful to keep in mind.

Remark 2. As we shall see, if H is infinite dimensional separable Hilbert space, the results analogous to properties eight and nine (the $T = UA$ and $S = A^*A$ factorization properties, and the idempotent property) hold intact for a nest algebra $\text{Alg}(\mathcal{N})$, where $\mathcal{N}$ is assumed to be a *complete* nest, *if and only if* $\mathcal{N}$ is *countable* as a set of subspaces. If $\mathcal{N}$ is uncountable, and in particular, if $\mathcal{N}$ is continuous, then there exists a nonzero idempotent P in $\text{Alg}(\mathcal{N})$ which fails to be similar to a projection via an operator in $\text{Alg}(\mathcal{N})$ with inverse also in $\text{Alg}(\mathcal{N})$. This fact comes out of, and is in fact equivalent to, the solution to John Ringrose's similarity problem in [84], and utilizes Niels Andersen's approximate equivalence theorem in [1].

A tenth property of the triangular algebra $\mathcal{A}$ is the specialization to this simple case of William Arveson's distance formula, (hyperreflexivity), which is valid for arbitrary nest algebras. This will be important in subsequent lectures. It is nontrivial even for algebras of this simple form. If $T \in B(H)$ is arbitrary, it states that the norm-distance from T to the algebra $\mathcal{A}$ is equal to the maximum of the norms of the *lower triangular blocks* of T of the form $(I - P_i)TP_i$, where P_i is the projection onto the subspace N_i, for $l \leq i \leq n$. That is,

$$d(T, \mathcal{A}) = \max_i \|P_i^{\perp} T P_i\|.$$

More will be said of this.

1. Some Roots of the Subject

1.1 Triangular Operator Algebras

It is generally understood that the first paper in the literature which was concerned entirely with the systematic investigation of properties of operator algebras that were not assumed to be selfadjoint ws the seminal work "Triangular Operator Algebras" by

R. V. Kadison and I. M. Singer [72]. We begin with a brief account of some of the aspects of that paper, maintaining most original terminology and notation, for perspective. In the context, certain of the later developments will also be briefly discussed.

Let $\mathcal{M}$ be a factor von Neumann algebra and $\mathcal{A}$ a maximal abelian self-adjoint subalgebra of $\mathcal{M}$. A subalgebra $\mathcal{T}$, of $\mathcal{M}$ is said to be *triangular in* $\mathcal{M}$ (or simply *triangular*, when $\mathcal{M}$ is the algebra of all bounded operators on the underlying Hilbert space) with diagonal $\mathcal{A}$ when

$$\mathcal{T} \cap \mathcal{T}^* = \mathcal{A}.$$

A triangular subalgebra of $\mathcal{M}$ is called *maximal* if it is not contained in any larger triangular subalgebra of $\mathcal{M}$. A simple argument using Zorn's Lemma shows that every triangular algebra is contained in a maximal triangular algebra.

The *hulls* of a triangular algebra $\mathcal{T}$ are the projections in $\mathcal{M}$ that are invariant under $\mathcal{T}$. Since the diagonal $\mathcal{A}$ of $\mathcal{T}$ is a m.a.s.a. and the hulls of $\mathcal{T}$ necessarily commute with $\mathcal{A}$, they are contained in $\mathcal{A}$. In particular, a triangular algebra contains its hulls. The *core* of $\mathcal{T}$ is the von Neumann algebra generated by the hulls of $\mathcal{T}$. So also,

$$\operatorname{core}(\mathcal{T}) \subseteq \mathcal{A}.$$

If H is a finite dimensional Hilbert space, and if $\mathcal{M} = B(H)$, then it is easily shown that every maximal triangular subalgebra of $\mathcal{M}$ is unitarily equivalent to the *elementary triangular algebra* discussed in §0. The defining property which abstracts the notion of *triangularity* in the Kadison-Singer theory is that $\operatorname{Diag} \mathcal{T}$ is required to be a m.a.s.a. This is property 2 of §0. Property 3 states that the set of hulls of the *elementary triangular algebra* is totally ordered by inclusion. A structurally revealing aspect of the Kadison-Singer theory is that the set of hulls of a maximal triangular subalgebra of a factor are also totally ordered. This points out that the Kadison-Singer defining property is a very natural condition, capturing the *essence* of the notion of triangularity. It is good to give a proof of this before continuing. The proof we give is structured somewhat differently from that in [72].

Lemma 1.1.1. (Kadison-Singer) *Let* $\mathcal{M}$ *be a factor von Neumann algebra, and let* $\mathcal{T}$ *be a maximal triangular subalgebra of* $\mathcal{M}$. *Then the hulls of* $\mathcal{T}$ *are totally ordered.*

Proof: Let the diagonal of $\mathcal{T}$ be

$$\mathcal{A} = \mathcal{T} \cap \mathcal{T}^*.$$

It will suffice to show that maximality of T implies that for each hull P of T we have

$$PMP^\perp \subseteq T.$$

For then, if P and Q are hulls of T, and if P fails to contain Q, or equivalently, if $P^\perp Q \neq 0$, then

$$QH \supseteq TQH$$
$$\supseteq PMP^\perp QH$$
$$= PH$$

so $Q \geq P$, where the final equality is obtained by utilizing the Comparison Theorem for factors.

So let P be a hull of T, let $S = PMP^\perp$, and let

$$T_0 = T + S \qquad \text{(no closure)}.$$

It will suffice to show that T_0 is closed under multiplication (so is an *algebra*), and that $T_0 \cap T_0^* = A$, for then maximality of T will imply that $T_0 = T$, and hence that T contains S, as desired.

Let $T_i \in T, S_i \in S, i = 1, 2$. Then $T_i P = PT_i P$, $P^\perp T_i = P^\perp T_i P^\perp$ and $S_i = PS_i P^\perp$. Compute

$$(T_1 + S_1)(T_2 + S_2) = T_1 T_2 + T_1 S_2 + S_1 T_2 + S_1 S_2.$$

We have $T_1 T_2 \in T$ and $S_1 S_2 = 0$. Also,

$$T_1 S_2 = T_1 P S_2 P^\perp$$
$$= PT_1 P S_2 P^\perp$$
$$\in S,$$

and

$$S_1 T_2 = PS_1 P^\perp T_2$$
$$= PS_1 P^\perp T_2 P^\perp$$
$$\in S,$$

so $(T_1 + S_1)(T_2 + S_2) \in T + S = T_0$, as desired.

To show that $T_0 \cap T_0^* = A$ it will suffice to consider selfadjoint elements of $T_0 \cap T_0^*$. Let B be a selfadjoint element of $T_0 \cap T_0^*$. Observe that P is invariant under T_0. Then P commutes with B. Decompose B as

$$B = T + S$$

with $S \in \mathcal{S}, T \in \mathcal{T}$. Since $PSP = 0 = P^{\perp}SP^{\perp}$, we then have

$$B = PBP + P^{\perp}BP^{\perp}$$
$$= P(T+S)P + P^{\perp}(T+S)P^{\perp}$$
$$= PTP + P^{\perp}TP^{\perp}.$$

Both PTP and $P^{\perp}TP^{\perp}$ are in $\mathcal{T}$, hence $B \in \mathcal{T}$. So since B is selfadjoint, $B \in \mathcal{A}$, as desired.

The proof is complete. ∎

We have established a formal structural analogy between the simple triangular algebra acting on a finite dimensional Hilbert space studied in §0 and an arbitrary maximal triangular subalgebra of a factor: the hulls of each are totally ordered. In the finite dimensional case the hulls generate the diagonal as a von Neumann algebra (i.e., core = diagonal), but here is where the analogy in the abstract setting can break down. There is no reason to expect that an arbitrary maximal triangular algebra need have any hulls other than O and I. Indeed, it need not have. This structural situation is frequently the case, and leads to the study of *irreducible* (that is, *transitive*) triangular algebras, an important subject to which many researchers have devoted much attention over the years. The core of such an algebra consists of scalar multiples of the identity operator. Initiating the investigation of such algebras, Kadison and Singer proved the following.

Theorem 1.1.2. (Kadison-Singer) *Let $\mathcal{A}$ be a m.a.s.a. in $B(H)$. If a unitary operator U acts ergodically on $\mathcal{A}$ (i.e., if $U\mathcal{A}U^* = \mathcal{A}$ and $\mathcal{A} \cap \text{Lat}\,(U) = \{0, I\}$) then the algebra generated (no closure) by $\mathcal{A}$ together with U is triangular, and is irreducible.*

The classic example in this vein is obtained by letting $H = L^2(C)$, where C is the unit circle, and where the measure is Lebesgue measure, and letting U be the unitary transformation of H induced by an irrational rotation of C. The action of U on the multiplication algebra $\mathcal{A}$ of $L^2(C)$ is ergodic, so the algebra $\mathcal{T}$ generated by $\mathcal{A}$ together with U is triangular and is transitive.

It is useful to observe that the condition

$$\mathcal{A} \cap \text{Lat}\,(U) = \{0, I\}$$

is equivalent to the condition: if $E \in \mathcal{A}$ is a projection with

$$UEU^* \leq E,$$

then $E = 0$ or $E = I$.

Also, it is good to note that the condition that $U\mathcal{A}U^* = \mathcal{A}$ implies that the algebra generated by $\mathcal{A}$ and U is the set of *polynomials*

$$\Sigma A_n U^n$$

in the operator U, with coefficients in $\mathcal{A}$, and thus, at least formally, has a *triangular form* independent of invariant projection considerations. Ergodicity of the action of U on $\mathcal{A}$ implies, in addition, that a *polynomial* $\Sigma A_n U^n$ will be 0 if and only if each coefficient A_n is 0.

In "A density theorem for operator algebras" [7], William Arveson proved that a weakly closed subalgebra of $B(H)$ which is transitive and which contains a m.a.s.a. is necessarily $B(H)$ itself. Moreover, *weakly closed* can be replaced by *ultra-weakly closed*, as shown in particular in [10]. Thus a transitive triangular subalgebra of $B(H)$ cannot be ultraweakly closed. Arveson has asked the question of whether the *norm* closure of a transitive triangular algebra is also triangular.

In "Operator algebras and measure preserving automorphisms" [8] by Arveson, and its sequel [9] by Arveson and Josephson, the norm closures of the algebras generated as in Theorem 1.1.2 were shown to be triangular. These algebras completely determine the associated actions of the group of integers on the m.a.s.a. $\mathcal{A}$ (via $n \in Z \to \mathrm{ad}(U^n)$) in that two such algebras are algebraically isomorphic if, and only if, the corresponding actions are conjugate. In "Ergodic automorphisms and linear spaces of operators" [63] Alan Hopenwasser obtained conjugacy equivalence in terms of *complete isometric* equivalence of the subspaces spanned by $\mathcal{A}$ and U. More recently, very interesting progress in the theory was made in "Irreducible triangular algebras" [116], by Baruch Solel. In "Radicals, crossed products, and flows" [94] Paul Muhly obtained interesting results on semi-simplicity of irreducible triangular algebras. Very recently, the paper "Coordinates for triangular algebras" [95] by Muhly, Saito, and Solel is an important contribution to the subject.

We return to further aspects of the Kadison-Singer paper [72]. In their work, the most incisive results were obtained for the class of triangular algebras of the extreme opposite to the irreducible class. A triangular subalgebra $\mathcal{T}$ of a factor $\mathcal{M}$ is called *hyperreducible* (or *hyperintransitive*) if the set of hulls of $\mathcal{T}$ is *thick* enough to *generate* the diagonal $\mathcal{A} = \mathcal{T} \cap \mathcal{T}^*$ as a von Neumann algebra. That is, $\mathcal{T}$ is hyperreducible if $\mathrm{core}\,(\mathcal{T}) = \mathrm{Diag}\,(\mathcal{T})$, as is the case for the maximal triangular subalgebras of $B(H)$

when H is finite dimensional (as in §0). We state a theorem of Kadison and Singer, in original terminology, which completely characterizes the hyperreducible maximal triangular algebras in the case where the factor in the definition of *triangular* is $B(H)$. (The notation and terminology of nest algebras, and of reflexive operator algebras, and lattices, is now usually used.)

Theorem 1.1.3. (Kadison-Singer) *If $\{E_\alpha\}$ is a totally ordered family of projections which generates the maximal abelian (selfadjoint) algebra $\mathcal{A}$ of $B(H)$), then $\mathcal{T}$, the set of all bounded linear operators (in $B(H)$) which leave each E_α invariant, is a maximal triangular algebra, with core and diagonal $\mathcal{A}$. If $\{E_\alpha\}$ is closed under unions and intersections (of arbitrary subfamilies) then it is the set of hulls of $\mathcal{T}$. Each hyperreducible maximal triangular subalgebra of $B(H)$ arises in this way.*

The class of hyperreducible triangular subalgebras of $B(H)$ is the prototype of the class of *nest algebras* of J. Ringrose, which we will consider at length in §2. A brief preview may be good: A nest is a chain of projections (or subspaces) as is $\{E_\alpha\}$ in Theorem 1.1.3., and the associated nest algebra is the set ($\mathcal{T}$, in this theorem) of all operators in $B(H)$ leaving every member of the nest invariant. The condition, in the next to the last sentence in the statement of the theorem, that $\{E_\alpha\}$ be closed under unions and intersections of arbitrary subfamilies, is simply the condition that $\{E_\alpha\}$, as a lattice of projections, be complete, or equivalently, in this case (since these projections commute), that $\{E_\alpha\}$ be closed in the strong operator topology. The assertion that, under this condition, $\{E_\alpha\}$ is the set of hulls of $\mathcal{T}$, means, in current terminology, that $\{E_\alpha\}$ is a reflexive lattice of projections. The requirement that $\{E_\alpha\}$ generate a m.a.s.a. means that the algebras $\mathcal{T}$ in Theorem 1.1.3. constitute a special class of nest algebras: those that have multiplicity one. (The term *multiplicity free* is also used for these.)

In [72], maximal triangular algebras which are hyperreducible are called *ordered bases*.

A m.a.s.a. is said to be totally atomic if it is generated by its minimal projections. It is not hard to show that a maximal triangular algebra with totally atomic diagonal is hyperreducible. (Of course, the factor in the definition of *triangular* must then be of type I if it is to contain a totally atomic m.a.s.a., so this factor may be assumed to be $B(H)$.) The first basic result concerning order isomorphisms between lattices of invariant projections of operator algebras is the following.

Theorem 1.1.4. (Kadison-Singer) *If $\mathcal{T}_1$ and $\mathcal{T}_2$ are maximal triangular algebras with*

132

totally-atomic diagonals and ϕ is an order isomorphism between their sets of hulls, then ϕ can be implemented by a unitary transformation which carries T_1 onto T_2.

The idea in Theorem 1.1.4. is simple. A proof sketch here is worthwhile. We may assume that the factor is $B(H)$. Thus the minimal projections in $\mathrm{Diag}\,(T_1)$ and $\mathrm{Diag}\,(T_2)$ are of rank one. The total ordering of the hulls of T_1 induces a natural total ordering "$<<$" of the minimal projections of $\mathrm{Diag}\,(T_1)$, where $P << P'$ means that the smallest hull of T_1 containing P' also contains P. The minimal projections of T_2 are similarly ordered. With this ordering, ϕ induces an order isomorphism $\tilde{\phi}$ between the minimal projections in $\mathrm{Diag}\,(T_1)$ and the minimal projections in $\mathrm{Diag}\,(T_2)$. So the set of minimal projections in $\mathrm{Diag}\,(T_1)$ can be written $\{P_\lambda : \lambda \in \Lambda\}$ for a totally ordered set Λ, so that $P_{\lambda_1}0 << P_{\lambda_2}$ if and only if $\lambda_1 < \lambda_2$, and the minimal projections in $\mathrm{Diag}\,(T_2)$ can be similarly indexed as $\{Q_\lambda : \lambda \in \Lambda\}$ for the same totally ordered set Λ. For each λ choose unit vectors e_λ and f_λ with span the ranges of P_λ and Q_λ, respectively. Then $\{e_\lambda : \lambda \in \Lambda\}$ and $\{f_\lambda : \lambda \in \Lambda\}$ are orthonormal bases for H, so there is a unitary operator $U\epsilon B(H)$ with $Ue_\lambda = f_\lambda, \lambda\epsilon\Lambda$. Then $UP_\lambda U^* = Q_\lambda$ for each λ. It follows easily that for any hull P of T_1 we have $\phi(P) = UPU^*$, and since T_1 contains every operator that leaves the set of hulls of T_1 invariant, and similarly for T_2, it follows that $T_2 = UT_1U^*$.

The next theorem is quite important. It completely characterizes, up to unitary equivalence, the hyperreducible maximal triangular subalgebras of $B(H)$ with nonatomic diagonal, for H a separable Hilbert space. (*Nonatomic* means that the diagonal contains no minimal projection.)

Theorem 1.1.5. (Kadison-Singer) *If T is a hyperreducible maximal triangular subalgebra of $B(H)$ with nonatomic diagonal, A, and with H a separable Hilbert space, then T is unitarily equivalent to T_0, the algebra of all bounded operators on $L_2(0,1)$ (Lebesgue measure) leaving each F_λ invariant, where F_λ is the projection due to multiplication by the characteristic function, $\chi_{_\lambda}$, of $[0, \lambda]$.*

Unlike the totally atomic diagonal case of Theorem 1.1.4, it was noted in [72] that *not* every order isomorphism between the sets of hulls of two hyperreducible maximal triangular subalgebras of $B(H)$ with nonatomic diagonals can be implemented by a unitary transformation. If $\{F_\lambda : 0 \le \lambda \le 1\}$ and T_0 are as in Theorem 1.1.5, and if f is any order isomorphism of $[0,1]$ onto $[0,1]$ which does not preserve the Borel sets of Lebesgue measure 0, then the order isomorphism from the hull set of T_0 onto itself given by $\phi(F_\lambda) = F_{f(\lambda)}, 0 \le \lambda \le 1$, cannot be implemented by a unitary transformation for measure-theoretic reasons. (For one example of such a function, we may let $f(\lambda) = (\lambda + g(\lambda))/2$, where g is the Cantor function.) For a given f, the corresponding order isomorphism ϕ can be implemented by a unitary transformation if and only if the measures μ and $\mu \circ f$ are mutually absolutely continuous, where μ is Legesgue measure. (It is now known that every such ϕ *can* be implemented by an invertible operator which is a small compact perturbation of a unitary transformation, where the implementation is in the sense that the invertible operator T maps the range space of F_λ onto the range space of $F_{f(\lambda)}$ for every λ (i.e., $TF_\lambda H = F_{f(\lambda)}H, 0 \le \lambda \le 1$) and we have $TT_0T^{-1} = T_0$. This belongs to the subject matter of similarity theory for nest algebras. It, together with other topics, will be discussed in the remaining chapters.) The algebra T_0 of Theorem 1.1.5 is important in nest algebra theory. It will be called the *Volterra* nest algebra, and its nest $\{F_\lambda\}$ will be called the Volterra nest.

Much of the balance of [72] deals with the classification, up to unitary equivalence, of the hyperreducible maximal triangular subalgebras of $B(H)$ with *mixed* (neither totally atomic nor nonatomic) diagonals. Dealing with the general classification problem is not simply a matter of separating the diagonal into its totally-atomic and non-atomic parts. The order involved places the atoms throughout the continuous portion of the diagonal in a manner which does not permit a separation consistant with the theory. The theory developed in [72] is useful, and is used in, and leads to, the multiplicity theory for general nests worked out in [35] and partially discussed in §2.3 of Chapter 2. Two aspects are important to note for our purpose. The first is that *every* hyperreducible maximal triangular subalgebra of $B(H)$ is unitarily equivalent to an algebra constructed as is the algebra T_0 in Theorem 1.5, with Lebesgue measure replaced with a perhaps different Borel measure on $[0,1]$, which may have atoms. Equivalently, every complete chain of projections in $B(H)$ which generates a m.a.s.a. is unitarily equivalent to a chain, such as $\{F_\lambda\}$, for some measure, where if the measure has atoms we may have to augment by including the additional projections corresponding to multiplication by the half-open intervals $[0, \lambda)$. The second aspect is that if μ and ν are Borel measures on $[0,1]$, and if $\{F_\lambda^\mu\}$ and $\{F_\lambda^\nu\}$ are the corresponding chains of projections constructed (as in Theorem 1.1.5) for μ and ν, respectively, then the order isomorphism $\phi(F_\lambda^\mu) = F_\lambda^\nu$ can be implemented by a unitary transformation if and only if the measures μ and ν are mutually absolutely continuous. As a special case, if μ is an arbitrary Borel measure and if f is an order isomorphism from $[0,1]$ onto $[0,1]$, then, just as discussed earlier

134

for Lebesgue measure, the order isomorphism $F_\lambda^\mu \to F_{f(\lambda)}^\mu$ can be implemented by a unitary transformation if and only if the measures μ and $\nu \circ f$ are mutually absolutely continuous. Now, if one is simply given two Borel measures μ and ν on $[0,1]$, and if one asks the question of whether the algebras T_0^μ and T_0^ν are unitarily equivalent, one has the problem of deciding when there is an order isomorphism f from $[0,1]$ onto $[0,1]$ such that ν and $\mu \circ f$ are mutually absoltuely continuous. Obvious technical considerations will arise.

Kadison and Singer posed several questions in [72], two of which are of particular interest here, because they stimulated much subsequent work by others. We state these, together with quoted comments from [72], for perspective.

Question A. (This is question 2.4.1 from [72].)

Is each bounded operator contained in some *hyperreducible* algebra?

"This would provide a *triangular form* for bounded operators. We are inclined to feel that there is little hope that this question has an affirmative answer. It raises, in a natural way, the following question."

Question B. (This is question 2.4.2 from [72].)

Is each bounded operator contained in some maximal triangular algebra?

"This is a much broader question, allowing, as it does, the possibility that the operator falls in an irreducible maximal triangular algebra. From this we would not conclude the existance of a single proper invariant subspace, although in the general sense of our theory, we would have the operator in *triangular form* and might gain knowledge about it from an analysis of the maximal triangular algebra in question."

1.2 Subdiagonal Operator Algebras

Another very important early paper, which has stimulated a great deal of work in the subject, was "Analyticity in operator algebras" by William B. Arveson [6], which appeared in 1967, and which introduced the class of *subdiagonal algebras*. We give a brief account of some of the points of this paper.

Let $\mathcal{B}$ be a von Neumann algebra, and let Φ be a faithful normal positive linear mapping of $\mathcal{B}$ into itself which is idempotent. [So $\Phi(A) \geq 0$ whenever $A \geq 0$ (positivity), $\Phi(A) > 0$ whenever $A > 0$ (faithfulness), Φ is ultraweakly continuous (normality), and $\Phi \circ \Phi = \Phi$ (the idempotent property).]

A subalgebra $\mathcal{A}$ of $\mathcal{B}$ is said to be *subdiagonal* (with respect to Φ) if it has the following properties:

(i) $\mathcal{A} + \mathcal{A}^*$ is ultraweakly dense in $\mathcal{B}$,

(ii) $\Phi(AB) = \Phi(A)\Phi(B), A, B \in \mathcal{A}$,

(iii) $\Phi(\mathcal{A}) \subseteq \mathcal{A} \cap \mathcal{A}^*$,

(iv) $(\mathcal{A} \cap \mathcal{A}^*)^2$ has trivial nullspace.

With this definition, the diagonal of $\mathcal{A}$ need not be abelian. The class of sub-diagonal algebras includes many of the triangular operator algebras with particularly interesting structural properties. It enlarges the set of algebras under consideration to include many that are not triangular in the Kadison-Singer sense.

From ultraweak continuity of Φ it follows that the ultraweak closure of a subdiagonal algebra is subdiagonal.

By an *expectation* from a von Neumann algebra $\mathcal{B}$ onto a von Neumann subalgebra $\mathcal{M}$ we mean, as usual, a positive linear map ψ (not necessarily faithful or normal) of $\mathcal{B}$ onto $\mathcal{M}$ which leaves the identity fixed and which satisfies $\psi(AX) = A\psi(X), A \in \mathcal{M}, X \in \mathcal{B}$. These conditions imply that the range of ψ is $\mathcal{M}$, that ψ is idempotent, and that $\psi(XA) = \psi(X)A, A \in \mathcal{M}, X \in \mathcal{B}$.

Now let $\mathcal{A}$ be subdiagonal in $\mathcal{B}$ with respect to Φ, let $\mathcal{D} = \mathcal{A} \cap \mathcal{A}^*$, and let $\mathcal{D}^-$ denote the ultraweak closure of $\mathcal{D}$.

Proposition 1.2.1. (Arveson) Φ *is an expectation from $\mathcal{B}$ onto $\mathcal{D}^-$.*

Now let $\mathcal{T} = \{T \in \mathcal{A} : \Phi(T) = 0\}$. From the defining property (ii), the restriction of Φ to $\mathcal{A}$ is an algebraic homomorphism, so $\mathcal{T} = \ker(\Phi|_\mathcal{A})$ is a 2-sided ideal in $\mathcal{A}$, and $\mathcal{A} = \mathcal{D} + \mathcal{T}$, and $\mathcal{T} \cap \mathcal{D} = (0)$ since Φ is an idempotent linear mapping.

Given $\mathcal{A}$ and Φ, there is a unique subdiagonal (with respect to Φ) subalgebra of $\mathcal{B}$ which is maximal with respect to the property that it contains $\mathcal{A}$ and is subdiagonal with respect to Φ. It is interesting that it has an algebraic characterization.

Theorem 1.2.2. (Arveson) *Let $\mathcal{A}_m$ be the set of $X \in \mathcal{B}$ with $\Phi(TXA) = \Phi(AXT) = 0$. Then $\mathcal{A}_m$ is a subdiagonal subalgebra of $\mathcal{B}$ (with respect to Φ), $\mathcal{A}_m$ contains $\mathcal{A}$, and $\mathcal{A}_m$ is maximal in the sense that it contains every subdiagonal algebra $\mathcal{A}_1$ with $\mathcal{A}_1 \geq \mathcal{A}$.*

Arveson analyzed several examples of concretely defined subdiagonal algebras, all of which have been relevant to subsequent work in the general subject. One of these is particularly relevant to these notes: algebras based on invariant subspaces. Others are algebras based on ordered groups, and algebras based on groups of measurable transformations. Although terminology and notation currently in use is different, we use the original, for perspective here.

Let $\mathcal{B}$ be a von Neumann algebra, and let $\mathcal{P}$ be a nonempty *abelian* family of projections in $\mathcal{B}$. Put

$$\mathcal{A} = \{X \in \mathcal{B} : EXE = XE \text{ for all } E \in \mathcal{P}\}$$

$$\mathcal{D} = \{X \in \mathcal{B} : EX = XE \text{ for all } E \in \mathcal{P}\}.$$

Then $\mathcal{A}$ is weakly closed, and $\mathcal{D} = \mathcal{A} \cap \mathcal{A}^* = P' \cap \mathcal{B}$. Assume $\mathcal{D}$ is *compatible with $\mathcal{B}$* in the sense that there is a faithful normal expectation of $\mathcal{B}$ onto $\mathcal{D}$. In this case (as is well known), Φ is unique.

Theorem 1.2.3. (Arveson) *If $\mathcal{D}$ is compatible with $\mathcal{B}$, then the unique expectation Φ from $\mathcal{B}$ onto $\mathcal{D}$ is multiplicative on $\mathcal{A}$. If $T \in \mathcal{A}$ and $\Phi(T) = 0$, then there is a uniformly bounded net $\{T_n\}$ in $\mathcal{A}$, of nilpotents, such that $\Phi(T_n) = 0$, and $T_n \to T$ in the strong operator topology. If $\mathcal{P}$ is linearly ordered, then every element of $\mathcal{B}$ is in the strong closure of a bounded subset of $\mathcal{A} + \mathcal{A}^*$. In particular, $\mathcal{A} + \mathcal{A}^*$ is ultraweakly dense in $\mathcal{B}$.*

Corollary 1.2.4. (Arveson) *If $\mathcal{P}$ is linearly ordered and $\mathcal{D}$ is compatible with $\mathcal{B}$, then $\mathcal{A}$ is maximal subdiagonal, with respect to the expectation Φ.*

The most incisive results obtained in [6] on subdiagonal algebras were for those which are *finite* in the sense that there is a faithful normal finite trace ϕ on $\mathcal{B}$ such that $\phi \circ \Phi = \phi$, and which satisfy a maximality condition. The elements of $\mathcal{B} \cap \mathcal{B}^{-1}$ are related to $\mathcal{A} \cap \mathcal{A}^{-1}$ by a factorization theorem, and this is used to study the relation between generalizations of some properties of bounded analytic functions in the open unit disk. "Jensen's" inequality, for instance, is shown to be valid in some examples, both hyperreducible, and irreducible, of triangular subalgebras of II_1 factors. An important factorization result is the following.

Theorem 1.2.5. (Arveson) *Let $\mathcal{A}$ be a finite maximal subdiagonal subalgebra of $\mathcal{B}$. Then every invertible operator in $\mathcal{B}$ admits a factorization UA, where U is unitary and $A \in \mathcal{A} \cap \mathcal{A}^{-1}$. If UA and VB are two such representations for the same operator, then there is a unitary W in the diagonal of $\mathcal{A}$ such that $V = UW^{-1}$ and $B = WA$.*

2. Nest Algebras

2.1 A Paper of Ringrose.

In the paper "On some algebras of operators" by John R. Ringrose [108] the subject of nest algebras of operators on Hilbert space was formally introduced. If we begin with the characterization of hyperreducible maximal triangular subalgebras of $B(H)$ obtained by Kadison and Singer in Theorem 1.1.3., and simply drop the condition that the family of projections $\{E_\alpha\}$ be required to generate a m.a.s.a., we have defined the class of nest algebras. This was *previewed* in the remark following the statement of Theorem 1.1.3. We begin with a formal account of some aspects of Ringrose's paper [108], maintaining most original terminology and notation.

Let H be a Hilbert space. A family $\mathcal{F}$ of subspaces of H will be termed a *nest* if it is totally ordered by inclusion. A nest $\mathcal{F}$ is *complete* if

   (i) $(0), H \in \mathcal{F}$,

and

   (ii) given any subset $\mathcal{F}_0 \subseteq \mathcal{F}$, the subspaces

$$\bigwedge_{L \in \mathcal{F}_0} L, \quad \bigvee_{L \in \mathcal{F}_0} L$$

   are both members of $\mathcal{F}$.

Given a complete nest $\mathcal{F}$, and a non-zero subspace M in $\mathcal{F}$, we define

$$M_- = \bigvee \{L : L \in \mathcal{F}, L \subset M\}.$$

Then $M_- \in \mathcal{F}$. If M has an immediate predecessor in $\mathcal{F}$ (with respect to inclusion), then this predecessr is M_-. If, on the other hand, M has no immediate predecessor, then $M_- = M$. (If $M_- \neq M$, then the subspace $M \ominus M_-$, or the projection onto this, is usually called a *gap* of the nest.)

A nest $\mathcal{F}$ is said to be *simple* if it is complete and satisfies the condition that given any M in $\mathcal{F}$ such that $M \neq (0)$, the subspace M_- has codimension at most 1 in M. A nest $\mathcal{F}$ is said to be *maximal* if it is not properly contained in any larger nest.

Lemma 2.1.1. (Ringrose) *A nest $\mathcal{F}$ is maximal if and only if it is simple.*

Henceforth, only the term *maximal* will be used.

A nest is a linearly ordered set so has *order intervals*. Let "$\subset$" denote *proper* inclusion. Given M and N in $\mathcal{F}$ with M containing N, we write

$$(M, N) = \{L \in \mathcal{F} : N \subset L \subset M\}$$
$$[M, N] = \{L \in \mathcal{F} : N \subseteq L \subseteq M\}$$
$$[M, N) = \{L \in \mathcal{F} : N \subseteq L \subset M\}$$
$$(M, N] = \{L \in \mathcal{F} : N \subset L \subseteq M\}.$$

An order interval of the first type is called *open*, and of the second type is called *closed*. A nest carries a natural topology called its *order topology* determined by taking as a subbase the set of open order intervals. The following lemma is useful.

Lemma 2.1.2. *A complete nest is compact in its order topology.*

Given a nest $\mathcal{F}$, the *nest algebra* associated with $\mathcal{F}$ is the class of all bounded linear operators from H into itself which leave invariant each subspace of $\mathcal{F}$. Ringrose's original notation for this is $\mathcal{N}_\mathcal{F}$. (The notation presently commonly used is Alg $(\mathcal{F})$.)

Given vectors e and f in H, the notation $e^* \otimes f$ will be used to denote the (rank-1) operator which assigns to each vector $U \in H$ the vector $< u, e > f$. (The notation used in [108] is $e \otimes f$. We utilize the *star* to avoid conflict with notation of others.) The following is simple, but useful.

Lemma 2.1.3. (Ringrose) *Let F be a complete nest, and let e and f be non-zero vectors in H. Then $e^* \otimes f \in N_\mathcal{F}$ if and only if there is a subspace M in $\mathcal{F}$ such that $f \in M$ and $e \in H \ominus M_-$.*

The next result in Ringrose's paper states that complete nests are *reflexive*, although the term, and the subject, of reflexivity of lattices of subspaces, had not yet been formally introduced. Basically, this, or rather the Kadison-Singer Theorem 1.1.3., is the first non-selfadjoint reflexivity theorem in the literature.

Theorem 2.1.4. *Let $\mathcal{F}$ be a complete nest, and let M be a subspace of H which is invariant under $\mathcal{N}_\mathcal{F}$. Then $M \in \mathcal{F}$.*

A nest algebra contains certain norm closed two–sided ideals termed the *diagonal ideals* by Ringrose. If L and M are contained in the nest $\mathcal{F}$ with $L \subset M$, let $P(L, M)$ denote the projection onto the difference space $M \ominus N$, and let $\rho_{L,M}$ denote the semi-norm defined on $\mathcal{N}_\mathcal{F}$ by $\rho_{L,M}(T) = \|P(L, M)TP(L, M)\|, T \in \mathcal{N}_\mathcal{F}$. It is easy to show that $\rho_{L,M}$ is multiplicative. (This follows from semi-invariance of the subspace $M \ominus N$.) Given $M \in \mathcal{F}$ with $M \neq (0)$, the diagonal ideal $\mathcal{J}_{M^-}$ is the set of all $T \in \mathcal{N}_\mathcal{F}$ with the following property: given any positive real number ϵ, there exists $L \in \mathcal{F}$ with $L \subset M$ and $\rho_{L,M}(T) < \epsilon$. (Since each $\rho_{L,M}$ is multiplicative, it is easy to show that this is indeed a closed two–sided ideal.) Similarly, given $M \in \mathcal{F}$ with $M \neq H$, the diagonal ideal $\mathcal{J}_{M^+}$ is the set of all $T \in \mathcal{N}_\mathcal{F}$ with the property that given $\epsilon > 0$ there exists $L \in \mathcal{F}$ with $M \subset L$ and $\rho_{M,L}(T) < \epsilon$. Recall, if $\mathcal{A}$ is a unital Banach algebra, then $\mathcal{A}$ contains a unique (necessarily norm closed) two–sided ideal maximal with respect to the property that all of its elements are quasinilpotent, and this coincides with the Jacobson radical of $\mathcal{A}$, which is abstractly defined to be the intersection of the kernels of all strictly irreducible (that is, strictly transitive) representations of $\mathcal{A}$. The following is proven.

Theorem 2.1.5. (Ringrose) *Let $\mathcal{F}$ be a complete nest of subspaces of a Hilbert space H. Then the radical $\mathcal{R}_\mathcal{F}$ of the nest algebra $\mathcal{N}_\mathcal{F}$ is the intersection of the diagonal ideals of $\mathcal{N}_\mathcal{F}$.*

Using Theorem 2.1.5. and compactness of $\mathcal{F}$ in its order topology, the following theorem is proven which is frequently called "The Ringrose Criterion."

Theorem 2.1.6. (Ringrose) *Let $\mathcal{F}$ be a complete nest of subspaces of a Hilbert space H, and let $T \in \mathcal{N}_\mathcal{F}$. Then, in order that $T \in \mathcal{R}_\mathcal{F}$, it is necessary and sufficient that the following condition be satisfied: given any positive real number ϵ, there exists a finite subnest $(L_0, L_1, \ldots, L_n)$ of $\mathcal{F}$ such that*

$$(0) = L_0 \subset L_1 \subset \ldots \subset L_n = H,$$

$$\|P(L_{j-1}, L_j)TP(L_{j-1}, L_j)\| < \epsilon \quad (j = 1, \ldots, n).$$

Much of the balance of [108] is devoted to investigation of properties of the *diagonal algebras*: the Banach algebras which are obtained as quotients of a nest algebra with its diagonal ideals. These are shown to be semi-simple. Also, using these, a functional representation is obtained for elements of a nest algebra modulo its radical.

2.2. Notation, Terminology, and Comments

Up to this point we have attempted to maintain original notation and terminology usage because we felt that it would be desirable to do so. Hereafter, the notation, and terminology, of reflexivity of operator algebras and of subspace lattices, as given in the introduction, will be standard in these lectures. Nests will be denoted by the symbols $\mathcal{L}, \mathcal{M}$, and $\mathcal{N}$. Given a nest $\mathcal{N}$, the corresponding nest algebra will be denoted by $\mathrm{Alg}\,\mathcal{N}$. The term *nest* will be used to denote either a chain of subspaces or of projections, the distinction determined by context. The Jacobson radical of $\mathrm{Alg}\,\mathcal{N}$ will simply be denoted by $\mathrm{rad}\,(\mathrm{Alg}\,\mathcal{N})$. The *diagonal* $\mathcal{A} \cap \mathcal{A}^*$ of an operator algebra will be denoted by $\mathrm{Diag}\,(\mathcal{A})$. The core of a nest algebra $\mathrm{Alg}\,\mathcal{N}$ (or of the nest $\mathcal{N}$) is the von Neuman algebra generatd by the projections in the nest $\mathcal{N}$, and is denoted by $\mathrm{core}\,(\mathcal{N})$. In the case where $\mathcal{N}$ generates a m.a.s.a., so $\mathrm{Alg}\,\mathcal{N}$ is a triangular operator algebra, this coincides with the definition of core given by Kadison and Singer in §2.1. Such a nest is said to be *multiplicity free* (or have *multiplicity one*.) A nest $\mathcal{N}$ is said to have *atomic core* if $\mathrm{core}(\mathcal{N})$ is a purely atomic von Neumann algebra. (In this case, $\mathcal{N}$ is frequently simply said to be an atomic nest.) A nest $\mathcal{N}$ is said to be *continuous* if no member of $\mathcal{N}$ has an immediate predecessor in $\mathcal{N}$. That is, if $N = N_-$ for every $N \in \mathcal{N}$. This is equivalent to the condition that core $(\mathcal{N})$ is a nonatomic (contains no minimal projection) von Neumann algebra. A complete continuous nest is maximal, by Lemma 2.1.1. If $\mathcal{N}$ is a nest, it is easily seen that $\mathrm{Diag}(\mathrm{Alg}\,\mathcal{N})$ is the commutant of $\mathrm{core}(\mathcal{N})$, and that $\mathrm{core}(\mathcal{N})$ is the center of $\mathrm{Diag}(\mathrm{Alg}\,\mathcal{N})$. If $\mathcal{N}$ is a nest of projections, an $\mathcal{N}$-*interval* is a projection of the form $E = M - N$, with $M, N \in \mathcal{N}$ and $N \le M$. The projections M and N are called the upper and, respectively, lower *endpoints* of E. It is easily verified that the endpoints of a nonzero $\mathcal{N}$-interval are well defined. A convenient, easily proven fact is that the set of minimal core projections coincides with the set of *gap* projections, and these are the minimal $\mathcal{N}$-intervals.

A *finite* $\mathcal{N}$-*partition* is a finite set $\{E_1, \ldots, E_n\}$ of mutually orthogonal $\mathcal{N}$ intervals with $E_1 + \ldots + E_n = I$. The criterion of Theorem 2.1.6 can now be stated in a convenient alternate form.

Theorem 2.2.1. (Ringrose Criterion) *Let $\mathcal{N}$ be a complete nest. If $T \in \mathrm{Alg}\,\mathcal{N}$, then $T \in \mathrm{rad}(\mathrm{Alg}\,\mathcal{N})$ if and only if for each $\epsilon > 0$ there is a finite $\mathcal{N}$-partition $\{E_i\}$ such that $\|E_i T E_i\| < \epsilon$ for each i.*

If P is a projection in $\mathcal{P}(H)$, and if $T \in B(H)$, then the operator $PT(I - P)$ leaves invariant every projection contained in P, and also every projection which contains P. Thus if $\mathcal{N}$ is a nest, and if $N \in \mathcal{N}$, then for each $T \in B(H)$ we have $NT(I - N) \in \mathrm{Alg}\,\mathcal{N}$. From this, the following corollary of Theorem 2.2.1, which is useful for perspective, is almost immediate.

Corollary 2.2.2. *Let $\mathcal{N}$ be a complete nest. Then $\mathrm{rad}(\mathrm{Alg}\,\mathcal{N})$ is the norm closed linear span of the set $\{NT(I - N) : N \in \mathcal{N}, T \in B(H)\}$.*

If $\mathcal{N}$ is a nest, and if E and F are nonzero $\mathcal{N}$-intervals, we will write $E << F$ if the upper endpoint of E is contained in the lower endpoint of F. Equivalently, if E and F are orthogonal $\mathcal{N}$-intervals, then $E << F$ if and only if $EB(H)F \subseteq \mathrm{Alg}\,\mathcal{N}$. The following corollary of Theorem 2.2.1. is also useful for perspective.

Corollary 2.2.3. *Let $\mathcal{N}$ be a complete nest. Then $T \in \mathrm{rad}(\mathrm{Alg}\,\mathcal{N})$ if and only if*

$T \in \text{Alg}\,\mathcal{N}$ and T is in the norm closed linear span of the set $\{ETF : E \text{ and } F \text{ are orthogonal } \mathcal{N}\text{-intervals with } E << F\}$.

Thus the radical of $\text{Alg}\,\mathcal{N}$ consists of those operators in $\text{Alg}\,\mathcal{N}$ which can be *synthesized* in the norm topology by their *strictly upper triangular blocks*. This is analogous to the case of the *elementary triangular algebra* mentioned in the description of property 6 in §0. However, the Wedderburn decomposition in property 6 does not hold for infinite nests. For an infinite nest $\mathcal{N}$, there are many *strictly upper triangular* operators in $\text{Alg}\,\mathcal{N}$ which are not quasinilpotent so cannot be contained in $\text{rad}(\text{Alg}\,\mathcal{N})$.

2.3. A Connection with Triangular Algebras

We give a brief account of an early paper of John Erdos, "Some results on triangular operator algebras" [34], which makes some contact between the study of nest algebras and one of the questions raised by Kadison and Singer for triangular operator algebras.

A maximal triangular subalgebra $\mathcal{T}$ of $B(H)$ is called *strongly reducible* if its nest $\mathcal{N}$, of hulls, is a maximal nest. Also, a nest $\mathcal{N}$ is called *quasi-maximal* if all of its *gaps* $N \ominus N_-, N \in \mathcal{N}$, have dimension 0, 1 or ∞. (So the nest $\{0, H\}$ is quasi-maximal, in particular.)

Theorem 2.3.1. (Erdos) *The nest of hulls of a maximal triangular subalgebra of $B(H)$ is quasi-maximal. Conversely, if $\mathcal{N}$ is any quasi-maximal nest of subspaces of a separable Hilbert space, then there exists a maximal triangular algebra having $\mathcal{N}$ as its set of hulls.*

If H is a finite dimensional Hilbert space, and if $\mathcal{N}$ is a maximal nest of subspaces of H, then it is easily seen that $\text{core}\,(\mathcal{N})$ is a m.a.s.a. so that $\text{Alg}\,(\mathcal{N})$ is triangular. However, if H is infinite dimensional there exist maximal nests $\mathcal{N}$ for which core $(\mathcal{N})$ is *not* a m.a.s.a. This makes the following result interesting.

Theorem 2.3.2. (Erdos) *Let $\mathcal{N}$ be a maximal nest, and let $\mathcal{K}$ denote the ideal of compact operators in $B(H)$. Let $\mathcal{D}$ be any m.a.s.a. in $B(H)$ which contains the projection onto each member of $\mathcal{N}$. Then the algebra generated by $\mathcal{D}$ and $(\text{Alg}\,\mathcal{N}) \cap \mathcal{K}$ is triangular.*

Remark. This algebra will be norm-closed, but not closed in the weak operator toplogy unless $\mathcal{N}$ is finite.

From this a positive answer *for compact operators* is easily obtained to the Kadison-Singer question B discussed at the end of §1.1.

142

Corollary 2.3.3. (Erdos) *Any compact operator acting on a Hilbert space is a member of some strongly reducible triangular algebra.*

Proof: It is well-known [c.f. 110] that if T is a compact operator then the invariant subspace lattice of T contains a maximal nest $\mathcal{N}$. Then simply let $\mathcal{D}$ be as in Theorem 2.3.2 for this nest.

Theorem 2.3.2 is actually a special case of a more general result.

Theorem 2.3.4. (Erdos) *Suppose that $\mathcal{N}$ is a complete nest. Let $\mathcal{D}$ be any m.a.s.a. in $B(H)$ which contains the projection onto each member of $\mathcal{N}$. Then the algebra generated by $\mathcal{D}$ and $\mathrm{rad}\,(\mathrm{Alg}\,\mathcal{N})$ is triangular.*

2.4. Multiplicity Theory for Nests

Extending work of Kadison and Singer [72] on hyperreducible maximal triangular subalgebras of $B(H)$, which are now referred to as nest algebras with multiplicity one (or multiplicity free) nests, in [35] John Erdos worked out a detailed multiplicity theory for arbitrary complete nests. Some of the results in [35] were used in this author's paper [84]. The theory was developed for Hilbert spaces which are not necessarily separable. It will not be possible to give a detailed account in this lecture. Instead, we will abstract some exposition, for the separable case, from Erdos's survey article [37].

Let Ω be any totally ordered set with extreme elements 0 and 1 which is complete in its order topology. Let $\{\mu_i\}$ be a sequence of positive Borel measures on Ω such that $\mathrm{supp}(\mu_1) = \Omega$ (that is, no open order interval is a null set with respect to the first measure μ_1), and $\mu_i > \mu_{i+1}$ for each $i = 1, 2, \ldots,$. For each $\omega \in \Omega$, let

$$\chi_{[0,\omega]} L^2(\Omega, \mu_i)$$

denote the subspace of the Hilbert space

$$L^2(\Omega, \mu_i)$$

consisting of all (equivalence classes of) functions in $L^2(\Omega, \mu_i)$ with $f(t) = 0, \mu_i$-almost everywhere, for all $t > \omega$.

The set

$$\mathcal{N} = \{\oplus_i \chi_{[0,\omega]} L^2(\Omega, \mu_i) : \omega \in \Omega\}$$

is a complete nest of subspaces of the direct sum Hilbert space

$$\oplus_i L^2(\Omega, \mu_i).$$

Any nest $\mathcal{N}$ (of subspaces of a separable Hilbert space) that is complete is *unitarily equivalent* to a nest of the above form $\mathcal{N}$.

The (integral) *multiplicity* of $\mathcal{N}$ is then the *cardinality* of the set of non-zero measures in the sequence $\{\mu_i\}$.

Integral multiplicity is not a *complete* unitary equivalence invariant for nests. For a complete invariant, one must pass to a concept of multiplicity function, just as one does in the spectral multiplicity theory of self-adjoint operators. We refer the reader to [35] for further details.

2.5. Some Examples of Nest Algebras

Example 1. The elementary triangular algebra whose properties were discussed in §0 is a nest algebra. Its nest is

$$\{0, [e_1], [e_1, e_2], \ldots, [e_1, \ldots e_{n-1}], H\}.$$

Its core is equal to its diagonal, which is the set of operators which are diagonal with respect to the basis $\{e_1, \ldots, e_n\}$. This nest has multiplicity one.

Example 2. Let H be 4-dimensional, with orthonormal basis $\{e_1, e_2, e_3, e_4\}$, and let A be the set of operators in $B(H)$ with matrix with respect to this basis of the form

$$\begin{pmatrix} * & * & * & * \\ * & * & * & * \\ 0 & 0 & * & * \\ 0 & 0 & * & * \end{pmatrix}$$

where "*" means arbitrary. Then $\mathcal{A}$ is a nest algebra with nest $\mathcal{N} = \{0, [e_1, e_2], H\}$, which has uniform multiplicity two. We have

$$\text{core}(\mathcal{N}) = \left\{ \begin{pmatrix} \lambda & 0 & 0 & 0 \\ 0 & \lambda & 0 & 0 \\ 0 & 0 & \mu & 0 \\ 0 & 0 & 0 & \mu \end{pmatrix} : \lambda, \mu \in \mathbb{C} \right\}$$

and

$$\text{Diag}(\text{Alg}\,\mathcal{N}) = \begin{pmatrix} * & * & 0 & 0 \\ * & * & 0 & 0 \\ 0 & 0 & * & * \\ 0 & 0 & * & * \end{pmatrix}.$$

Example 3. Let H again be 4-dimensional, with $\{e_1, e_2, e_3, e_4\}$ an orthonormal basis, and let $\mathcal{A}$ be the set of operators in $B(H)$ with matrix with respect to this basis of the

144

form

$$\begin{pmatrix} * & * & * & * \\ * & * & * & * \\ 0 & 0 & * & * \\ 0 & 0 & 0 & * \end{pmatrix}.$$

Then $\mathcal{A}$ is a nest algebra with nest

$$\mathcal{N} = \{0, [e_1, e_2], [e_1, e_2, e_3], H\},$$

and has *mixed* multiplicity according to the definition in [35]. Its *integral* multiplicity is two. We have

$$\text{core}\,(N) = \left\{ \begin{pmatrix} \lambda & 0 & 0 & 0 \\ 0 & \lambda & 0 & 0 \\ 0 & 0 & \mu & 0 \\ 0 & 0 & 0 & \xi \end{pmatrix} : \lambda, \mu, \xi \in \mathbf{C} \right\}$$

and

$$\text{Diag}\,(\text{Alg}\,\mathcal{N}) = \begin{pmatrix} * & * & 0 & 0 \\ * & * & 0 & 0 \\ 0 & 0 & * & 0 \\ 0 & 0 & 0 & * \end{pmatrix}$$

Example 4. Let μ be a Borel measure on the unit interval $[0,1]$ and let H be the Hilbert space $L^2([0,1]; \mu)$. For each $0 \le t \le 1$, let $N_t(\mu) = L^2([0,t]; \mu)$, regarded as a subspace of H, where we use the same notation for μ as for its restriction measure. If μ is a nonatomic measure, then $\mathcal{N}(\mu) = \{N_t(\mu) : 0 \le t \le 1\}$ is a *complete* nest. (The subspaces $N_t(\mu)$ are not necessarily distinct; the distinct members form a complete nest.) If μ is a measure with atoms then $\{N_t(\mu) : 0 \le t \le 1\}$ may not be complete. The completion, which we denote by $\mathcal{N}(\mu)$, may contain certain additional subspaces of the form $L^2([0,t); \mu)$. In either case, core($\mathcal{N}(\mu)$) is the multiplication algebra of $L^2([0,1], \mu) : \{M_f : f \in L^\infty(\mu)\}$, where $M_f g = fg$, $g \in L^2(\mu)$. Since this is a m.a.s.a., $\mathcal{N}(\mu)$ has multiplicity one, and core$\,(\mathcal{N}(\mu)) = \text{Diag}\,(\text{Alg}\,\mathcal{N}(\mu))$.

Example 4 is more than an example: it is a *characterization* of multiplicity one separably acting nests which was obtained by Kadison and Singer in [72]. This was discussed in §2.1. We recapitulate:

Theorem 2.5.1. *Every complete multiplicity one nest acting on a separable Hilbert space is unitarily equivalent to a nest of the form $\mathcal{N}(\mu)$ for some finite Borel measure μ on $[0,1]$.*

If we extend the scope of Example 4 to *projection-valued* Borel measures μ on $[0,1]$ taking values in $B(\mathcal{K})$ for $\mathcal{K}$ a separable Hilbert space, the results of [35] yield a characterization of arbitrary nests.

145

Theorem 2.5.2. *Every complete separably acting nest is unitarily equivalent to a nest of the form $\mathcal{N}(\mu)$ for some projection-valued Borel measure μ on $[0,1]$.*

Example 5. If H is a finite dimensional Hilbert space, then every maximal nest of subspaces of H has multiplicity one, and the corresponding nest algebra is simply the full algebra of upper triangular operators with respect to an orthonormal basis of H. This implication need not hold if H is infinite dimensional.

Let $I = [0,1]$, let $\mu \times \mu$ be product Lebesgue measure on $I \times I$, and let $H = L^2(I \times I, \mu \times \mu)$. For $0 \le t \le 1$, let $N_t = L^2([0,t] \times I; \mu \times \mu)$, regarded as a subspace of H, where again we use the same notation for $\mu \times \mu$ as for its restriction measure. Then $\mathcal{N}$ is a complete nest, and is continuous, so is maximal. But $\mathcal{N}$ does not have multiplicity one. To see this, let E be the subset $I \times [0,1/2]$ of $I \times I$, and let P be the projection on $L^2(I \times I, \mu \times \mu)$ determined by multiplication of elements by the characteristic function of E. Then P commutes with the projections onto the subspaces in $\mathcal{N}$, but a simple computation shows that P is not in the weakly closed algebra generated by the projections onto the members of $\mathcal{N}$ (i.e. $\operatorname{core}(\mathcal{N})$). Thus $\operatorname{core}(\mathcal{N})$ is *not* maximal abelian in $B(H)$. Using definitions in [35], $\mathcal{N}$ can be shown to have uniform infinite multiplicity. The diagonal $(\operatorname{Alg}\mathcal{N}) \cap (\operatorname{Alg}\mathcal{N})^* = (\mathcal{C}_\mathcal{N})'$ is a von Neumann algebra of type I_∞.

Example 6. Let $\mathcal{N}$ be the nest of Example 4, for some measure μ. Define a nest algebra $\mathcal{A}$ on the direct sum $L^2([0,1]; \mu) \oplus L^2([0,1]; \mu)$ as follows:

$$\mathcal{A} = \operatorname{alg}\{N_t \oplus N_t : 0 \le t \le 1\}.$$

Let $\mathcal{M}$ be the nest $\{M_t : 0 \le t \le 1\}$, where $M_t = N_t \oplus N_t$. The nest $\mathcal{M}$ is of uniform multiplicity two, and is maximal if and only if the measure μ is nonatomic. (If μ is nonatomic then $\mathcal{M}$ will be continuous; conversely, if μ has an atom then $\mathcal{N}$ will have a 1-dimensional gap, so $\mathcal{M}$ will have a 2-dimensional gap, and hence will fail to be maximal.) The diagonal $(\operatorname{Alg}\mathcal{M}) \cap (\operatorname{Alg}\mathcal{M})^*$ is a (noncommutative) von Neumann algebra of type I_2. In particular, the nests $\mathcal{N}$ and $\mathcal{M}$ cannot be unitarily equivalent (except, of course, in the trivial case $\mu = 0$), but as we will see in Chapters 4 and 5, they are similar if μ is nonatomic: each is similar to the nest of Example 5.

Example 7. Let H be an infinite dimensional separable Hilbert space, and let Λ be a *countable* totally ordered set. Let $\{e_\lambda : \lambda \in \Lambda\}$ be an orthonormal basis for H indexed by Λ. For each $\lambda \in \Lambda$, define closed subspaces

$$N_\lambda = \overline{\mathrm{span}}\{e_\mu : \mu \leq \lambda\}$$
$$N_\lambda^- = \overline{\mathrm{span}}\{e_\mu : \mu < \lambda\},$$

and let

$$\mathcal{N} = \{N_\lambda : \lambda \in \Lambda\} \cup \{N_\lambda^- : \lambda \in \Lambda\}.$$

Then $\mathcal{N}$ is a complete multiplicity one nest. The gaps of $\mathcal{N}$ are precisely the subspaces $N_\lambda \ominus N_\lambda^- = \mathbb{C}\, e_\lambda, \lambda \in \Lambda$. $\mathcal{N}$ has atomic core, the minimal projections being the rank-1 projections $e_\lambda^* \otimes e_\lambda, \lambda \in \Lambda$.

Of course, this example can also be obtained via the construction in Example 4 by embedding Λ into [0,1] with its usual order, and defining μ by

$$\mu(E) = \mathrm{cardinality}\,(E \cap \Lambda)$$

for Borel subsets E of [0,1]. The vantage point given by the orthonormal basis construction is sometimes more natural, and can be useful for perspective.

Two special cases of Example 7 are noteworthy.

Case I. Let Λ denote the set of positive integers with usual order. Then $\mathrm{Alg}\,\mathcal{N}$ is the algebra of operators which have infinite upper triangular form with respect to the basis $\{e_1, e_2, \ldots\}$.

Case II. Let $\Lambda = Q$, the set of rational numbers, with usual order. This case has the curious property that

$$\{N_\lambda : \lambda \in \Lambda\} \cap \{N_\lambda^- : \lambda \in \Lambda\} = \phi.$$

This nest has tomic core, yet is uncountable. It has proven useful in similarity theory, and was called the *model* nest $\mathcal{N}(Q)$ in [84], and the *Cantor* nest in [29]. This nest is order isoorphic to the Cantor set (with usual total order as a subset of [0,1] with its usual order).

3. Several Topics

3.1. Some Notes on Similarity Transformations

Let H and $\mathcal{K}$ be Hilbert spaces, and let $T : H \to \mathcal{K}$ be a bounded invertible linear transformation. If $\mathcal{N}$ is a nest of closed subspaces of H, we let

$$T\mathcal{N} = \{TN : N \in \mathcal{N}\}$$

denote the image nest of closed subspaces of $\mathcal{K}$. A few pieces of information are easily deduced:

(i) If $\mathcal{N}$ is a complete nest, then the image nest $T\mathcal{N}$ is a complete nest in $\mathcal{K}$.

(ii) If $\mathcal{N}$ is a maximal nest in H, then $T\mathcal{N}$ is maximal in $\mathcal{K}$.

(iii) If $\mathcal{N}$ is a continuous nest, then $T\mathcal{N}$ is also continuous.

(iv) If $N \in \mathcal{N}$ and $N \neq N_-$, then $TN \neq (TN)_-$ in the nest $T\mathcal{N}$.

We have $(TN)_- = T(N_-)$. More precisely, if $0 \leq n \leq \aleph_0$, and if N_- has codimension n in N, then a dimension argument, using invertibility of T, shows that $(TN)_-$ has codimension n in TN. So the *positions* and *dimensions* of any gaps *match up* under the action of T. Nothing more can be said without more information about the operator T and the Nest $\mathcal{N}$.

The following question was posed by Ringrose and arose from his work in [105]: If $\mathcal{T}$ is a hyperreducible maximal triangular algebra, so is an ordered basis, and if T is an invertible bounded linear transformation, is the similarity transform algebra

$$T\mathcal{T}T^{-1} = \{TAT^{-1} : A \in \mathcal{T}\}$$

also an ordered basis?

We now know that the answer can be negative. However, unless $\mathcal{T}$ satisfies the very special property that $\mathrm{Lat}\,(\mathcal{T})$ is countable, in which the answer is always positive, there is no known characterization of those operators T which do yield a positive answer. Also, there has been no actual *construction* of an operator T which yields a negative answer. Proofs have been of *existence* type. Because of this situation, a preliminary exposition of alternate forms, and strengthenings of this question, for general $\mathcal{T}$, specific $\mathcal{T}$, general T and specific T, is in order, and serves as a means of introducing some of the relevant ideas.

If $\mathcal{L}$ is a family of closed subspaces of a Hilbert space H, and if T is an invertible bounded linear transformation from H onto a Hilbert space $\mathcal{K}$, and if we write $T\mathcal{L} = \{TL : L \in: \mathcal{L}\}$, it is easily deduced that

$$\mathrm{Alg}\,(T\mathcal{L}) = T(\mathrm{Alg}\,\mathcal{L})T^{-1} = \{TAT^{-1} : A \in \mathrm{Alg}\,\mathcal{L}).$$

Thus an equivalent formulation of the question above is: If $\mathcal{N}$ is a complete nest of multiplicity one, is the image nest $T\mathcal{N}$ of multiplicity one?

If $\mathcal{A} \subseteq B(H)$ is an algebra, and if T is an invertible bounded linear transformation, the map $A \to TAT^{-1}, A \in \mathcal{A}$, is, as usual, called the similarity transformation of $\mathcal{A}$ induced by T. If $\mathcal{L}$ is a lattice of subspaces, then, since this transformation applied to $\mathrm{Alg}\,\mathcal{L}$ yields $\mathrm{Alg}\,(T\mathcal{L})$, it is natural to refer to the map $L \to TL, L \in \mathcal{L}$, as a similarity transformation. The lattices $\mathcal{L}$ and $T\mathcal{L}$ are said to be *similar*. If T is a unitary operator, then $\mathcal{L}$ and $T\mathcal{L}$ are said to be *unitarily equivalent*. In this case the algebras $\mathrm{Alg}\,\mathcal{L}$ and $\mathrm{Alg}\,(T\mathcal{L})$ are unitarily equivalent.

If $\mathcal{L}$ is a lattice of projections, by $T\mathcal{L}$ we mean the lattice of orthogonal projections onto the subspaces $\{TLH : L \in \mathcal{L}\}$, so the system of notation is consistent. If T is a *unitary* transformation, then the projection onto TLH is simply TLT^*, so in this case we do have $T\mathcal{L} = \{TLT^* : L \in \mathcal{L}\}$, and so the map $L \to \mathrm{proj}\,(TLH)$ *is* a unitary transformation between the sets of projections in the usual sense. In general, when T is not unitary, there is no such simple algebraic relation between a projection L and its image projection under the similarity transformation induced by T. The correspondence is geometric, rather than algebraic.

Questions somewhat stronger than the one stated above would be whether multiplicity (in the sense of §2.4) was preserved under similarity, or whether $T\mathcal{N}$ was unitarily equivalent to $\mathcal{N}$. (Two nests can be order isomorphic and both may have, say, multiplicity one, and yet they may not be unitarily equivalent. One may have atomic core and the other a continuous part, for example. That is, there may be a difference in *measure type*. The strongest question in this vein would be whether, given an indexing of a nest $\mathcal{N}$ by a totally ordered set λ, so $\mathcal{N} = \{N_\lambda : \lambda \in \Lambda\}$, there exists a unitary operator U such that $TN_\lambda = UN_\lambda$ for each λ. That is, not only require that the *sets*

$$\{TN : N \in \mathcal{N}\} \text{ and } \{UN : N \in \mathcal{N}\}$$

be equal, as would be the definition of unitary equivalence, but also require that the specific order isomorphism

$$N_\lambda \to TN_\lambda$$

be implemented by the unitary transformation. (For example, it is now known that if $\mathcal{N}$ is, say, a continuous multiplicity one nest there exist bounded inveretible operators T which map $\mathcal{N}$ onto itself, in the sense of similarity of lattices, so $\mathcal{N}$ and $T\mathcal{N}$ are the same nest, but for which the order isomorphism $N \to TN$ cannot be implemented by a unitary transformation for the measure theoretic reasons discussed in §2.1.)

We note that if $\mathcal{N} = \{N_\lambda\}$ is a nest, and if T is an operator for which there exists a unitary operator U such that

$$TN_\lambda = UN_\lambda \text{ for all } \lambda,$$

then the operator U^*T maps N_λ onto N_λ for each λ, hence so does its inverse $(U^*T)^{-1}$, so both U^*T and its inverse lie in alg $\mathcal{N}$. Thus T admits a factorization $T = UA$, with U unitary and

$$A = U^*T \in (\text{Alg}\,\mathcal{N}) \cap (\text{Alg}\,\mathcal{N})^{-1}.$$

The converse is also true: if T factors thus, it is clear that the *action* of T on $\mathcal{N}$ can be implemented by the unitary U.

This $T = UA$ factorization of a bounded invertible operator T relative to a nest $\mathcal{N}$ is equivalent to the factorization of the positive invertible operator T^*T as

$$T^*T = A^*A, \text{for } A \in (\text{Alg}\,\mathcal{N}) \cap (\text{Alg}\,\mathcal{N})^{-1}.$$

Indeed, if $T = UA$ then $T^*T = A^*U^*UA = A^*A$ as desired. Conversely, if $T^*T = A^*A$ for some invertible operator $A \in \text{Alg}\,\mathcal{N}$ with $A^{-1} \in \text{Alg}\,\mathcal{N}$ also, let $T = V|T|$ and $A = W|A|$ be the corresponding polar decompositions, and noting that $|T| = |A|$, we then have $T = UA$ where $U = WV^*$, as desired.

We recapitulate the above facts in a lemma:

Lemma 3.1.1. *If $\mathcal{N}$ is a nest and T an invertible operator the following properties for T are equivalent:*

   (i) There exists a unitary U such that $TN = UN, N \in \mathcal{N}$.
   (ii) There exists a unitary U such that $UT \in (\text{Alg}\,\mathcal{N}) \cap (\text{Alg}\,\mathcal{N})^{-1}$.
   (iii) $T^*T = A^*A$ for some $A \in (\text{Alg}\,\mathcal{N}) \cap (\text{Alg}\,\mathcal{N})^{-1}$.

A second lemma is useful.

Lemma 3.1.2. *Let $\mathcal{N}$ be a nest and T a positive invertible operator. Suppose $T = A^*B$ with $A, B \in (\text{Alg}\,\mathcal{N}) \cap (\text{Alg}\,\mathcal{N})^{-1}$. Then $T = C^*C$ for some $C \in (\text{Alg}\,\mathcal{N}) \cap (\text{Alg}\,\mathcal{N})^{-1}$.*

Proof: We have $T = T^*$ so $A^*B = B^*A$, so $(A^*)^{-1}B^* = BA^{-1}$, hence $BA^{-1} \in \mathcal{D}_\mathcal{N}$. Let $D = BA^{-1}$. Then $B = DA$, so $T = A^*DA$, so $D = (A^{-1})^*T(A^{-1})$, a positive operator. Now let $C = D^{1/2}A$.

150

We note that if $\mathcal{N}$ is a complete nest of projections in a finite factor von Neumann algebra $\mathcal{B}$, then the nest subalgebra $\mathrm{Alg}_\mathcal{B}(\mathcal{N}) = (\mathrm{Alg}\,\mathcal{N}) \cap \mathcal{B}$ is a finite subdiagonal algebra, and so Theorem 1.2.5 (Arveson) shows that the *action* of any invertible operator $T \in \mathcal{B}$ on $\mathcal{N}$ can be implemented by a unitary operator in $\mathcal{B}$. So if $\mathcal{N}$ is a nest of projections in $B(H)$ which happens to be contained in a finite subfactor of $B(H)$, this shows that factorization holds relative to $\mathcal{N}$ for a large class of operators.

A positive invertible operator T is said to admit *universal factorization* if it factors in the sense of (iii) of Lemma 3.1.1 with respect to every nest of closed subspaces of H. With this definition, the theory developed by Brodskii, Gohberg, Krein, Macaev, and others for abstract Volterra operators yields the most general universal factorization result to date. We present the following theorem of Gohberg-Krein [53, Chapter 4, Theorem 6.2] for the special case in which T is positive, together with corollaries relevant to these notes. The notation $C_p, 1 \leq p < \infty$ will denote the Schatten p-classes and C_ω the Macaev ideal. We recall that C_ω is a symmetrically normed ideal which contains $C_p, 1 \leq p < \infty$, and which has the property that for any maximal nest $\mathcal{N}$ the integral of triangular truncation for each element of C_ω with respect to $\mathcal{N}$ converges.

Theorem 3.1.3. (Gohberg-Krein). *Let $\mathcal{N}$ be a maximal nest and let T be a positive invertible operator with $T - I \in C_\omega$. Then T^{-1} factors $T^{-1} = (I + X_+)D(I + X_-)$ with X_+, X_- quasinilpotents in $\mathrm{Alg}\,\mathcal{N}$, $\mathrm{Alg}\,\mathcal{N}^\perp$, respectively and D an invertible operator in $\mathrm{Diag}\,(\mathrm{Alg}\,\mathcal{N})$. [Here $\mathcal{N}^\perp$ denotes the dual nest consisting of the orthogonal complements of elements of $\mathcal{N}$. We have $\mathrm{Alg}\,(\mathcal{N}^\perp) = (\mathrm{Alg}\,\mathcal{N})^*$].*

Corollary 3.1.4. *Let $\mathcal{N}$ be an arbitrary nest and let T be a positive invertible operator with $T - I \in C_\omega$. Then T factors $T = A^*A$ for $A \in (\mathrm{Alg}\,\mathcal{N}) \cap (\mathrm{Alg}\,\mathcal{N})^{-1}$ and T satisfies (i), (ii), (iii) of Lemma 3.1.1.*

Proof: Imbed $\mathcal{N}$ in a maximal nest $\mathcal{M}$ if necessary, noting that $\mathrm{alg}\,\mathcal{M} \subset \mathrm{alg}\,\mathcal{N}$. Apply Theorem 3.2.2 and let $B = I + X_+, C = D(I + X_-)$ obtaining $T^{-1} = BC$ so $T = C^{-1}B^{-1}$ with $C^*, B \in (\mathrm{Alg}\,\mathcal{N}) \cap (\mathrm{Alg}\,\mathcal{N})^{-1}$. Now apply Lemma 3.1.2 obtaining $T = A^*A$, $A \in (\mathrm{Alg}\,\mathcal{N}) \cap (\mathrm{Alg}\,\mathcal{N})^{-1}$. Now note that $T^2 - I \in C_\omega$, hence also T satisfies (i), (ii), (iii) of lemma 3.1.1.

Corollary 3.1.5. *Let $\mathcal{N}$ be an arbitrary nest and suppose T is an invertible operator of the form $T = U + K$ with U unitary and $K \in C_\omega$. Then T satisfies (i), (ii), (iii) of Lemma 3.1.1 relative to $\mathcal{N}$.*

151

Proof: Then $T^*T - I \in C_\omega$.

We will continue with similarity theory in Chapters 4 and 5. Before doing this, we will give an account of some of the work leading to this. We begin with an account of Arveson's work on commutative subspace lattices.

3.2. Commutative Subspace Lattices

An important paper, which stimulated much of the work accomplished in recent years on reflexivity, was "Operator algebras and invariant subspaces" by William B. Arveson [10]. We give a brief account of some of its aspects.

A lattice $\mathcal{L}$ of projections in $B(H)$ is called a *subspace lattice* if it is closed in the strong operator topology. $\mathcal{L}$ is said to be *commutative* if the elements of $\mathcal{L}$ commute. Alan Hopenwasser has suggested that the abbreviation "CSL" be used, and we adopt this notation. So a CSL is a commutative subspace lattice, and a CSL-algebra is a reflexive operator algebra whose lattice of invariant projections is commutative. Equivalently, a CSL-algebra is a subalgebra of $B(H)$ of the form Alg $\mathcal{L}$, with $\mathcal{L}$ a CSL.

If $\mathcal{L}$ is an arbitrary subspace lattice, then every operator A which commutes with $\mathcal{L}$ has the property that both A and A^* leave every member of $\mathcal{L}$ invariant, so A is contained in the diagonal $(\text{Alg}\,\mathcal{L}) \cap (\text{Alg}\,\mathcal{L})^*$. The converse is also true: if $A \in \text{Diag}\,(\text{Alg}\,\mathcal{L})$ then $A \in (\mathcal{L})'$. So in general, $\text{Diag}\,(\mathcal{A}) = (\mathcal{L})'$.

If $\mathcal{L}$ is a CSL, then $\mathcal{L} \subset \text{Alg}\,(\mathcal{L})$, and in fact this inclusion characterizes those subspace lattices which are commutative. Also, since $\mathcal{L}$ is commutative, the commutant of $\mathcal{L}$ contains a m.a.s.a., so in particular, Alg $\mathcal{L}$ contains a m.a.s.a. The converse is also true here: if $\mathcal{A}$ is an operator algebra which contains a m.a.s.a., then every projection in Lat $(\mathcal{A})$ is invariant under the m.a.s.a. so commutes with the m.a.s.a. and hence is contained in the m.a.s.a., and thus Lat $(\mathcal{A})$ is commutative. So $\mathcal{L}$ is a CSL if and only if Alg $(\mathcal{L})$ contains a m.a.s.a. if and only if $\mathcal{L} \subseteq \text{Alg}\,(\mathcal{L})$.

In Chapter 1 of [10], Arveson proved that separably acting commutative subspace lattices are reflexive. That is, if H is a separable Hilbert space and if $\mathcal{L}$ is a CSL acting on H, then $\mathcal{L} = \text{Lat Alg}\,\mathcal{L}$. The proof of this theorem utilizes a *spectral representation theorem* for commutative subspace lattices which has played an important role in much of the subsequent work that has been accomplished in the literature concerning reflexive operator algebras.

Chapter II of [10] concerns operator algeras (not necessarily reflexive) which contain a m.a.s.a. It is shown that every such algebra $\mathcal{A}$ which is closed in the weak operator topology is *pre-reflexive* in the sense that $\mathcal{A}$ and Alg Lat $(\mathcal{A})$ have the same diagonal. A reflexive lattice $\mathcal{L}$ is defined to be *synthetic* if the only ultraweakly closed pre-reflexive algebra $\mathcal{A}$ satisfying the relation Lat $\mathcal{A} = \mathcal{L}$ is the algebra $\mathcal{A} = $ Alg $\mathcal{L}$. Arveson showed that if $\mathcal{L}$ is commutative there is a *smallest* pre-reflexive algebra, called $\mathcal{A}_{\min}$, for which Lat $(\mathcal{A}_{\min}) = \mathcal{L}$. This is generated by the class of *pseudo-integral operators* for $\mathcal{L}$, defined by Arveson. We will not have space to consider this here. A subspace lattice $\mathcal{L}$ is said to have *finite width* if there is a finite set of nests $\{C_1, C_2, \ldots, C_n\}$ such that

$$\mathcal{L} = C_1 \vee C_2 \vee \ldots \vee C_n.$$

(Here, the *join* of a family of subspace lattices is defined to be the smallest subspace lattice containing every lattice in the family.) The *width* of $\mathcal{L}$ is defined as the smallest such integer n, when $\mathcal{L}$ has finite width, and is defined as ∞ otherwise. Arveson proved that every commutative subspace lattice of finite width is synthetic.

In Chapter III, Arveson dealt with aspects of the general problem of classifying (infinite) complete distributive lattices. This is closely connected with the problem of classifying certain algebras with respect to similarity. (If two operator algebras $\mathcal{A}$ and $\mathcal{B}$ have non-isomorphic lattices, then they cannot be similar.)

We next give an account, with a bit of detail, of the spectral representation theory for commutative subspace lattices.

A *partially ordered Borel space* is a pair $(X, \leq)$ consisting of a Borel space X (a measurable space; the measurable sets are called Borel sets) and a relation $\leq$ in X which is transitive and symmetric. (It is allowed that perhaps $x \leq y$ and $y \leq x$ for distinct x and y in X.) A partially ordered Borel space $(X, \leq)$ is called *standard* if X is standard as a Borel space, and if there exists a sequence $\{f_1, f_2, \ldots\}$ of real-valued Borel functions on X such that, for all $x, y \in X$,

$$x \leq y \text{ if and only if } f_n(x) \leq f_n(y), n \geq 1.$$

(Recall that a Borel space X is defined to be standard if it is Borel-isomorphic to a Borel subset of some separable complete metric space in its relative Borel structure.) A *partially ordered measure space* is defined to be a partially ordered Borel space $(X, \leq)$ together with a σ-finite positive Borel measure m on X.

Given a partially ordered measure space $(X, \leq, m)$, we construct a commutataive subspace lattice acting on the Hilbert space $L^2(X, m)$ in the following fashion:

Define a Borel subset E of X to be *almost increasing* if there is a null set N (with respect to m) such that whenever x and y are points in the complement of N with $x \in E$ and with $x \leq y$, then $y \in E$ also. Let $L_m(X, \leq)$ denote the family of all such almost increasing sets, and observe that this family is closed under unions and intersections, so it is a lattice of sets. Now for each $E \in L_m(X, \leq)$ let P_E denote the projection in $B(H)$ corresponding to multiplication by the characteristic function, χ_E, on $L^2(X, m)$. Now define

$$\mathcal{L}(X, \leq, m) = \{P_E : E \in L_m(X, \leq)\}.$$

This is a commuting lattice of projections. It is closed in the strong operator topology. Hence it is a CSL. Arveson's representation theorem states that up to unitary equivalence, every CSL has this form.

In the following theorem, a partial order on a topological space X is said to be *closed* if $\{(x, y) : y \leq x\}$ is a closed subset of the Cartesian product $X \times X$.

Theorem 3.2.1. (Arveson) *For every separably acting commutative subspace lattice $\mathcal{L}$, there is a compact metric space X, a closed partial order $\leq$ on X, and a finite Borel measure m on X such that $\mathcal{L}$ is unitarily equivalent to $\mathcal{L}(X \leq, m)$.*

Two examples are noteworthy.

Example 3.2.2. Let G be a separable locally compact abelian group equipped with Haar measure m, and let Σ be a closed subset of G containing the identity with $\Sigma + \Sigma \subseteq \Sigma$. Define $x \leq y$ to mean $y - x \in \Sigma$. Then $(X, \leq, m)$ satisfies our requirements.

Example 3.2.3. Let I be the unit interval $[0,1]$ equipped with Lebesgue measure μ, and let $X = I \times I$ be the unit square with product measure $\mu \times \mu$. Let $\leq$ be the usual product order on X defined by: $(x_1, y_1) \leq (x_2, y_2)$ if $x_1 \leq x_2$ and $y_1 \leq y_2$. Then $(X, \leq, \mu)$ satisfies our requirements. It can be shown that the CSL given by $\mathcal{L}(X, \leq, \mu \times \mu)$ is unitarily equivalent to the tensor product of two copies of the invariant subspace (or projection) lattice of the Volterra integral operator.

3.3. Quasitriangular Operator Algebras

An operator $T \in B(H)$ was defined by Paul Halmos [56] to be *quasitriangular* if there is an increasing sequence $\{P_n\}$ of finite rank projections which converges in the strong operator topology to the identity operator I such that the sequence $\{(I - P_n)T P_n\}$ converges to 0 in norm. Every quasitriangular operator is a compact perturbation of a triangular operator, and the set QT of quasitriangular operators is norm closed in $B(H)$. A famous theorem of C. Apostol, C. Foias, and D. Voiculescu states that if A is a Hilbert space operator for which the semi-Fredholm index of $(A-\lambda I)$ s nonnegative for all complex numbers for which $(A - \lambda I)$ is semi-Fredholm, then A is quasitriangular. This yields that any non-quasitriangular operator must have a non-trivial invariant subspace.

Let $\mathcal{N}$ be a multiplicity one nest order isomorphic to the positive extended integers. Such a nest can be constructed by letting $\{e_1, \ldots, e_n\}$ be an orthonormal basis for H, defining $N_n = \text{span}\{e_1, \ldots, e_n\}, 1 \le n < \infty$, and letting $\mathcal{N} = \{0, H\} \cup \{N_n : 1 \le n < \infty\}$. Every multiplicity one nest order isomorphic to the positive extended integers is, of course, unitarily equivalent to this nest. We will denote this by $\mathcal{N}(\mathbb{Z}_+)$. Every quasitriangular operator is unitarily equivalent to an operator in $\text{Alg}\,\mathcal{N}(\mathbb{Z}_+) + \mathcal{K}$.

One of the results that Arveson obtained in the paper [11] is that the compactly perturbed algebra $\text{Alg}\,\mathcal{N}(\mathbb{Z}_+) + \mathcal{K}$ is norm closed. An important ingredient in this paper was a formula for the norm distance of an operator in $B(H)$ to an arbitrary nest algebra. In [80], C. Lance obtained this formula independently, utilizing different techniques. S. Power [100] has obtained an elegant proof of Arveson's formula. We refer the reader to the excellent lecture notes [101].

Arveson's formula states the if $\mathcal{N}$ is an arbitrary complete nest of projections in $B(H)$, and if $T \in B(H)$ is arbitrary, then

$$\text{distance}\,(T, \text{alg}\,\mathcal{N}) = \sup\{\|(I - P)TP\| : P \in \mathcal{N}\}.$$

Thus, descriptively, the distance from an operator to a nest algebra is the supremum of the norms of the *strictly lower triangular blocks*, with respect to the nest, of the operator. More will be said about Arveson's formula later, and about *hyperreflexivity* in general.

We define a *quasitriangular operator algebra* to be an algebra of the form

$$\mathcal{A} + \mathcal{K} = \{A + K : A \in \mathcal{A}, K \in \mathcal{K}\}$$

where $\mathcal{A}$ is a nest algebra.

In the paper [40], T. Fall, W. Arveson, and P. Muhly extended some known quasitriangularity results to this setting. It was proven that an arbitrary quasitriangular algebra is norm closed, generalizing Arveson's result for $\mathcal{N}(\mathbb{Z}_+)$. A criterion was obtained for membership in a quasitriangular algebra: If $T \in B(H)$, and if $\mathcal{N}$ is a complete nest of projections in $\mathcal{P}(H)$, then $T \in \operatorname{Alg}\mathcal{N} + \mathcal{K}$ if and only if two conditions are satisfied. The first is that $(I - P)TP$ must be compact for every projection P in the nest $\mathcal{N}$, and the second is that the map $P \to (I - P)TP$ must be continuous, as a map from $\mathcal{N}$ into $B(H)$, for the (relative) strong operator topology on $\mathcal{N}$ and the norm topology on $B(H)$.

For a nest $\mathcal{N}$, define $\mathcal{QT}(\mathcal{N})$ to be the set of all operators $T \in B(H)$ for which there is a unitary operator $U \in B(H)$ with $UTU^* \in \operatorname{Alg}\mathcal{N} + \mathcal{K}$. Answering a question of Arveson, in [60] Domingo Herrero showed that $\mathcal{QT}(\mathcal{N}) = B(H)$ for *most* nests $\mathcal{N}$, the only possible obstructions being index obstructions. Thus for *most* nests $\operatorname{Alg}\mathcal{N} + \mathcal{K}$ provides a *universal model* for Hilbert space operators. This is true if $\mathcal{N}$ is continuous, in particular. Herrero's proof utilized a result of Niels Andersen in [1].

The paper "Compact perturbations of reflexive algebras" [1] by Niels Toft Andersen, has proven to be very important to several developments, especially to those which will be discussed in Chapter 4 and 5. We describe some of the main results.

If $\mathcal{L}$ is a commutative subspace lattice, then the commutant, $\mathcal{L}'$, of $\mathcal{L}$ is the diagonal of the associated algebra $\operatorname{Alg}\mathcal{L}$. If P is a projection in $\mathcal{L}'$, then the set of projections $\{P, I\} \cup \{(I - P)L : L \in \mathcal{L}\}$ is also a commutative subspace lattice in $\mathcal{P}(H)$ which we denote by $\mathcal{L}^P$. We have

$$\operatorname{Alg}(\mathcal{L}^P) = \{PBP + (I - P)A(I - P) : B \in B(H), A \in \operatorname{Alg}\mathcal{L}\}.$$

Two algebras $\mathcal{A}$ and $\mathcal{B}$ are said to have the same compact perturbation if $\mathcal{A} + \mathcal{K} = \mathcal{B} + \mathcal{K}$. If P is a finite rank projection in $\mathcal{L}'$ then $\operatorname{Alg}\mathcal{L}$ and $\operatorname{Alg}(\mathcal{L}^P)$ have the same compact perturbation.

Definition 3.3.1. (Andersen) Let $\mathcal{L}_1$ and $\mathcal{L}_2$ be two commutative subspace lattices. We say that they are *compactly perturbed* if there are finite-dimensional projections $P_0 \in \mathcal{L}'_1$ and $Q_0 \in \mathcal{L}'_2$, and a bijection θ of $\mathcal{L}_1^{P_0}$ onto $\mathcal{L}_2^{Q_0}$, so that for $\epsilon > 0$ there exists a unitary operator U with $U - I \in \mathcal{K}$ such that

$$\|(U \cdot U^* - \theta)|_{\mathcal{L}_1 P_0}\| < \epsilon.$$

(Here, $\|(U \cdot U^* - \theta)|_{\mathcal{L}_1 P_0}\|$ means $\sup\{\|UQU^* - \theta(Q)\| : Q \in \mathcal{L}_1^{P_0}\}$.)

Theorem 3.3.2. (Andersen) Let $\mathcal{L}_1$ and $\mathcal{L}_2$ be commutative subspace lattices in $\mathcal{P}(H)$. Assume that $\operatorname{Alg}\mathcal{L}_1$ and $\operatorname{Alg}\mathcal{L}_2$ have the same compact perturbations. Then $\mathcal{L}_1$ and $\mathcal{L}_2$ are compactly perturbed lattices.

If $\mathcal{L}_1$ and $\mathcal{L}_2$ are nests (or more generally, if they are hyperreflexive) then the converse of Theorem 3.3.2 is valid.

Theorem 3.3.3. (Andersen) *Let $\mathcal{L}_1$ and $\mathcal{L}_2$ be two compactly perturbed lattices in $P(H)$ which are nests. Then $\operatorname{Alg} \mathcal{L}_1$ and $\operatorname{Alg} \mathcal{L}_2$ have the same compact perturbation.*

The most important results of [1] were those for continuous nests. If $\mathcal{N}$ is a complete continuous nest of projections in $B(H)$, for H separable, then there is a bijection between the elements of $\mathcal{N}$ and the points in the closed unit interval [0,1] which is an order isomorphism, and which is a homeomorphism for the usual topology on [0,1] and the (relative) strong operator topology on $\mathcal{N}$ (equvalently, the order topology on $\mathcal{N}$). If $\mathcal{N} = \{P_\alpha : \alpha \in [0,1]\}$ is a nest so indexed, we will simply use the notation "P_α" to denote the nest.

If P_α and Q_α are continuous nests in $P(H)$, we write

$$P_\alpha \simeq Q_\alpha$$

if

 (i) $P_\alpha - Q_\alpha$ is compact, $0 \leq \alpha \leq 1$, and
 (ii) the map $\alpha \to P_\alpha - Q_\alpha$ is continuous for the usual topology on [0,1] and the norm topology on $B(H)$.

Since [0,1] is compact, (ii) implies that the set of differences $\{P_\alpha - Q_\alpha : \alpha \in [0,1]\}$ is norm compact.

The following is an important theorem which culminates a difficult argument in [1].

Theorem 3.3.4. (Andersen) *For any two continuous nests P_α and Q_α, there is a unitary operator $U \in B(H)$ such that $U^* P_\alpha U \simeq Q_\alpha$.*

Results of [40] then yield the following, which is the starting point for work presented in Chapter 4.

Corollary 3.3.5. (Andersen) *For any two continuous nests P_α and Q_α, the quasitriangular algebras $\operatorname{Alg}(P_\alpha) + \mathcal{K}$ and $\operatorname{Alg}(Q_\alpha) + \mathcal{K}$ are unitarily equivalent.*

If $\mathcal{L}$ is any set of projections and U is a unitary operator then, since $U(\operatorname{Alg} \mathcal{L})U^* = \operatorname{Alg}(U\mathcal{L}U^*)$, where $U\mathcal{L}U^* = \{ULU^* : L \in \mathcal{L}\}$, and since $U\mathcal{K}U^* = \mathcal{K}$, we have $U(\operatorname{Alg} \mathcal{L} + \mathcal{K})U^* = \operatorname{Alg}(U\mathcal{L}U^*) + \mathcal{K}$. Thus by Corollary 3.3.5, given any two continuous nests $\mathcal{N}$ and $\mathcal{M}$ there is a unitary operator U such that $\operatorname{Alg}\mathcal{M}+\mathcal{K}=\operatorname{Alg}(U\mathcal{L}U^*)+\mathcal{K}$. So if $\mathcal{N}_1 = U\mathcal{N}U^*$, then $\operatorname{Alg}\mathcal{M}$ and $\operatorname{Alg}\mathcal{N}_1$ have the same compact perturbation, and hence $\mathcal{M}$ and $\mathcal{N}_1$ are compactly perturbed lattices. Since $\mathcal{M}$ and $\mathcal{N}_1$ are continuous nests and thus have nonatomic core, their commutants contain no nonzero finite rank operators, and so the projections P_0 and Q_0 appearing in Definition 3.3.1 are zero. (That is, $\mathcal{N}_1 = \mathcal{N}_1^{P_0}$ and $\mathcal{M} = \mathcal{M}^{Q_0}$.) Thus we have *approximate unitary equivalence of continuous nests*.

Other interesting papers on quasitriangular algebras include "Derivations of quasitriangular algebras" by Bruce Wagner [121], and "Automorphisms of quasitriangular algebras" by Kenneth Davidson and Bruce Wagner [32].

4. On Similarity Theory for Nests

In this chapter we describe our work in [83] and [84] on similarity transformations for nests. An expanded proof is given of a result in [83], providing a negative answer to the question of Ringrose discussed in §3.1. Then we briefly sketch the steps in [84] leading to a general factorization result for positive invertible operators along nests. Other proofs, along with subsequent work, will be discussed in Chapter 5.

The proof of Theorem 4.1, below, was given in condensed form in [83], with several technical details left to the reader. Since these lecture notes are written for beginning graduate students in the field, we expand this proof.

An improvement of Theorem 4.1 was needed for the first step of work in [84]. This is Theorem 4.3 below. The proof was similar to that of Theroem 4.1, but was necessarily different in enough respects to make it less intuitive. After proving Theorem 4.1, and a corollary, we will state the improved version, noting some differences in proof, and then sketch remaining steps in [84].

Theorem 4.1. *Let $\mathcal{N}$ be a continuous nest of multiplicity one. Then there exists a positive invertible operator T such that $T\mathcal{N} = \{TN : N \in \mathcal{N}\}$ fails to have multiplicity one.*

Before giving the proof, we will briefly mention the *program*. The starting point is the result of Andersen that arbitrary quasitriangular algebras based on continuous nests are unitarily equivalent. The basic idea, which may have wider application, is that if $\mathcal{A}$ and $\mathcal{B}$ are subalgebras of $B(H)$ for which $\mathcal{A} + \mathcal{K} = \mathcal{B} + \mathcal{K}$, then it follows, via the second isomorphism theorem of classical ring theory, that, since $\mathcal{K}$ is a two-sided ideal in $B(H)$, the quotient algebras $\mathcal{A}/(\mathcal{A} \cap \mathcal{K})$ and $\mathcal{B}/(\mathcal{B} \cap \mathcal{K})$ are algebraically isomorphic. We have

$$\mathcal{A}/(\mathcal{A} \cap \mathcal{K}) \approx (\mathcal{A} + \mathcal{K})/\mathcal{K}$$

and

$$\mathcal{B}/(\mathcal{B} \cap \mathcal{K} \approx (\mathcal{B} + \mathcal{K})/\mathcal{K}.$$

So if $\mathcal{B}$ contains structural information that does not disappear under application of the Calkin homomorphism $\Pi : B(H) \to B(H)/\mathcal{K}$, then the quotient $\mathcal{A}/(\mathcal{A} \cap \mathcal{K})$ also contains this information. If we are fortunate, we may be able to *lift* this information to corresponding information about $\mathcal{A}$, and then utilize this.

We note that, in case $\mathcal{A}$ and $\mathcal{B}$ are norm closed, the isomorphism $\mathcal{A}/(\mathcal{A} \cap \mathcal{K}) \approx \mathcal{B}/(\mathcal{B} \cap \mathcal{K})$ is bounded, with bounded inverse.

Proof of Theorem 4.1. By Corollary 3.3.5 (Andersen) there exists a continuous nest $\mathcal{M}$, not of multiplicity one, such that, as described above, we have the diagram:

$$
\begin{array}{ccc}
\text{Alg}\,\mathcal{N} + \mathcal{K} & = & \text{Alg}\,\mathcal{M} + \mathcal{K} \\
\downarrow \Pi & & \downarrow \Pi \\
(\text{Alg}\,\mathcal{N} + \mathcal{K})/\mathcal{K} & = & (\text{Alg}\,\mathcal{M} + \mathcal{K})/\mathcal{K} \\
\downarrow \phi & & \downarrow \psi \\
\text{Alg}\,\mathcal{N}/(\text{Alg}\,\mathcal{N}) \cap \mathcal{K} & \overset{\theta}{\leftarrow} & \text{Alg}\,\mathcal{M}/(\text{Alg}\mathcal{M}) \cap \mathcal{K}
\end{array}
$$

where ϕ, ψ and θ are algebra isomorphisms.

Since $\mathcal{M}$ does not have multiplicity one, the diagonal $\mathcal{D}_{\mathcal{M}}$ is a noncommutative von Neumann algebra, so $\mathcal{D}_{\mathcal{M}}$ contains a nonzero partial isometry V with the property that its initial projectin V^*V is orthogonal to its final projection VV^*. Now let

$$\widetilde{S} = (I - VV^* - V^*V) + V + V^*.$$

Then $\widetilde{S}$ is a self-adjoint unitary (a symmetry). We have $\widetilde{S}^2 = I, \widetilde{S}V\widetilde{S} = V^*, \widetilde{S}V^*\widetilde{S} = V$. Let $\widetilde{P} = VV^*$. Then $\widetilde{P} \neq 0$, and $\widetilde{P}\widetilde{S}\widetilde{P} = 0$. A computation shows that the group generated by $\widetilde{S}$ together with the symmetry $(I - 2\widetilde{P})$ is noncommutative and has order eight: (It is isomorphic to D_4–the dihedral group on four elements.)

Since $\mathcal{M}$ is continuous, the core $\mathcal{C}_{\mathcal{M}}$ is a nonatomic abelian von Neumann algebra, hence $\mathcal{C}_{\mathcal{M}}$ has no nonzero compact operator in its commutant. So $\mathcal{D}_{\mathcal{M}} \cap \mathcal{K} = (0)$, hence $\Pi|_{\mathcal{D}_{\mathcal{M}}}$ is injective, and thus $\{\Pi(\widetilde{S}), \Pi(\widetilde{P})\}$ posses the same algebraic properties possessed by the pair $\{\widetilde{S}, \widetilde{P}\}$. Let

$$\widehat{P} = (\theta \circ \psi \circ \Pi)(\widetilde{P})$$

and

$$\widehat{S} = (\theta \circ \psi \circ \Pi)(\widetilde{S}).$$

Now choose A and B in $\mathrm{Alg}\,(\mathcal{N})$ with $\widehat{P} = (\theta \circ \Pi)(A)$ and $\widehat{S} = (\phi \circ \Pi)(B)$. Since $\widehat{P}^2 = \widehat{P} \neq 0, (\widehat{S})^2 = I$, and $\widehat{P}\widehat{S}\widehat{P} = 0$, it follows that the operators $A^2 - A, B^2 - I$, and ABA are contained in $(\mathrm{Alg}\,\mathcal{N}) \cap \mathcal{K}$. Also, since $\mathcal{N}$ is continuous and hence has no gaps, every compact operator contained in $\mathrm{Alg}\,\mathcal{N}$ is quasinilpotent [110]. Thus $(\mathrm{Alg}\,\mathcal{N}) \cap \mathcal{K}$ is a topologically nil ideal in $\mathrm{Alg}\,\mathcal{N}$, hence is contained in the Jacobson radical of $\mathrm{Alg}\,\mathcal{N}$. So A is an idempotent modulo the radical, and hence decomposes as $A = P + R$ where P is an idempotent in $\mathrm{Alg}\,\mathcal{N}$ and $R \in \mathrm{rad}\,(\mathrm{Alg}\,\mathcal{N})$. (The fact that an element of a Banach algebra which is idempotent modulo the radical differs from an idempotent by an element of the radical can be found as Theorem 2.3.9 in [104].) Note that $P \neq 0$, since otherwise A would be in the radical, hence quasi-nilpotent, so $\widehat{P}$ would be a quasinilpotent idempotent. But a quasinilpotent idempotent is zero, and we earlier obtained that $\widehat{P} \neq 0$.

We must modify B. Since $ABA \in \mathrm{rad}\,(\mathrm{Alg}\,\mathcal{N})$ and $A - P \in \mathrm{rad}\,(\mathrm{Alg}\,\mathcal{N})$ we have $PBP \in \mathrm{rad}\,(\mathrm{Alg}\,\mathcal{N})$. Also, since $B^2 - I \in \mathrm{rad}\,(\mathrm{Alg}\,\mathcal{N})$ we have $\sigma(B) \subset \{-1, 1\}$, so B is invertible in $\mathrm{Alg}\,(\mathcal{N})$. Let $B_1 = B - PBP$. Then $PB_1P = 0$, and since addition of an element of the radical to an element of a Banach algebra does not affect spectrum, B_1 is also invertible in $\mathrm{Alg}\,\mathcal{N}$. Now let

$$S = B_1 P + PB_1^{-1}(I - P) + I - P - B_1 PB_1^{-1}(I - P).$$

The operator S will be the required modification of B. (From the definitions of $\widetilde{S}$ and $\widetilde{P}$ near the beginning of this proof it is easily computed that $\widetilde{S} = \widetilde{S}\widetilde{P} + \widetilde{P}\widetilde{S}(I - \widetilde{P}) + I - \widetilde{P} - \widetilde{S}\widetilde{P}\widetilde{S}(I - \widetilde{P})$. It is also true that $B_1 - B_1^{-1} \in \mathrm{rad}\,(\mathrm{Alg}\,\mathcal{N})$. From this it follows that $S - B \in \mathrm{rad}(\mathrm{Alg}\,\mathcal{N})$. So S is a radical perturbation of B, although this fact will not be used in the remaider of the proof.)

It is clear that $PSP = 0$. It is also true that $S^2 = I$. To obtain this latter property, let

$$\alpha = B_1 P + PB_1^{-1}(I - P)$$

and

160

$$\beta = I - P - B_1 P B_1^{-1}(I - P)$$

so $S = \alpha + \beta$, and $S^2 = \alpha^2 + \alpha\beta + \beta\alpha + \beta^2$. It will suffice to show that $\beta^2 = \beta, \alpha\beta = \beta\alpha = 0$ and $\alpha^2 = I - \beta$.

We have

$$\begin{aligned}
\alpha^2 &= [B_1 P + P B_1^{-1}(I - P)]^2 \\
&= B_1 P B_1 P + B_1 P B_1^{-1}(I - P) \\
&\quad + P B_1^{-1}(I - P)B_1 P + P B_1^{-1}(I - P)P B_1^{-1}(I - P).
\end{aligned}$$

The first and fourth terms are zero because $PB_1 P = 0$ and $(I-P)P = 0$, and the third term reduces to P since $PB_1 P = 0$. Thus

$$\alpha^2 = P + B_1 P B_1^{-1}(I - P) = I - \beta,$$

as desired.

Next, compute

$$\begin{aligned}
\beta^2 &= [(I - P) - B_1 P B_1^{-1}(I - P)]^2 \\
&= (I - P) - (I - P)B_1 P B_1^{-1}(I - P) \\
&\quad - B_1 P B_1^{-1}(I - P) + B_1 P B_1^{-1}(I - P)B_1 P B_1^{-1}(I - P).
\end{aligned}$$

Since $PB_1 P = 0$ we have $(I - P)B_1 P = B_1 P$, so the second term reduces to

$$-B_1 P B_1^{-1}(I - P),$$

and the fourth term to

$$+B_1 P B_1^{-1}(I - P),$$

thus cancelling. This gives

$$\beta^2 = (I - P) - B_1 P B_1^{-1}(I - P) = \beta,$$

as required.

For $\alpha\beta$, we have

$$\begin{aligned}
\alpha\beta &= [B_1 P + P B_1^{-1}(I - P)][(I - P) - B_1 P B_1^{-1}(I - P)] \\
&= B_1 P(I - P) - B_1 P B_1 P B_1^{-1}(I - P) \\
&\quad + P B_1^{-1}(I - P) - P B_1^{-1}(I - P)B_1 P B_1^{-1}(I - P).
\end{aligned}$$

The first and second terms are 0 since $P(I - P) = 0$ and $PB_1P = 0$. The fourth term reduces to $-PB_1^{-1}(I - P)$, using the fact that $(I - P)B_1P = B_1P$, thus cancelling the third term and yielding $\alpha\beta = 0$.

For $\beta\alpha$, we have

$$\beta\alpha = [(I - P) - B_1PB_1^{-1}(I - P)][B_1P + PB_1^{-1}(I - P)]$$
$$= (I - P)B_1P + (I - P)PB_1^{-1}(I - P)$$
$$- B_1PB_1^{-1}(I - P)B_1P - B_1PB_1^{-1}(I - P)PB_1^{-1}(I - P).$$

As above, the second and fourth terms are zero, the first reduces to B_1P, and the third to $-B_1P$, cancelling the first, thus yielding $\beta\alpha = 0$, as required.

We have succeeded in showing that $\text{Alg}\,\mathcal{N}$ contains a pair $\{P, S\}$ satisfying the properties: $P^2 = P \neq 0, S^2 = I, PSP = 0$. The operators P and S do not commute, since otherwise we would have $P = S^2P = SPSP = 0$, a contradiction. Now let $R = I - 2P$.

We have $SR \neq RS, S^2 = I$ and $R^2 \neq I$. Also, since $PSP = 0$ we have

$$PSR = PS(I - 2P) = PS,$$

and

$$RSP = (I - 2P)SP = SP,$$

so

$$PSRS = PS^2 = P,$$

and

$$SRSP = S^2P = P.$$

This shows that P commutes with SRS, hence also that R commutes with SRS. It follows that the set

$$G = \{I, S, R, RS, SR, SRS, RSR, SRSR\}$$

consists of distinct elements and is closed under multiplication. (It is isomorphic to D_4.)

We have shown that $\text{Alg}\,(\mathcal{N})$ contains a finite nonabelian group. A finite group is amenable (averaging yields the invariant mean), and it is well known that a bounded amenable subgroup of $B(H)$ is similar, via an invertible operator in $B(H)$, to a group of unitary operators. In this case, since G is finite, we may simply take the operator to be

$$T = \left(\sum g^*g : g \in G\right)^{1/2}.$$

For then $h^*T^2h = T^2$ for every $h \in G$, so $(ThT^{-1})^*(ThT^{-1}) = (T^{-1}h^*T)(ThT^{-1}) = T^{-1}h^*T^2hT^{-1} = T^{-1}T^2T^{-1} = I$. Since ThT^{-1} is invertible, this shows that it is a unitary operator. Since $h \in G$ was arbitrary, TGT^{-1} is a unitary group. It is nonabelian since G is nonabelian. The important feature used here is that it is a noncommutative subset of $B(H)$ which is closed under the adjoint operation. Let $\mathcal{U} = TGT^{-1}$.

The operator T is positive and invertible. Let $\mathcal{L}$ be the image nest of $\mathcal{N}$ under the similarity transformation induced by T. So $\mathcal{L} = T\mathcal{N} = \{TN : N \in \mathcal{N}\}$, and $\text{Alg}(\mathcal{L}) = T(\text{Alg}\,\mathcal{N})T^{-1} = \{TCT^{-1} : C \in \text{Alg}\,\mathcal{N}\}$. Since $G \subseteq \text{Alg}\,\mathcal{N}$ we have $\mathcal{U} \subseteq \text{Alg}\,\mathcal{L}$. Since $\mathcal{U}$ is selfadjoint, we have $\mathcal{U} \subseteq (\text{Alg}\,\mathcal{L}) \cap (\text{Alg}\,\mathcal{L})^* = \text{Diag}\,(\text{Alg}\,\mathcal{L})$. This shows that the diagonal of $\text{Alg}\,(\mathcal{L})$ is nonabelian, and hence its center, the core of $\text{Alg}\,\mathcal{L}$, cannot be maximal abelian in $B(H)$. This means that $\mathcal{L}$ fails to have multiplicity one. The proof is complete. ■

A corollary of this is that there exists a non-hyperintransitive compact operator. (That is, a compact operator which fails to be contained in any hyperreducible triangular algebra. The terms *hyperintransitive* and *hyperreducible* mean the same.) The fact that this would be a corollary of Theorem 4.1 was known before the proof. The argument was shown to be by William Arveson, and I believe was first obtained by John Erdos.) It answers negatively the Question 2.4.1 in [72] (Question A in 1.1 of these notes) and also a question of Gohberg-Krein in [53], which arose independently.

Corollary 4.2. *There exists a non-hyperintransitive compact operator.*

Proof: Let V be the Volterra integral operator defined by

$$(Vf)(t) = \int_0^t f(s)ds, \quad f \in L^2([0,1]; \mu),$$

where μ is Lebesgue measure. It is well known that the invariant subspace lattice for V is the continuous multiplicity one nest consisting of all subspaces of $L^2([0,1]; \mu)$ of the form $N_t = \{\chi_{[t,1]}f : f \in L^2([0,1]; \mu)\}$. By Theorem 4.1 there is an invertible operator T such that $\{TN_t : 0 \le t \le 1\}$ is a nest not of multiplicity one, which is continuous since $\{N_t : 0 \le t \le 1\}$ is continuous, so is maximal. Then the similarity TVT^{-1} leaves $\{TN_t : 0 \le t \le 1\}$ invariant, so is in the nest algebra for this nest. The invariant subspace lattice for TVT^{-1} is precsely $\{TN_t : 0 \leq t \leq 1\}$, so Lat (TVT^{-1}) cannot contain a multiplicity one nest. That is, TVT^{-1} is not hyperintransitive. ∎

The following is Theorem 2.2 from [84].

Theorem 4.3. *Let $\mathcal{N}$ be a continuous nest of multiplicity one. Then given $\epsilon > 0$ there exists a positive invertible operator T such that $T - I$ is compact, $\|T - I\| < \epsilon$, and $T\mathcal{N}$ fails to have multiplicity one.*

For the proof of Theorem 4.3, idempotent lifting modulo the radical, as in Theorem 4.1, is replaced with lifting modulo the smaller ideal $\mathrm{Alg}\,\mathcal{N} \cap \mathcal{K}$. The main parts of the proofs, up to construction of T, are much alike except that the intuitive program used in (4.1) cannot be retained intact. The operators R and S, as in (4.1), are obtained as compact perturbations of unitary operators, and it follows that $T - I = K$ is compact. To make this difference small, we now apply the Gohberg-Krein factorization theory as follows: Let K_0 be a finite-rank operator with $\|K - K_0\|$ sufficiently small. Then since $K_0 \in C_\omega$, and also $(I + K_0)^{-1} - I \in C_\omega$, by Corollary 3.1.4, the operator $(I + K_0)^{-1}$ preserves multiplicity of the nest $T\mathcal{N}$. So $[(I + K_0)^{-1}T]\mathcal{N}$ fails to have multiplicity one. Now let $T_0 = [(I + K_0)^{-1}T]$, and let $T_0 = V|T_0|$ be the polar decomposition. Then $\| |T_0| - I \|$ is small, $|T_0| - I$ is compact, $|T_0|$ is positive, and, since V^* is unitary, $|T_0| = V^*T_0$ fails to preserve multiplicity of $\mathcal{N}$. Now replace T with $|T_0|$.

For the next step in [84], Theorem 4.3 is again refined, to show that if $\mathcal{N}$ is any continuous nest, T can be obtained, as in Theorem 4.3, also satisfying the additional property that for some $B \in \mathrm{Alg}\,\mathcal{N}$ the similarity transform $W = TBT^{-1}$ is a partial isometry with $WW^* + W^*W = I$, and with W and W^* in $\mathrm{Diag}\,(\mathrm{Alg}\,(T\mathcal{N}))$.

The next step is a proof that arbitrary continuous nests are similar. This is accomplished by showing that any continuous nest is similar to a nest of uniform infinite multiplicity, and then invoking [35] which yields that two continuous nests of uniform infinite multiplicity are unitarily equivalent. It follows from [35] that a continuous nest $\mathcal{N}$ has uniform infinite multiplicity if and only if $\mathrm{Diag}\,(\mathrm{Alg}\,\mathcal{N})$ contains an infinite sequence of mutually orthogonal projections each with central support I. So given $\mathcal{N}$, the idea is to construct a norm convergent sequence of invertible operators $\{T_n\}$, such that for each n, $\mathrm{Diag}\,(\mathrm{Alg}\,(T_n\mathcal{N}))$ contains a family of n such projections, taking care so that at each step the new family contains the old, and then to take a limit. In practice, it turns out to be convenient to work with systems of partial isometries (incomplete systems of matrix units) rather than sets of projections, and to construct $\{T_n\}$ as the

164

sequence of partial products of a convergent infinite product. The method is to use the refinement of Theorem 4.3 discussed in the preceeding paragraph in an inductive argument.

Let $\{W_i\}$ be a sequence of partial isometries. We call $< W_1, W_2, \ldots, >$ a *directed system* if

$$W_i W_i^* + W_i^* W_i = W_{i-1} W_{i-1}^*$$

and

$$W_i W_i^* \neq 0, \quad i = 1, \ldots \ .$$

It is called *proper* if $W_1 W_1^* + W_1^* W_1 = I$. If the diagonal of a nest algebra contains an infinite proper directed system, then it contains an infinite sequence of mutually orthogonal projections, each with central support I, and thus the nest has uniform infinite multiplicity.

If $< W_1, \ldots, W_n >$ is the first n terms of a directed system, let

$$P_n = W_n W_n^* \text{ and } P_0 = W_1 W_1^* + W_1^* W_1.$$

If $S \in B(H)$ with $S = P_n S P_n$, and if $S \neq 0$, then S cannot commute with the set $\{W_1, \ldots, W_n\}$. However, S extends to an operator $\widetilde{S}$, supported on P_0, which does commute with $\{W_1, \ldots, W_n\}$. (The operator $\widetilde{S}$ is a direct sum of 2^n copies of $P_n S|_{P_n H}$, the equivalences between the direct summand spaces being determined by the W_i.) The mapping $S \to \widetilde{S}$, viewed as a map from $B(P_n H)$ into $B(H)$, is isometric, positive, linear, and maps compact operators to compact operators.

The following lemma captures our approach.

Lemma 4.4. *The $\mathcal{N}$ be a continuous nest and let $\epsilon > 0$ be given. Then there exists a sequence $\{S_n\}$ of positive invertible operators, and a sequence of partial isometries $\{W_n\}$, such that:*

(i) $S_n - I \in \mathcal{K}, \quad \|S_n - I\| < \epsilon/2^n, \quad n \geq 1.$

(ii) $< W_1, \ldots >$ *is an infinite proper directed system.*

(iii) *For each $1 \leq n < \infty$, the set $\{W_1, \ldots, W_n\}$ is contained in the diagonal of the nest algebra $\mathrm{Alg}\,(S_n S_{n-1} \ldots S_1 \mathcal{N})$.*

(iv) *For each $1 \leq n < \infty$, S_{n+1} commutes with $\{W_1, \ldots, W_n\}$.*

The idea in Lemma 4.4 is to apply, at the nth step, the refinement of Theorem 4.3 to the compression of the nest

$$S_n S_{n-1} \dots S_1 \mathcal{N}$$

to the projection $W_{n+1} W_{n+1}^*$, obtaining a positive invertible operator S for that compression, and then extending, via the map "$\sim$", obtaining an operator $S_{n+1} \in B(H)$ satisfying our requirements.

Item (i) implies that the infinite product

$$\text{Lim}_n (S_n S_{n-1} \dots S_1)$$

converges in norm to an operator $S \in B(H)$, and $\|S - I\| < \exp(\epsilon) - 1$. The difference $S - I$ is compact since each $S_n - I \in \mathcal{K}$. For suitable ϵ we have $\|S - I\| < 1$, so S will be invertible. Items (iii) and (iv) imply that, for $1 \le n < \infty$,

$$W_n \in \text{Diag Alg} (S_n S_{n-1} \dots S_1 \mathcal{N}),$$

and that

$$(\dots S_{n+2} S_{n+1})$$

commutes with W_n. Since

$$S = (\dots S_{n+2} S_{n+1}) S_n S_{n-1} \dots S_1,$$

this implies that $W_n \in \text{Diag Alg} (S\mathcal{N})$. So $\text{Diag} (\text{Alg} (S\mathcal{N}))$ contains an infinite directed system, hence $S\mathcal{N}$ has uniform infinite multiplicity.

It follows from above that arbitrary continuous nests $\mathcal{N}$ and $\mathcal{M}$ are similar: given $\epsilon > 0$ there is a unitary operator U and a compact operator K of norm less than ϵ such that $\mathcal{M} = (U + K)\mathcal{N}$. A corollary is that every maximal nest is similar to a nest of multiplicity one: apply the above to modify the continuous part of the nest. Since every compact operator has a maximal nest of invariant subspaces, it follows that every compact operator is *similar* to a hyperintransitive operator.

For a general factorization result it was necessary to deal with the atomic core parts of nests. Absolute continuity machinery was developed for this.

Let H, K be Hilbert spaces with $T \in B(H, K)$ invertible. Let $\mathcal{N}$ be a complete nest of closed subspaces of H and let $\widetilde{\mathcal{N}} = T\mathcal{N}$. If E is an $\mathcal{N}$-interval with endpoints M, N let $\widetilde{E}$ denote the $\widetilde{\mathcal{N}}$-interval with endpoints TM and TN. Then $E \leftrightarrow \widetilde{E}$ is a bijection between the family of $\mathcal{N}$-intervals and the family of $\widetilde{\mathcal{N}}$-intervals. Let $\mathcal{E}(\cdot)$ and $\widetilde{\mathcal{E}}(\cdot)$ denote the *natural* projection valued measures on $\mathcal{N}$ and $\widetilde{\mathcal{N}}$, respectively [35]. Let $\phi_T(\cdot)$ denote the order isomorphism $N \to TN$. T acts *absolutely continuously* on $\mathcal{N}$ if the composition $\widetilde{\mathcal{E}}(\phi_T(\cdot))$ is absolutely continuous with respect to $\mathcal{E}(\cdot)$.

Given a nest $\mathcal{N}$, an (infinite) $\mathcal{N}$-partition is a sequence $\{E_n\}$ of mutually orthogonal $\mathcal{N}$-intervals with $\sum E_n = I$, the sum converging in the strong operator topology. With this definition, it is easy to see that T acts absolutely continuously on $\mathcal{N}$ if and only if whenever $\{E_n\}$ is an $\mathcal{N}$-partition, then $\{\widetilde{E}_n\}$ is an $\widetilde{\mathcal{N}}$-partition. If T is unitary, or more generally, if T factors as $T = UA$ with U unitary and $A \in (\mathrm{Alg}\,\mathcal{N}) \cap (\mathrm{Alg}\,\mathcal{N})^{-1}$, then T acts absolutely continuously on $\mathcal{N}$. If $\mathcal{N}$ has atomic core, then the set of minimal core projections is an $\mathcal{N}$-partition, and it follows that T acts absolutely continuously on $\mathcal{N}$ if and only if $T\mathcal{N}$ also has atomic core. In this case, it is easily shown that T must factor as above. It $\mathcal{N}$ is a complete nest which is *countable*, then, since a complete countable nest has atomic core, this condition is automatically satisfied. The plan is to show that a complete uncountable nest with atomic core admits an invertible operator that does not act absolutely continuously. Then a synthesis of this and the similarity result for continuous nests yields the general factorization result. Analysis of absolute continuity is accomplished via a certain two-sided ideal, $\mathcal{R}_{\mathcal{N}}^{\infty}$, which is defined below. This gives us a means of transfer of information from a continuous special case to the atomic core case, and then to the general case. The information transfered concerns the idempotent structure of the ideal $\mathcal{R}_{\mathcal{N}}^{\infty}$.

Definition 4.5. Let $\mathcal{N}$ be a complete nest. Then $\mathcal{R}_{\mathcal{N}}^{\infty}$ is the set of all operators $A \in \mathrm{Alg}\,\mathcal{N}$ for which, given $\epsilon > 0$ there exists a, perhaps infinite, $\mathcal{N}$-partition $\{E_n\}$ for which $\|E_n A E_n\| < \epsilon$ for each n. (Equivalently, $A \in \mathcal{R}_{\mathcal{N}}^{\infty}$ iff given $\epsilon > 0$ there exists a complete countable subnest $\mathcal{N}_0$ such that the diagonal of A with respect to $\mathcal{N}_0$ has norm less than ϵ.)

Formally, the definition of $\mathcal{R}_{\mathcal{N}}^{\infty}$ can be obtained from the statement of the Ringrose criterion for containment in the radical $\mathcal{R}_{\mathcal{N}} = \mathrm{rad}\,(\mathrm{Alg}\,\mathcal{N})$ by replacing *finite partition* in Theorem 2.2.1 with *partition*.

The following is a statement of some basic properties of $\mathcal{R}_{\mathcal{N}}^{\infty}$.

Proposition 4.6. $\mathcal{R}_{\mathcal{N}}^{\infty}$ is a norm-closed two-sided ideal in $\mathrm{Alg}\,\mathcal{N}$ containing $\mathcal{R}_{\mathcal{N}}$, with equality if and only if $\mathcal{N}$ has only a finite number of members. $\mathcal{R}_{\mathcal{N}}^{\infty}$ is contained in the strong closure of $\mathcal{R}_{\mathcal{N}}$. $\mathcal{R}_{\mathcal{N}}^{\infty}$ is diagonal-disjoint in the sense that $\mathcal{R}_{\mathcal{N}}^{\infty} \cap \mathcal{D}_{\mathcal{N}} = \{0\}$. The algebraic sum $\mathcal{R}_{\mathcal{N}}^{\infty} + \mathcal{D}_{\mathcal{N}}$ is a norm-closed subalgebra of Alg $\mathcal{N}$ with the property that for every $D \in \mathcal{D}_{\mathcal{N}}, R \in \mathcal{R}_{\mathcal{N}}^{\infty}$, we have $\|D + R\| \geq \|D\|$. All expectations from $B(H)$ onto $\mathcal{D}_{\mathcal{N}}$ annihilate $\mathcal{R}_{\mathcal{N}}^{\infty}$ and so agree on $\mathcal{R}_{\mathcal{N}}^{\infty} + \mathcal{D}_{\mathcal{N}}$; the common restriction to $\mathcal{R}_{\mathcal{N}}^{\infty} + \mathcal{D}_{\mathcal{N}}$ is a homomorphism If $\mathcal{N} \supset \mathcal{M}$ are nests, then $\mathcal{R}_{\mathcal{N}} \supset \mathcal{R}_{\mathcal{M}}$ and $\mathcal{R}_{\mathcal{N}}^{\infty} \supset \mathcal{R}_{\mathcal{M}}^{\infty}$.

If P is a nonzero idempotent contained in $\mathcal{R}_{\mathcal{N}}^{\infty}$, then there is no invertible operator A in $(\text{Alg}\,\mathcal{N}) \cap (\text{Alg}\,(\mathcal{N})^{-1})$ such that APA^{-1} is selfadjoint. This follows from the property that $\mathcal{R}_{\mathcal{N}}^{\infty}$ is a diagonal-disjoint two-sided ideal in Alg $\mathcal{N}$. So if $T \in B(H)$ is an invertible operator which *normalizes* P in the sense that TPT^{-1} is selfadjoint (for instance, we may take $T = [P^*P + (I - P^*)(I - P)]^{1/2}$), then T cannot factor as $T = UA$ with U unitary and $A \in (\text{Alg}\,\mathcal{N}) \cap (\text{Alg}\,\mathcal{N})^{-1}$, because if T factored thus,then APA^{-1} would also be selfadjoint, an impossibility. The following gives the *means of transfer* mentioned above. In particular, a nest $\mathcal{N}$ admits an invertible operator which acts *discontinuously* if and only if $\mathcal{R}_{\mathcal{N}}^{\infty}$ contains a nonzero idempotent.

Proposition 4.7. *Let $\mathcal{N}$ be a complete nest and let T be an invertible operator. The following are equivalent:*

(i) T *acts absolutely continuously on $\mathcal{N}$.*

(ii) T *acts absolutely continuously on every subnest of $\mathcal{N}$.*

(iii) T *acts absolutely continuously on every subnest of $\mathcal{N}$ with purely atomic core.*

(iv) $\mathcal{R}_{\mathcal{N}}^{\infty}$ *does not contain a nonzero idempotent P which is normalized by T.*

(v) $\mathcal{R}_{T\mathcal{N}}^{\infty} \supseteq T\mathcal{R}_{\mathcal{N}}^{\infty}T^{-1}$.

The next step is to prove that for *some* nest $\mathcal{N}$ the ideal $\mathcal{R}_{\mathcal{N}}^{\infty}$ contains a nonzero idempotent. Andersen's theorem is used in the proof.

Theorem 4.8. *Let $\mathcal{N}$ be a continuous nest of multiplicity one. Then for each $\epsilon > 0$, $\mathcal{R}_{\mathcal{N}}^{\infty}$ contains a nonzero idempotent Q with $Q - Q^*$ compact such that $\|Q - Q^*\| < \epsilon$.*

Next, from Theorem 4.8, together with Proposition 4.7, it follows that for *some* nest $\mathcal{M}$ with atomic core (a subnest of the nest of Theorem 4.7) the ideal $\mathcal{R}_{\mathcal{M}}^{\infty}$ contains a nonzero idempotent. The nest $\mathcal{M}$ cannot be countable. From this, via a compression argument, it follows that the *model nest* $\mathcal{N}(\mathcal{Q})$ of §2.5, example 7, case II, (the Cantor nest) has the property that $\mathcal{R}_{\mathcal{N}(\mathcal{Q})}^{\infty}$ contains a nonzero idempotent. From this property of $\mathcal{N}(\mathcal{Q})$, compression and embedding techniques yield that *any* complete nest $\mathcal{N}$ with purely atomic core which is *uncountable* has the property that $\mathcal{R}_{\mathcal{N}}^{\infty}$ contins a nonzero idempotent. It follows easily that for a complete nest $\mathcal{N}$, the ideal $\mathcal{R}_{\mathcal{N}}^{\infty}$ contains a nonzero idempotent if and only if $\mathcal{N}$ is uncountable. So $\mathcal{N}$ admits an invertible operator which acts discontinuously if and only if $\mathcal{N}$ is uncountable. From this, together with the easily proven result that complete countable nests admit universal factorization, it follows that a complete nest $\mathcal{N}$ admits universal factorization of positive invertible operators (see the exposition in §3.1) if and only if $\mathcal{N}$ is countable. Equivalently, a complete nest $\mathcal{N}$ has the property that for every invertible bounded linear operator T there is a unitary operator U such that

$$UT \in (\operatorname{Alg} \mathcal{N}) \cap (\operatorname{Alg} \mathcal{N})^{-1}$$

if and only if $\mathcal{N}$ is countable. Utilizing the full statement of Theorem 4.8, taking care computationally, yields stronger forms of these results:

Theorem 4.9. *Let $\mathcal{N}$ be a complete nest. Then $\mathcal{R}_\mathcal{N}^\infty$ contains a nonzero idempotent if and only if $\mathcal{N}$ is uncountable. If $\mathcal{N}$ is a complete uncountable nest, then for each $\epsilon > 0, \mathcal{R}_\mathcal{N}^\infty$ contains a nonzero idempotent Q with $Q - Q^*$ compact and $\|Q - Q^*\| < \epsilon$.*

Theroem 4.10. *A complete nest has the factorization property if and only if it is countable. If a complete nest $\mathcal{N}$ is uncountable then for each $\epsilon > 0$ there exists a positive invertible operator T with $T - I$ compact and with $\|T - I\| < \epsilon$ such that $T^{1/2}$ fails to act absolutely continuously on $\mathcal{N}$ so T does not equal A^*A for any invertible $A \in \operatorname{Alg} \mathcal{N}$ with $A^{-1} \in \operatorname{Alg} \mathcal{N}$ also.*

5. Further Developments

5.1. Background

In §4 we gave an account of our own work in the solution of John Ringrose's question and further related developments. Other solutions to this question, which are technically simpler than our own, and which have some dramatic features, have been obtained by Niels Andersen [2] and by Kenneth Davidson [29]. The perturbation theoretic techniques, and results, of E. Christopher Lance [80] played an important role in [2], and the essential aspects of these techniques, developed independently, with some additional features, were important in [29]. A self-contained solution, developing and using these techniques, is contained in Arveson's notes [14]. All of these proofs take as starting point Andersen's results from [1].

Recapitulating briefly, in §4, a solution was given to a question that had been raised by Ringrose, continuous nests were shown to be similar, and it was shown that if $\mathcal{N}$ is a complete nest which is not countable as a set of subspaces of H, there exists an invertible operator T such that $T\mathcal{N}$ is not unitarily equivalent to $\mathcal{N}$. This was captured in Theorem 4.10 which states that a complete nest $\mathcal{N}$ admits a factorization $S = A^*A$, $A \in (\text{Alg}\,\mathcal{N}) \cap (\text{Alg}\,\mathcal{N})^{-1}$, for every positive invertible operator S, if and only if $\mathcal{N}$ is countable. To prove this, absolute continuity machinery was developed, centering on the ideal $\mathcal{R}_\mathcal{N}^\infty$. Proofs of the existence of T, implementing such a similarity transformation, were not constructive, and depended heavily on a result of Andersen [1]. All that is known about T, implementing a given transformation, is that it can be taken to be a small compact perturbation of a unitary operator, and that the compact operator cannot lie in the Macaev ideal.

Subsequently, Davidson [29] (independent from Andersen's work [2]), utilizing a quite different method of proof from that in §4, again based on Andersen's Theorem [1], improved and amplified some of our main results to show that arbitrary order isomorphic nests are similar, provided that the dimensions of the gaps match up under the order isomorphism, thus obtaining complete similarity invariants for nests, and that such order isomorphisms θ, mapping $\mathcal{N} \to \mathcal{M}$, can be precisely implemented in the sense that there exists an invertible operator T such that $T N = \theta(N)$ for each $N \in \mathcal{N}$. (This is the strong form of similarity, discussed in §3.1 in the context of strong form of unitary equivalence.) Again, T can be taken to be a small compact perturbation of a unitary operator, and again, proofs are not constructive. One feature of Davidson's proof is that it made very clear the role, in similarity theory, of Arveson's formula [11] for the distance from an operator to a nest algebra. Andersen (independent from Davidson's work [29]) also utilized Arveson's formula in obtaining a yet different proof of similarity of continuous nests. His proof, in [2], used the approximate equivalence result in [1] together with the results of Lance in [80] in a direct fashion, and also yielded the strong form of similarity. In essence, Lance proved that two *close* nest algebras are similar. Andersen's results yielded that two nest algebras can be *close* yet not unitarily equivalent. This proof pointed out the usefulness of Lance's work, and ideas, in similarity theory. Lance had independently proven the distance formula for nest algebras in [80], and developed Hochschild cohomology theory for nest algebras, as necessary tools needed in obtaining perturbation results. Lance's proof of the distance formula was of duality-type, and some of his ideas motivated in part the study of duality-reflexivity-hyperreflexivity connections we began in [82], and continued, in several papers.

In [13] Arveson obtained an alternate proof of Andersen's approximate equivalence result in [1] by extending the methods of his earlier paper [12] to yield a version of a theorem of Voiculescu [119] valid for certain non-separable C^*-algebras. Arveson's proof generalizes Andersen's result to yield the corresponding approximate equivalence result valid for a *wide* class of CSL algebras. It is not known whether these algebras are similar.

We have used the term *hyperreflexive* several times during the course of these lectures. We will now define the term properly.

In [11], as discussed in §3.3, W. Arveson showed that if $\mathcal{A}$ is a nest algebra, then the distance from the arbitrary operator T in $B(H)$ to $\mathcal{A}$ is given by the formula $d(T, \mathcal{A}) = \sup\{\|P^\perp TP\| : P \in \operatorname{Lat} \mathcal{A}\}$. As indicated above, this formula has been very useful in investigating problems involving compact perturbations and similarity theory for nests. Certain other reflexive algebras have been shown to possess the property that there exists a constant K such that $d(T, \mathcal{A}) \leq K \sup\{\|P^\perp TP\| : P \in \operatorname{Lat} \mathcal{A}\}$ for all $T \in B(H)$. A simple example (unpublished) due to M. D. Choi showed that this constant need not be one. A question which arose was whether every reflexive algebra has such a distance estimate for some constant K. Arveson conjectured that the answer to this question is no. He coined the term *hyperreflexive* to denote reflexive subalgebras of $B(H)$ for which there is a finite constant K, as above.

In [75] the authors constructed a non-hyperreflexive reflexive algebra, verifying this conjecture. The idea employed was to use duality theory for reflexivity to construct a linear subspace of $B(H)$, for two-dimensional H, which is reflexive in the Loginov-Sulman sense [92], yet with the property that the reflexive algebra

$$A = \begin{pmatrix} \mathbb{C} & S \\ 0 & \mathbb{C} \end{pmatrix}$$

has large distance constant. The idea of hyperreflexivity was extended to subspaces by defining $\alpha(T, \mathcal{S}) = \sup\{\|P^\perp TQ\| : PSQ = 0\}$, and it was shown (suggested by Arveson) that, if $K(T, \mathcal{S}) = d(T, \mathcal{S})/\alpha(T, \mathcal{S})$, and $K(\mathcal{S}) = \sup\{K(T, \mathcal{S}) : T \notin \mathcal{S}\}$, then $K(\mathcal{A}) \geq K(\mathcal{S})$, where $\mathcal{A}$ is the algebra defined above. A direct sum of these, with increasing constants, is then non-hyperreflexive.

The example of [75] was not a CSL (commutative subspace lattice) algebra. The construction by Davidson and Power [30] of a non-hyperreflexive CSL algebra was the next development. The technique used in [75] could not be used for this. In [30], a sequence of subspaces of $B(H)$ was constructed inductively, from first principles, with H increasing in dimension, such that each subspace was a bimodule over the corresponding diagonal algebra, and such that the constants of the algebras

$$\begin{pmatrix} \mathcal{D} & \mathcal{S} \\ 0 & \mathcal{D} \end{pmatrix}$$

formed an unbounded set.

In [85] it was shown that the key construction in [30] could be interpreted as a special case of a general *dual-product* subspace construction. Given subspace $\mathcal{S} \subseteq B(H)$ and $\mathcal{T} \subseteq B(K)$, denote by $\mathcal{S} * \mathcal{T}$ the subspace of $B(H) \otimes B(K)$ determined by first taking the tensor product of the preannihilators $\mathcal{S}_\perp \otimes \mathcal{T}_\perp$, regarded as a trace class norm-closed linear subspace of the ideal of trace-class operators in $B(H) \otimes B(K)$, and then taking the annihilator of $\mathcal{S}_\perp \otimes \mathcal{T}_\perp$ in $B(H) \otimes B(K)$. That is,

$$\mathcal{S} * \mathcal{T} = (\mathcal{S}_\perp \otimes \mathcal{T}_\perp)^\perp.$$

This will be reflexive if both $\mathcal{S}$ and $\mathcal{T}$ are reflexive. It was shown that if $\mathcal{S}$ is a proper subspace of $B(H)$, and if $\mathcal{T}$ is a proper subspace of $B(K)$, then the hyperreflexivity constant for $\mathcal{S} * \mathcal{T}$ is no smaller than the product of the constants for $\mathcal{S}$ and for $\mathcal{T}$. That is,

$$K(\mathcal{S} * \mathcal{T}) \geq K(\mathcal{S}) \cdot K(\mathcal{T}).$$

The [30] construction can be realized by taking $\mathcal{S} = \mathcal{D}$, the 3×3 diagonal matrices, and repeating this n-times. So the n-th subspace becomes $\mathcal{S}_n = \mathcal{D} * \mathcal{D} * \ldots * \mathcal{D}$ (n-times) with

$$K(\mathcal{S}_n) \geq (K(\mathcal{D}))^n.$$

A proof of M. D. Choi gives that $K(\mathcal{D}) \geq 3/(2\sqrt{2})$ and so $K(\mathcal{A}_n) \geq (3/(2\sqrt{2}))^n$, where $\mathcal{A}_n$ is the algebra

$$\mathcal{A}_n = \begin{pmatrix} \mathcal{D} \otimes \ldots \otimes \mathcal{D} & \mathcal{D} * \ldots * \mathcal{D} \\ 0 & \mathcal{D} \otimes \ldots \otimes \mathcal{D} \end{pmatrix}.$$

Very little concerning hyperreflexivity is presently known. Evidence seems to suggest that perhaps *most* reflexive algebras are not hyperreflexive. It is not known whether every von Neumann algebra is hyperreflexive. It follows from work of Eric Christensen [26] that if $\mathcal{A}$ is a C^*-algebra in $B(H)$, then every derivation from $\mathcal{A}$ into $B(H)$ is inner in $B(H)$ if and only if the commutant von Neumann algebra $\mathcal{A}'$ is hyperreflexive. Also, it is not known whether the algebras in *Arveson's class* are hyperreflexive. One question, which we feel is a very good *test* question for future progress, is due to Arveson [14]: If $\mathcal{A} = \text{Alg}\,\mathcal{N}(\mathbf{Z}_+)$, is the tensor product $\mathcal{A} \otimes \mathcal{A}$ hyperreflexive?

The reason that hyperreflexivity questions are relevant to similarity theory will be made clear in the next section.

5.2. On Hyperreflexivity and Similarity of Subspace Lattices

The purpose of this section is to point out as precisely as possible the role of hyperreflexivity in similarity theory. It is semi-expository in the sense that many of the ideas presented are at least implicitly contained in the work of [2], [29], and [80] for nest algebras. Much of this material was presented without great detail as part of a talk given by this author at the NSF-CBMS Regional Conference with William Arveson as principal lecturer, August, 1983, in Lubbock, Texas. We will try to show clearly how similarity theory for general subspace lattices, at least from one point of view, *breaks cleanly* into two distinct classes of subproblems: approximate equivalence problems and distance estimate (hyperreflexivity and variations thereof) problems.

Penetrating results have been obtained on approximate equivalence by William Arveson [13] for a class of CSL algebras much wider than the class of nests. Also, Kenneth Davidson [31] has recently obtained a very nice proof, shorter and more easily understood than Niels Andersen's original, that order isomorphisms between continuous nests can be implemented by approximate equivalences. This is useful for understanding. Many problems remain, however.

Distance estimate problems have been studied by a number of authors [1, 14, 21, 26, 28, 30, 44, 75, 76, 80, 82, and 85]. Interest has increased since the importance of the formula for nest algebras, as relating to similarity, was demonstrated. A number of results have been obtained. However, at present, the topic seems to be in even greater need of further development than the topic of approximate equivalence. Further results of a *positive* nature must be obtained to complement (hopefully) known approximate equivalence results. Although this is an opinion, it is hoped that the presentation we give will bear this feeling out to some degree.

We will develop the connection between approximate equivalence and hyperreflexivity in a setting without compact perturbation considerations. This is in reality a bit misleading, because the only proofs to date concerning nontrivial approximate equivalence results strongly utilize compact perturbation results and techniques. However, it is hoped that our approach here of *starting from the middle of the subject and working towards both ends* will enhance perspective and may suggest some new directions for research. The demonstration of the apparent *clean break* is the goal here.

By a hyperreflexive lattice, we mean a reflexive lattice $\mathcal{L}$ for which $\operatorname{Alg} \mathcal{L}$ is a hyperreflexive algebra.

Let $\mathcal{L}$ and $\mathcal{M}$ be subspace lattices acting on a Hilbert space $\mathcal{H}$ which are *order isomorphic* in the sense that there is a 1-1 map θ from $\mathcal{L}$ onto $\mathcal{M}$ which preserves the lattice operations. If there is an invertible operator $T \in B(H)$ such that for each $L \in \mathcal{L}$, $\theta(L)$ is the projection onto the closed linear span $[TLH]$, we say that θ is spatially implemented and write $\theta = \theta_T$. In this case $\operatorname{Alg} \mathcal{M} = T(\operatorname{Alg} \mathcal{L})T^{-1}$, as is easily checked, so $\operatorname{Alg} \mathcal{M}$ and $\operatorname{Alg} \mathcal{L}$ are similar algebras in the usual sense, and we say that the map θ_T is a similarity transformation of subspace lattices. If T can be taken to be unitary, then $\theta = \theta_T$ is called an equivalence.

We note that if an isomorphism θ is spatially implemented, the operator T for which $\theta = \theta_T$ is highly nonunique. Indeed, if A is any invertible operator in $\operatorname{Alg} \mathcal{L}$ for which also $A^{-1} \in \operatorname{Alg} \mathcal{L}$, then for any $L \in \mathcal{L}$ we have $ALH = LH$. Hence, if $S = TA$, then $SLH = TLH$ for each $L \in \mathcal{L}$, so $\theta_S = \theta_T$. The converse of this is also true: if S, T are invertible operators such that for the lattice $\mathcal{L}$, we have $\theta_S = \theta_T$, then $SLH = TLH$ for each $L \in \mathcal{L}$, so if $A = T^{-1}S$ then $ALH = LH$, $A^{-1}LH = LH$, $L \in \mathcal{L}$, hence $A \in (\operatorname{Alg} \mathcal{L}) \cap (\operatorname{Alg} \mathcal{L})^{-1}$, and S factors $S = TA$, as above.

Similarity is one type of equivalence relation among subspace lattices. Another is the notion of *approximate equivalence* introduced by Niels Andersen. (We will consider a somewhat weaker form of this.) Here, we will say that subspace lattices $\mathcal{L}$ and $\mathcal{M}$ are approximately equivalent if there is an order isomorphism $\theta : \mathcal{L} \to \mathcal{M}$ and a sequence of unitary operators $\{U_i\}_{i=1}^{\infty}$ with the property that given $\epsilon > 0$ there exists $N > 0$ such that $\|\theta(L) - U_i L U_i^*\| < \epsilon$ for all $L \in \mathcal{L}$ whenever $i \geq N$. Noting that $U_i L U_i^* = \operatorname{proj} [U_i LH] = \theta_{U_i}(L)$, this just says that the sequence of maps θ_{U_i} when considered as a sequence of projection valued functions from $\mathcal{L}$ into $B(H)$ with the norm topology, converges uniformly on $\mathcal{L}$ to the map θ.

The situation for subspace lattices which are nests and which act on a separable Hilbert space can be simply stated: any order isomorphism between separably acting nests, for which the necessary condition that the dimensions of the *gaps* between projections in the nests match up under the isomorphism, can be implemented by an approximate equivalence. Andersen obtained this for continuous nests in [1], and Davidson

174

extended this to arbitrary nests (i.e. having perhaps an atomic core part) in [29], via a proof that every such isomorphism for nests is spatially implemented. (This approach to similarity was different than that used in [84].)

If one wishes proofs of the strongest possible results in similarity theory it is essential that at least the portion of Andersen's Theorem which states that the differences $\theta(L) - U_i \, L \, U^*$ can be taken compact be fully utilized. A distilling of the central ideas involved, however, yields the following proof, valid for arbitrary subspace lattices, which we feel is reasonably transparent, that approximate equivalence alone (without compact perturbation considerations), together with hyperreflexivity of either lattice, does yield similarity, at the least.

We break the proof into two pieces. The first is proposition A, which isolates the application of hyperreflexivity in the construction. The proof of this consists of fitting together several *general principles* lemmas. Theorem B then gives the construction, and utilizes proposition A in one key position.

Proposition A. *Let $\mathcal{L}$ be a hyperreflexive lattice with constant K, and let $0 < \epsilon < 1/4$. If U is a unitary operator such that*

$$\sup_{L \in \mathcal{L}} \|ULU^* - L\| < \epsilon/K,$$

then there is an invertible operator $T \in B(H)$ with $\|T - I\| < \epsilon$ such that $ULU^* = $ proj $[TLH]$ for every $L \in \mathcal{L}$.

Remark. An example of Lance [80] shows that even when $\mathcal{L}$ is a nest T cannot always be taken to be unitary.

For the proof of Proposition A we require the Lemmas A1, A2, and A3 below. For the first, let $\mathcal{P}$ be an arbitrary subspace lattice acting on H, and let $V \in B(H)$ be a unitary operator. Let

$$\alpha(V, \mathcal{P}) = \sup_{P \in \mathcal{P}} \|P^\perp V P\|$$

denote the Arveson estimate for the distance from V to the algebra Alg $(\mathcal{P})$. Let

$$\mathcal{P}' = \{VPV^* : P \in \mathcal{P}\}$$

denote the image lattice, so Alg $(\mathcal{P}') = V(\text{Alg } \mathcal{P})V^*$. Let $\theta_V : \mathcal{P} \to \mathcal{P}'$ denote the order isomorphism defined by $\theta_V(P) = VPV^*$, $P \in \mathcal{P}$. Let

$$\|(\theta_V - \text{id})|_\mathcal{P}\| = \sup_{P \in \mathcal{P}} \|\theta_V(P) - P\|.$$

175

Lemma A1. *With the above notation,*

$$\|(\theta_V - \mathrm{id})|_{\mathcal{P}}\| = \max\{\alpha(V, \mathcal{P}), \alpha(V^*, \mathcal{P})\}.$$

Proof: Fix $P \in \mathcal{P}$, and let $Q = VPV^*$. Then

$$\|P^\perp Q\| = \|P^\perp VPV^*\| = \|P^\perp VP\|,$$

and

$$\|PQ^\perp\| = \|PVP^\perp V^*\| = \|PVP^\perp\| = \|P^\perp V^* P\|.$$

Thus

$$\|P - Q\| = \|P^\perp Q - PQ^\perp\|$$
$$= \max\{\|P^\perp Q\|, \|PQ^\perp\|\}$$
$$= \max\{\|P^\perp VP\|, \|P^\perp V^* P\|\}.$$

The taking of suprema now completes the proof. $\blacksquare$

Remark. Lemma (A1) shows that, when applied to a unitary operator and its adjoint in this manner, the distance estimate actually yields a *formula* for the distance a unitary operator *moves* an arbitrary subspace lattice. More generally, if T is an invertible operator which is not unitary, then

$$\max\{\alpha(T, \mathcal{P}), \alpha(T^{-1}, \mathcal{P})\}$$

yields an estimate for $\|(\theta_T - \mathrm{id})|_{\mathcal{P}}\|$, and vice-versa, with constant depending only on the product of norms $\|T\| \cdot \|T^{-1}\|$. These considerations do not depend on any assumed hyperreflexivity properties of $\mathcal{P}$. Hyperreflexivity enters into similarity computations in the following way: A unitary U is assumed to have the property that $\|(\theta_U - \mathrm{id})|_{\mathcal{P}}\|$ is small. Hyperreflexivity of $\mathcal{P}$ then implies, via Lemma (A1), that both U and U^* are close to alg $(\mathcal{P})$. This implies, via the *general principle* Lemma (A3) below, (which does not involve hyprreflexivity), that U^* is close to the group of invertible elements of Alg $(\mathcal{P})$. If A is a nearby element of $(\mathrm{Alg}\,(\mathcal{P})) \cap (\mathrm{Alg}\,(\mathcal{P}))^{-1}$, then the operator $T = UA$ is close to the identity operator and on $\mathcal{P}$ satisfies $\theta_T = \theta_U$. This is, informally, the content of Proposition A. Given an approximate equivalence, this device of replacing a unitary with an invertible operator close to the identity operator is then used repeatedly in a certain manner (the proof of Theorem B) to construct an invertible operator implementing the order isomorphism as an infinite product of operators, each close to I.

176

For the proof of Lemma (A3) we utilize the following standard inequality. The proof is short, so we include it for completness.

Lemma A2. Let $A, U \in B(H)$ with U a unitary operator. If $\|A - U\| < 1$, then A is invertible, and
$$\|A^{-1} - U^*\| \le \|A - U\|/(1 - \|A - U\|).$$

Proof: We have $\|U^*A - I\| = \|A - U\| < 1$ so U^*A is invertible, and hence A is invertible. Write
$$A = U + (A - U) = U[I + U^*(A - U)].$$

Then $A^{-1} = [I + U^*(A - U)]^{-1}U^*$, so

$$A^{-1} - U^* = [[I + U^*(A - U)]^{-1} - I]U^*.$$

Let $b = U^*(A - U)$. Then
$$\|b\| = \|A - U\| < \epsilon < 1,$$

so $I + b$ has inverse given by the Neumann series

$$\sum_{\ell=0}^{\infty}(-1)^{\ell}b^{\ell}.$$

We have
$$\|A^{-1} - U^*\| = \|(I + b)^{-1} - I\|$$
$$\le \sum_{\ell=1}^{\infty}\|b\|^{\ell}$$
$$= \|b\|/(1 - \|b\|)$$
$$= \|A - U\|/(1 - \|A - U\|). \quad \blacksquare$$

Lemma A3. Let A be an arbitrary norm closed subalgebra of $B(H)$ containing I, and let U be a unitary operator. If $0 < \epsilon < 1/4$, and if $d(U, A) < \epsilon$ and $d(U^*, A) < \epsilon$, then every element $A \in A$ with $\|A - U\| < \epsilon$ is in $A \cap A^{-1}$. In particular, $d(U, A \cap A^{-1}) < \epsilon$.

Proof: By Lemma (A2), A is invertible and $\|U^* - A^{-1}\| < \epsilon/(1 - \epsilon)$. We also have $\|A\| < 1 + \epsilon$. Choose $B \in A$ with $\|U^* - B\| < \epsilon$. Then

$$\|A^{-1} - B\| < \epsilon + \epsilon/(1 - \epsilon).$$

So

$$\|AB - I\| \le \|A\|\,\|B - A^{-1}\|$$
$$< (1 + \epsilon)[\epsilon + \epsilon/(1 - \epsilon)]$$
$$< (1 + \frac{1}{4})[\frac{1}{4} + \frac{1}{4}/(1 - \frac{1}{4})]$$
$$= \frac{35}{48} < 1.$$

Hence AB is invertible with inverse in $\mathcal{A}$. Then $A^{-1} = B(AB)^{-1} \in \mathcal{A}$, as desired.

Proof of Proposition A.

In the notation of Lemma (A1) we have $\|\theta_U - \mathrm{id}\| < \epsilon/K$ so $\alpha(U, \mathcal{L}) < \epsilon/K$ and $\alpha(U^*, \mathcal{L}) < \epsilon/K$ and hence $d(U, \mathrm{Alg}\,\mathcal{L}) < \epsilon$ and $d(U^*, \mathrm{Alg}\,\mathcal{L}) < \epsilon$ by hyperreflexivity of $\mathcal{L}$. Thus by Lemma (A3), we have

$$d(U^*, (\mathrm{Alg}\,\mathcal{L}) \cap (\mathrm{alg}\,\mathcal{L})^{-1}) < \epsilon.$$

Let $A \in (\mathrm{Alg}\,\mathcal{L}) \cap (\mathrm{Alg}\,\mathcal{L})^{-1}$ with $\|U^* - A\| < \epsilon$, and let $T = UA$. Then

$$\|T - I\| = \|UA^{-1} - I\|$$
$$= \|A - U^*\| < \epsilon.$$

Since $A \in (\mathrm{Alg}\,\mathcal{L}) \cap (\mathrm{Alg}\,\mathcal{L})^{-1}$, it follows that $ALH = LH, L \in \mathcal{L}$, and hence that $TLH = ULH$. So

$$\mathrm{proj}\,[TLH] = \mathrm{proj}\,[ULH] = ULU^*$$

for each $L \in \mathcal{L}$ as desired. ∎

Theorem B. *Let $\mathcal{L}$ and $\mathcal{M}$ be subspace lattices, and suppose $\theta : \mathcal{L} \to \mathcal{M}$ is an order isomorphism. Suppose there is a sequence $\{U_i\}$ of unitary operators satisfying the property that given $\epsilon > 0$ there exists $N > 0$ such that for each $i \ge N$ we have*

$$\|\theta(L) - U_i L U_i^*\| < \epsilon \text{ for all } L \in \mathcal{L}.$$

(That is, we assume $\mathcal{L}$ and $\mathcal{M}$ are approximately equivalent.) If either $\mathcal{L}$ or $\mathcal{M}$ is hyperreflexive, then so is the other with the same hyperreflexivity constant. In this event there is an invertible operator T such that $\theta(L) = \mathrm{proj}\,(TLH)$ for each $L \in \mathcal{L}$. (That is, $\mathcal{L}$ and $\mathcal{M}$ are similar, and $\theta = \theta_T$.) Moreover, given any $\eta > 0$ there is an

178

invertible operator T satisfying this property $\theta = \theta_T$ and a unitary operator W such that $\|T - W\| < \eta$.

Proof: Let $\theta : \mathcal{L} \to \mathcal{M}$ be an order isomorphism and $\{U_i\}$ a sequence of unitary operators satisfying the hypothesis. First, we will assume that $\mathcal{L}$ is hyperreflexive with constant K. Let $0 < \eta < 1/2$ be given arbitrarily, and let $\beta = ln(1 + \eta) < 1/2$.

For each i, let $\theta_i = \theta_{U_i}$, so

$$\theta_i(L) = U_i L U_i^* = \text{proj}\,[U_i L H], L \in \mathcal{L},$$

and let $\mathcal{L}_i = \{\theta_i(L) : L \in \mathcal{L}\}$ be the image lattice. So $\{\mathcal{L}_i\}$ is a sequence of subspace lattices unitarily equivalent to $\mathcal{L}$, and $\{\theta_i\}$ is a sequence of order isomorphisms having common domain $\mathcal{L}$ and ranges the $\mathcal{L}_i$.

Define

$$\|\theta - \theta_i\| = \sup_{L \in \mathcal{L}} \|\theta(L) - \theta_i\|.$$

Then

$$\text{Lim}_i \|\theta - \theta_i\| = 0.$$

Rapid convergence of $\{\theta_i\}$ will be required, so replace the sequence $\{\theta_i\}$, if necessary, with a subsequence, to insure that

$$\sum_i \|\theta - \theta_i\| < \beta/4K.$$

Using the notation

$$\|\theta_i - \theta_{i+1}\| = \sup_{L \in \mathcal{L}} \|\theta_i(L) - \theta_{i+1}(L)\| \quad,$$

since then

$$\|\theta_i - \theta_{i+1}\| \leq \|\theta_i - \theta\| + \|\theta - \theta_{i+1}\| \quad,$$

we have

$$\sum_i \|\theta_i - \theta_{i+1}\| < \beta/2K.$$

Define *difference* isomorphisms

$$\phi_i : \mathcal{L}_i \to \mathcal{L}_{i+1}, \text{ by } \phi_i = \theta_{i+1} \circ \theta_i^{-1}.$$

Denoting

$$\|\phi_i - \text{id}\| = \sup_{P \in \mathcal{L}_i} \|\phi_i(P) - P\|,$$

179

we have

$$\|\phi_i - \mathrm{id}\| = \|\theta_i - \theta_{i+1}\|,$$

so

$$\sum \|\phi_i - \mathrm{id}\| < \beta/2K.$$

Note that ϕ_i is implementd by the unitary $V_i = U_{i+1}U_i^*$.

Now observe that each lattice $\mathcal{L}_i$ is hyperreflexive with constant K since it is unitarily equivalent to $\mathcal{L}$. So *by proposition A*, corresponding to the sequence $\{V_i\}$ there is a sequence $\{T_i\}$ of invertible operators, with $\|T_i - I\| < 2K\|\phi_i - \mathrm{id}\|$, such that ϕ_i is implemented by T_i in the sense that for each i and for each $P \in \mathcal{L}_i$ we have

$$\mathrm{proj}\,[T_i P H] = \mathrm{proj}\,[V_i P H] = V_i P V_i^*.$$

Since $\sum \|\phi_i - \mathrm{id}\| < \beta/2K$, and $\|T_i - I\| < 2K\|\phi_i - \mathrm{id}\|$ for each i, we have

$$\sum \|T_i - I\| < \beta.$$

A standard argument shows that since $\sum \|T_i - I\|$ is finite, the sequence of partial products $S_n = T_n T_{n-1} \ldots T_1$ is norm-Cauchy and hence the infinite product

$$\ldots\, T_n T_{n-1} \ldots\, T_1$$

converges in norm to an operator S. An estimate shows that

$$\|S - I\| < e^\beta - 1 = \eta.$$

The operator S is invertible because $\|S - I\| < 1$.

Another standard argument shows that if a sequence of invertible operators $\{S_n\}$ converges in norm to an invertible operator S, then for every closed subspace M of H the sequence of projections $\{\mathrm{proj}\,[S_n M]\}$ converges in norm to the projection $\mathrm{proj}\,[SM]$.

We now let $T = SU_1$, where U_1 is the first unitary operator in the (perhaps reduced) sequence. For each i, the order isomorphism θ_i is implemented by $S_i U_1$. So, for each $L \in \mathcal{L}$ we have

$$\mathrm{proj}\,[TLH] = \mathrm{Lim}_i \, \mathrm{proj}\,[S_i U_1 L H] = \mathrm{Lim}_i \, \theta_i(L) = \theta(L).$$

Hence, T implements θ, as required. If we let $W = U_1$, then

$$\|T - W\| = \|SU_1 - U_1\| = \|S - I\| < \eta.$$

It is easily shown that any lattice which is similar to a hyperreflexive lattice is hyperreflexive, and an estimate yields

$$K(\mathcal{M}) \leq \|T\|^2 \cdot \|T^{-1}\|^2 \cdot K(\mathcal{L})$$
$$\leq \|T\|^4 \cdot \|T^{-1}\|^4 \cdot K(\mathcal{M}).$$

So, since T can be taken arbitrarily close to a unitary operator, it follows that $K(\mathcal{M}) = K(\mathcal{L})$.

We have shown that the hyperreflexivity hypothesis on $\mathcal{L}$ yields similarity. If instead $\mathcal{M}$ were assumed hyperreflexive, it is easily shown that the inverse order isomorphism $\theta^{-1} : \mathcal{M} \to \mathcal{L}$ and the sequence of adjoints $\{U_i^*\}$ also yields an approximate equivalence.

For each i we have

$$\sup_{P \in \mathcal{L}} \|U_i P U_i^* - \theta(P)\| = \sup_{P \in \mathcal{L}} \|P - U_i^* \theta(P) U_i\|$$
$$= \sup_{Q \in \mathcal{M}} \|\theta^{-1}(Q) - U_i^* Q U_i\|.$$

Hence, the above argument gives an invertible operator T which implements θ^{-1}. Hence, T^{-1} implements θ. ■

References

1. Andersen, N. T., Compact perturbations of reflexive algebras, J. Funct. Anal. **38** (1980), 366-400.

2. ———, Similarity of continuous nests, Bull. London Math. Soc., **15** (1983), 131-132.

3. Apostol, C., Ultra weakly closed operator algebras, J. Operator Theory **2** (1979), 49-67.

4. ———, L. Fialkow, D. Herrero, and D. Voiculescu, *Approximation of Hilbert Space Operators, Vol. II*, Pitman.

5. ———, F. Gilfeather, Isomorphisms modulo the compact operators of nest algebras, Pac. J. Math., **122** (1986), 263-286.

6. Arveson, W. B., Analyticity in operator algebras, Amer. J. Math., **89** (1967), 578-642.

7. ———, A density theorem for operator algebras, Duke Math. J., **23** (1967), 635-648.

8. _____, Operator algebras and measure preserving automorphisms, Acta Math., **118** (1967), 95-108.

9. _____ and K. Josephson, Operator algebras and measure preserving automorphisms II, J. Funct. Anal. **2** (1969), 100-134.

10. _____, Operator algebras and invariant subspaces, Ann. of Math. **3** (1974), 433-532.

11. _____, Interpolation problems in nest algebras, J. Funct. Anal. **20** (1975), 208-233.

12. _____, Notes on extensions of C^*-algebras, Duke Math. J. **44** (1977), 329-355.

13. _____, Perturbation theory for groups and lattices, J. Funct. Anal. **53** (1983), 22-73.

14. _____, Ten lectures on operator algebras, CBMS Regional Conf. Series, No. 55, A.M.S., Providence, 1984.

15. Azoff, E. A., K-reflexivity in finite dimensional spaces, Duke Math. J. **40** (1973), 821-830.

16. _____, C. K. Fong, and F. Gilfeather, A reduction theory for non-self-adjoint operator algerbras. Trans. A.M.S. **224** (1976), 351-366.

17. _____, On finite rank operators and preannhilators, Memoirs A.M.S., vol. 64, number 357.

18. Brown, S., B. Chevreau, and C. Pearcy, Contractions with rich spectrum, have invariant subspaces, J. Operator Theory **1** (1979), 123-136.

19. Bercovici, H., C. Foias, and C. Pearcy, Invariant subspaces, dilation theory, and dual algebras, CBMS Regional Conference Series, No. 56, A.M.S., Providence, 1985.

20. Chevreau, B. and C. Pearcy, Some new criteria for membershp in $\mathcal{A}_{\mathcal{N}_0}$ with applications to invariant subspaces, J. Funct. Anal., to appear.

21. Choi, M. D. and K. R. Davidson, Perturbations of matrix algebras, Mich. Math. J. **33** (1986), 273-287.

22. Christensen, E., Derivations of nest algebras, Math. Ann., **229** (1977), 155-161.

23. _____ and C. Peligrad, Commutants of nest algebras modulo the compact operators, Invent. Math. **56** (1980), 113-116.

24. _____, Perturbations of operator algebras II, Ind. U. Math. J. **26** (1977), 891-904.

25. _____, Extensions of derivations, J. Funct. Anal. **27** (1978), 234-247.

26. _____, Extensions of derivations II, Math. Scand. **50** (1982), 111-122.

27. Conway, J. B., A complete Boolean algebra of subspaces which is not reflexive, Bull. Am. Math. Soc. **79** (1973), 720-722.

28. Davidson, K. R., Commutative subspace lattices, Ind. U. Math. J. **27** (1978), 479-490.

29. ———, Similarity and compact perturbations of nest algebras, J. Reine Agnew. Math. **348** (1984), 72-87.

30. ——— and S. C. Power, Failure of the distance formula, J. London Math. Soc. **32** (1985), 157-165.

31. ———, Approximate unitary equivalence of continuous nests, Proc. Am. Math. Soc. **97** (1986), 655-660.

32. ——— and B. Wagner, Automorphisms of quasitriangular algebras, J. Funct. Anal. **59** (1984), 612-627.

33. Deddens, J. A., Another description of nest algebras, Lect. Notes Math. 693, Springer-Verlag 77-86.

34. Erdos, J. A., Some results on triangular operator algebras, Amer. J. Math. **89** (1967), 85-93.

35. ———, Unitary invariants for nests, Pacific J. Math. **23** (1967), 229-256.

36. ———, Some questions concerning triangular operator algebras, Proc. R. Irish Acad. **74A** (1974), 223-232.

37. ———, Non-selfadjoint operator algebras, Proc. R. Irish Acad. **81A** (1981), 127-145.

38. ——— and S. C. Power, Weakly closed ideals of nest algebras, J. Operator Theory **7** (1982), 219-235.

39. ———, Reflexivity for subspace maps and linear spaces of operators, Proc. London Math. Soc. **52** (1986), 582-600.

40. Fall, T., W. B. Arveson, and P. S. Muhly, Perturbations of nest algebras, J. Operator Theory **1** (1979), 137-150.

41. Feintuch, A. and R. Saeks, *System Theory: A Hilbert Space Approach*, Academic Press, New York, 1982.

42. Fillmore, P., On invariant linear manifolds, Proc. Amer. Math. Soc. **41** (1973), 501-505.

43. Gilfeather, F. and D. Larson, Nest subalgebras of von Neumann algebras, Advances in Math. **46** (1982), 176-199.

44. _____ and D. Larson, Nest subalgebras of von Neumann algebras: commutants modulo compacts and distance estimates, J. of Operator Theory **7** (1982), 279-302.

45. _____ and D. Larson, Nest subalgebras of von Neumann algebras: Commutants modulo the Jacobson radical, J. of Operator Theory **10** (1983), 95-118.

46. _____ and D. Larson, Structure in reflexive subspace lattices, J. London Math. Soc. **26** (1982), 117-131.

47. _____ and D. Larson, Commutants modulo the compact operators of certain CSL algebras. *Topics in Modern Operator Theory: Advances and Applications*, Birkhauser **2** (1981), 105-120.

48. _____ and D. Larson, Commutants modulo the compact operators of certain CSL algebras II, Integral Eq. and Operator Theory **6** (1983), 345-356.

49. _____, A. Hopenwasser, and D. Larson, Reflexive algebras with finite width lattices: Tensor products, cohomology and compact perturbations, J. of Funct. Anal. **55** (1984), 176-199.

50. _____, Derivations on certain CSL algebras, J. Operator Theory **11** (1984), 145-156.

51. _____ and R. Moore, Isomorphisms of certain CSL algebras J. Funct. Anal.**67** (1986), 264-291.

52. Gohberg, I. C. and M. G. Krein, *Introduction to the Theory of Linear Non-selfadjoint Operators*, Transl. Math. Monographs **18**, Amer. Math. Soc., R.I., 1969.

53. _____ and M. G. Krein, *Theory and Application of Volterra Operators in Hilbert Space*, Transl. Math. Monographs **24**, Amer. Math. Soc., R.I., 1970.

54. Hadwin, D. W., E. A. Nordgren, Subalgebras of reflexive algebras, J. Operator Theory **7** (1982), 3-23.

55. _____, Algebraically reflexive linear transformations, Linear and Multilinear Algebra **14** (1983), 225-233.

56. Halmos, P. R., Quasitriangular operators, Acta. Sci. Math., Szeged, **29** (1968), 283-293.

57. _____, Reflexive lattices of subspaces, J. Lond. Math. Soc. **4** (1971), 257-263.

58. Harrison, K. J., Certain distributive lattices of subspaces are reflexive, J. London Math. Soc., 51-56.

59. Herrero, D. A., Approximation of Hilbert Space Operators, vol. I, Pitman, 1982.

60. _____, Compact perturbations of nest algebras, index obstructions and a problem of Arveson, J. Funct. Anal. **55** (1984), 78-109.

184

61. _____, Compact perturbations of continuous nest algebras, J. London Math. Soc.(2) **27** (1983), 339-344.

62. _____, The diagonal entries in the formula: Quasitriangular − compact = triangular, and restrictions of quasitriangularity, Trans. A.M.S. **298** (1986), 1-42.

63. Hopenwasser, A., Ergodic automorphisms and linear spaces of operators, Duke Math. J. **41** (1974), 747-757.

64. _____, Completely isometric maps and triangular operator algebras, Proc. London Math. Soc. **25** (1972), 96-114.

65. _____, Isometries on irreducible triangular operator algebras, Math. Scand **30** (1972), 136-140.

66. _____, The radical of a reflexive operator algebra, Pac. J. Math. **65** (1976), 375-392.

67. _____ and J. Plastiras, Isometries of quasitriangular operator algebras, Proc. Am. Math. Soc. **65** (1977), 242-244.

68. _____ and D. Larson, The carrier space of a reflexive operator algebra, Pacific J. Math. **81** (1979), 417-434.

69. _____, C. Laurie, and R. Moore, Reflexive algebras with completely distributive subspace lattices, J. Operator Theory, II (1984), 91-108.

70. Johnson, B., Cohomohology in Banach algebras, Mem. A.M.S. **127** (1972).

71. _____ and S. K. Parrott, Operators commuting with a von Neumann algebra modulo the set of compact operators, J. Funct. Anal. **11** (1972), 39-61.

72. Kadison, R. V. and I. M. Singer, Triangular operator algebras, Amer. J. Math. **82** (1960), 227-259.

73. _____ and D. Kastler, Perturbations of von Neumann algebras I, Am. J. Math. **94** (1972), 38-54.

74. Kraus, J., Tensor products of reflexive algebras, J. London Math. Soc. (2) **28** (1983), 35-358.

75. _____ and D. Larson, Some applications of a technique for constructing reflexive operator algebras, J. Operator Theory **13** (1985), 227-236.

76. _____ and D. Larson, Reflexivity and distance formulae, Proc. London Math. Soc.**53** (1986), 340-356.

77. Lambrou, M. S., Complete atomic Boolean lattices, J. London Math. Soc. **15** (1977), 387-390.

78. _____ and W. E. Longstaff, Abelian algebras and reflexive lattices, Bull. London Math. Soc. **12** (1980), 165-168.

79. Lance, E. C., Some properties of nest algebras, Proc. London Math. Soc. **19** (1969), 45-68.

80. _____, Cohomology and perturbations of nest algebras, Proc. London Math. Soc. (3) **43** (1981), 334-356.

81. Larson, D. R., On the structure of certain reflexive algebras, J. Funct. Anal. **31** (1979), 275-292.

82. _____, Annihilators of operator algebras, *Topics in Modern Operator Theory*, Vol. **6** (1982), 119-130.

83. _____, A solution to a problem of J. R. Ringrose, Bull. (New Series) A.M.S. **7** (1982), 243-246.

84. _____, Nest algebras and similarity transformations, Annals of Math. **121** (1985), 409-427.

85. _____, Hyperreflexivity and a dual product construction, Transactions A.M.S. **294** (1986), 79-88.

86. _____ and B. Solel, Nests and inner flows, J. Operator Theory **16** (1986), 157-164.

87. _____, Reflexivity, algebraic reflexivity, and linear interpolation, Amer. J. Math., to appear.

88. Laurie, C., Invariant subspace lattices and compact operators, Pac. J. Math. **89** (1980), 351-365.

89. _____, On density of compact operators in reflexive algebras, Indiana Univ. Math. J. **30** (1981), 1-16.

90. _____ and W. Longstaff, A note on rank-one operators in reflexive algebras, Proc. Amer. Math. Soc. **89** (1983), 293-297.

91. Loebl, R. I. and P. S. Muhly, Analyticity and flows in von Neumann algebras, J. Funct. Anal. **29** (1978), 214-252.

92. Loginov, A. and V. Sulman, Hereditary and intermediate reflexivity of W^*-algebras, Izv. Akad. Nauk SSSR **39** (1975), 1260-1273; USSR-Isv. **9** (1975), 1189-1201.

93. Longstaff, W. E., Generators of nest algebras, Can. J. Math. **26** (1974), 565-575.

94. Muhly, P. S., Radicals, crossed products, and flows, Ann. Polon. Math. **XLII** (1983), 35-42.

95. ———, K. S. Saito, and B. Solel, Coordinates for triangular operator algebras, Annals of Math., to appear.

96. Narcowich, F. J., J. D. Ward, and D. A. Legg, Best approximation from stepped subspaces, Approximation theory and its applications **1** (1985), 29-49.

97. Nordgren, E., H. Radjavi, and P. Rosenthal, On Arveson's characterization of hyperreducible triangular algebras, Indiana Univ. Math. J. **26** (1977), 179-182.

98. Olin, R. and J. Thomson, Algebras of subnormal operators, J. Funct. Anal. **37** (1980), 271-301.

99. Plastiras, J., Quasitriangular operator algebras, Pac. J. Math. **64** (1976), 543-550.

100. Power, S. C., The distance to upper triangular operators, Math. Proc. Camb. Phil. Soc. **88** (1980), 827-829.

101. ———, Analysis in nest algebras, Lecture notes (1986).

102. Radjavi, H. and P. Rosenthal, On invariant subspaces and reflexive algebras, Am. J. Math. **91** (1969), 683-692.

103. ——— and P. Rosenthal, *Invariant Subspaces*, Springer-Verlag, New York, 1973.

104. Rickart, C., *General Theory of Banach Algebras*, Van Nostrand, New York, 1960.

105. Ringrose, J. R., Algebraic isomorphisms between ordered bases, Am. J. Math. **83** (1961), 463-478.

106. ———, Super-diagonal forms for compact linear operators, Proc. London Math. Soc. **12** (1962), 367-384.

107. ———, On the triangular representation of integral operators, Proc. London Math. Soc. **12** (1962), 385-399.

108. ———, On some algebras of operators, Proc. London Math. Soc., (3) **15** (1965), 61-83.

109. ———, On some algebras of operators II, Proc. London Math. Soc., (3) **16** (1966), 385-402.

110. ———, *Compact Non-selfadjoint Operators*, Van Nostrand Reinhold, 1971.

111. Rosenthal, P., Weakly closed maximal triangular algebras are hyperreducible, Proc. Am. Math. Soc. **24** (1970), 220.

112. ———, Applications of Lomonosov's lemma to non-selfadjoint operator algebras, Proc. R. Ir. Acad. **74** (1974), 271-281.

113. Saeks, R., Causality in Hilbert space, SIAM Rev. **12** (1970), 357-383.

114. Sarason, D., Invariant subspaces and unstarred operator algebras, Pacific J. Math. **17** (1966), 511-517.

115. Schue, J. R., The structure of hyperreducible triangular algebras, Proc. Am. Math. Soc. **15** (1964), 766-772.

116. Solel, B., Irreducible triangular algebras, Ph.D. Dissertation, U. of Pennsylvania, 1981.

117. ———, Algebras of analytic operators associated with a periodic flow on a von Neumann algebra, Can. J. Math. **37** (1985), 405-429.

118. Tsuji, K., Annihilators of operator algebras, Mem. Fac. Sci., Kochi Univ. (Math), Vol. 4.

119. Voiculescu, D., A non-commutative Weyl-von Neumann theorem, Rev. Roum. Math. Pures et Appl. **21** (1976), 97-113.

120. Wagner, B., Automorphisms and derivations of certain operator algebras, Ph.D. Dissertation, UC-Berkeley, 1982.

121. ———, Derivations of quasitriangular algebras, Pacific J. Math. **114** (1984), 243-255.

David Larson

Department of Mathematics

Texas A&M University

College Station, Texas 77843

U.S.A.

Analysis in Nest Algebras

by

Stephen Power

Introduction

We present some of the foundational theory of nest algebras together with applications and related topics. To make these notes of working use to students, a pedagogical dichotomy has been used. Almost all discussion and historical commentary has been relegated to the notes, and the mathematics is given, with full proofs, in a pure unencumbered atmosphere.

A recurring theme which is deliberately emphasised here is the close parallel that exists between nest algebra theory and the theory of analytic functions, particularly H^p-spaces. Indeed, it is now common to refer to non-self-adjoint operator algebras as analytic or asymmetric operator algebras. After some preliminaries, we develop four sections under the following headings:

1. **Distance:** Arveson's fundamental distance formula is proved here in two ways (neither proof is Arveson's) and applications are given that include

   (i) best approximation by upper triangular compact operators,
   (ii) Nehari's theorem for Hankel operators,
   (iii) an integral operator variant of Hardy's inequality for H^1 functions.

2. **Density:** The important Erdos density theorem is given two proofs (neither proof is Erdos') and a number of applications are given, including a classical tricky theorem of Lidskii concerning the equality of trace and spectral trace for a trace class operator.

3. **Factorisation:** A Cholesky factorisation method is developed in Hilbert space and used to obtain the Arveson inner-outer factorisation of operators relative to certain nest algebras. We also discuss Szego type factorisations, Riesz factorisation, weak factorisation, and Hankel forms.

4. **Beyond total order:** We present a key example of a reflexive operator algebra, with commutative subspace lattice, that nevertheless fails to possess a distance formula.

I would like to record here thanks to John Conway for organising such an exciting special year in Operator Theory, and to Alan Hopenwasser and Ken Davidson for conversations that improved these notes.

Preliminaries

Let $\mathcal{H}$ be a complex Hilbert space. We refer to the closed linear subspaces of $\mathcal{H}$ simply a subspaces. A *nest* in $\mathcal{H}$ is a family of subspaces that includes $\{0\}$ and $\mathcal{H}$ and is totally ordered with respect to inclusion. A *complete nest* is a nest that is closed under the formation of closed unions and arbitrary intersections. Clearly, to each nest there is a unique minimal complete nest containing it, called the completion. We can restrict our attention to complete nests since a nest and its completion determine identical nest algebras. The *nest algebra* associated with a nest is the algebra of operators that leave each element of the nest invariant.

Let Ω be a totally ordered set and suppose that $\mathcal{H}$ has an orthonormal basis indexed by Ω, namely $\{e_\omega : \omega \in \Omega\}$. Then the subspaces

$$N_\omega = \overline{\text{span}}\{e_\sigma : \sigma \leq \omega\},$$

together with $\{0\}$ and $\mathcal{H}$ form a nest and determine a nest algebra that we write as $T(\Omega)$. Of course we are usually indifferent to the particular spatial realisation of a nest algebra and are primarily interested in the unitary equivalence class. Of particular interest are the algebras $T(\mathbf{N}), T(\mathbf{Q})$ and $T(Z)$. Notice that the nest associated with Ω is not usually complete and that its completion is often quite unrelated to the nest associated with the order completion of Ω. (Take $\Omega = \mathbf{Q}$ to illustrate this.)

If $\mathcal{A}$ is an algebra of operators on $\mathcal{H}$ then we write Lat$\mathcal{A}$ for the lattice of invariant self-adjoint projections of $\mathcal{A}$;

$$\text{Lat}\mathcal{A} = \{E : (I - E)AE = 0 \text{ for all } A \text{ in } \mathcal{A}\}.$$

Taking the dual viewpoint, if $\mathcal{E}$ is a collection of self adjoint projections, then we write Alg$\mathcal{E}$ for the algebra of operators that leave invariant all the projections in $\mathcal{E}$;

$$\text{Alg}\mathcal{E} = \{A : (I - E)AE = 0 \text{ for all } E \text{ in } \mathcal{E}\}.$$

Nest algebras are precisely the operator algebras $\mathcal{A} = \text{Alg}\mathcal{E}$ associated with a totally ordered family of projections $\mathcal{E}$ (under the usual ordering of positive operators). In

190

particular, they are *reflexive algebras* since they satisfy the duality condition

$$\mathcal{A} = \text{Alg Lat} \mathcal{A}.$$

Here is some standard terminology concerning totally ordered families $\mathcal{E}$ of self adjoint projections. It will prove conveninet to fix attention to projection nests rather than the corresponding nest of subspaces. $\mathcal{E}$ is a (complete) *projection nest* if it contains O and I and is closed in any of the weak operator topologies. This is equivalent to the completeness of the subspace nest associated with $\mathcal{E}$. If $E, F \in \mathcal{E}$ and $E < F$, then the (nonzero) projection $F - E$ is called an *interval* of $\mathcal{E}$. An *atom* of $\mathcal{E}$ is an interval that does not properly contain any other interval. If $E < I$ then the projection

$$E_+ = \inf\{F : E < F, F \in \mathcal{E}\}$$

is well-defined in a projection nest $\mathcal{E}$. If $E < E_+$ then E_+ is called the immediate successor of E. If $E \neq E_+$ for every E in $\mathcal{E}$, then the nest is said to be *well ordered*. Similarly, if $E > 0$, we can define E_- and the notion of immediate predecessor in a projection nest. The atoms of $\mathcal{E}$ are precisely the intervals of the form $E_+ - E$. A projection nest is *purely atomic* it if is generated by the atoms Q in the sense that

$$\mathcal{H} = \sum_{\text{atoms } Q} \oplus Q\mathcal{H}.$$

In particular, if $\mathcal{H}$ is separable, then every projection E of $\mathcal{E}$ is the weak operator topology sum of any series formed from the set of atoms dominated by E. If $\mathcal{E}$ possesses no atoms, then it is said to be a continuous nest.

The canonical continuous nest algebras are those associated with the Volterra projection nests on $L^2(\mathbf{R})$ and $L^2[0,1]$;

$$\mathcal{E}_\mathbf{R} = \{P_t : \text{ran} P_t = L^2(-\infty, t), t \in \mathbf{R}\} \cup \{0, I\};$$
$$\mathcal{E}_{[0,1]} = \{P_t : \text{ran} P_t = L^2[o, t], 0 \le t \le 1\}.$$

We abuse our earlier notation and write $T(\mathbf{R})$ and $T([0,1])$ for $\text{Alg}\mathcal{E}_\mathbf{R}$ and $\text{Alg}\mathcal{E}_{[0,1]}$ respectively. This should not lead to confusion as we will have no cause to consider the nonseparable nest algebras associated with orthonormal bases indexed by $\mathbf{R}$ or by $[0,1]$. Note that $T(\mathbf{R})$ and $T([0,1])$ are unitarily equivalent. These nest algebras, together with $T(\mathbf{N}), T(\mathbf{Z})$ and $T(\mathbf{Q})$ have the property that the diagonal algebra $\text{Alg}\mathcal{E} \cap (\text{Alg}\mathcal{E})^*$ is maximal abelian.

The rank one operators in a nest algebra $\text{Alg}\mathcal{E}$ form an important and useful class and are characterised in the following way. Write $e \otimes f$ for the rank one operator

$$(e \otimes f)x = (x, f)e$$

so that $e \neq \otimes f$ is linear in the first entry, conjugate linear in the second entry, and $(e \otimes f)(g \otimes h) = (g, f)e \otimes h$. The rank one operator $e \otimes f$ belongs to the nest algebra $\text{Alg}\mathcal{E}$ if and only if there exists a projection E in $\mathcal{E}$ such that $Ee = e$ and $E_- f = 0$. Exercise: Prove this and use it to show that $\mathcal{E} = \text{LatAlg}\mathcal{E}$.

We write $\mathcal{C}_p(\mathcal{H})$ for the pth von Neumann Schatten class of operators on $\mathcal{H}$, and $\mathcal{K}$ for the ideal of compact operators.

1. Distance

Our first proof of the Arveson distance formula (Theorem 1.3) is of a constructive nature, and its essential ingredient is the following lemma. An important feature of the lemma, not made explicit in the statement, is that it applies to general 2×2 operator matrices with respect to any pairs of orthogonal decompositions of domain and range. It is therefore beautifully poised for inductive use.

Lemma 1.1. *The minimum operator norm of the operator matrices*

$$\begin{bmatrix} A & B \\ C & X \end{bmatrix},$$

for variable X, is attained at an operator of the form $X_1 = C_1 A^* B_1$. This minimum is equal to the maximum of the norm of the operators

$$\begin{bmatrix} A & B \\ 0 & 0 \end{bmatrix}, \begin{bmatrix} A & 0 \\ C & 0 \end{bmatrix}.$$

Proof: There is no loss of generality in the assumption that the maximum in the lemma is no greater than unity. Equivalently, we may assume that

$$AA^* + BB^* \leq Q, A^*A + C^*C \leq P,$$

where P (resp. Q) is the orthogonal projection onto the first summand in the decomposition of the domain (resp. range). Since $BB^* \leq Q - AA^*$ and $C^*C \leq P - A^*A$, there exists contractions B_1, C_1 such that

$$B^* = B_1(Q - AA^*)^{1/2}, C = C_1(P - A^*)^{1/2}.$$

192

In particular, if $X_o = -C_1 A^* B_1$, we see that

$$\begin{bmatrix} A & B \\ C & X_o \end{bmatrix} = \begin{bmatrix} Q & O \\ O & C_1 \end{bmatrix} \begin{bmatrix} A & (Q - AA^*)^{1/2} \\ (P - A^* A)^{1/2} & -A^* \end{bmatrix} \begin{bmatrix} P & O \\ O & B_1 \end{bmatrix}.$$

The middle term of the right hand side is a unitary operator. This is an immediate consequence of the identity $A(P - A^* A)^{1/2} = (Q - AA^*)^{1/2} A$, and so the proof os complete. ∎

Note that if A lies in a particular ideal of $\mathcal{L}(\mathcal{H})$, then the minimising operator may be chosen from the same ideal.

Lemma 1.2. (W. B. Arveson) *The minimum operator norm, α say, of the operator matrices*

$$\begin{pmatrix} X_{11} & X_{12} & \cdots & X_{1n} \\ Y_{21} & X_{22} & \cdots & X_{2n} \\ Y_{31} & Y_{32} & X_{33} & X_{3n} \\ \vdots & \vdots & \ddots & \vdots \\ Y_{n1} & Y_{n2} & \cdots & X_{nn} \end{pmatrix}$$

where the X_{ij} are variable and the Y_{ij} are fixed, is achieved and equal to the maximum of the operator norms of the lower triangular block matrices. That is, $\alpha = \beta$ where

$$\beta = \max_{2 \leq k \leq n} \left\| \begin{pmatrix} Y_{k1} & \cdots & Y_{kk-1} \\ \vdots & & \vdots \\ Y_{n1} & \cdots & Y_{nk-1} \end{pmatrix} \right\|.$$

Remark. In the case where domain and range receive the same decomposition into n summands the lemma expresses the Arveson distance formula for the case of a nest with $n + 1$ elements. It is clear that the minimum can be achieved with the first row and last column identically zero.

Proof: (S. C. Power) Set $X_{11}, X_{12}, \ldots, X_{1n}$ and $X_{2n}, X_{3n}, \ldots, X_{nn}$ to be the zero operators on the appropriate spaces. Choose X_{22} by means of Lemma 1.1, so that the operator norm of the submatrix

$$\begin{bmatrix} Y_{21} & X_{22} \\ Y_{31} & Y_{32} \\ \vdots & \vdots \\ Y_{n1} & Y_{n2} \end{bmatrix}$$

is no greater than β. Now, using Lemma 1.2 again, choose X_{33} in a similar way for the submatrix

$$\begin{bmatrix} Y_{31} & Y_{32} & X_{33} \\ Y_{41} & Y_{42} & Y_{43} \\ \vdots & \vdots & \\ Y_{n1} & & Y_{n3} \end{bmatrix}$$

In this way construct $X_{22}, X_{33}, \ldots, X_{n-1,n-1}$, and it is clear that we can similarly construct successive diagonals until all the X_{ij} are defined and the resulting operator has norm no greater than, and hence equal to, β. ■

Theorem 1.3. (W. B. Arveson) *Let* Alg$\mathcal{E}$ *be the nest algebra associated with the projection nest* $\mathcal{E}$ *in the Hilbert space* $\mathcal{H}$. *Then for each operator* X *in* $\mathcal{L}(\mathcal{H})$

$$\inf_{A \in \text{Alg}\mathcal{E}} \|X - A\| = \sup_{E \in \mathcal{E}} \|(I - E)XE\|.$$

Proof: Note that Alg$\mathcal{E}$ is the intersection of the nest algebras Alg$\mathcal{F}$ taken over all finite subsets $\mathcal{F}$ of $\mathcal{E}$. Moreover, $\text{dist}(X, \mathcal{F}) = \sup\{\|(I - E)XE\| : E \in \mathcal{F}\}$ by Lemma 1.2. It is sufficient then to show that

$$\text{dist}(X, \text{Alg}\mathcal{E}) = \sup\{\text{dist}(X, \text{Alg}\mathcal{F}) : \mathcal{F} \subset \mathcal{E}, \mathcal{F} \text{ finite}\}.$$

Let σ denote the supremum and let $\epsilon > 0$. The collection $C_{\mathcal{F}}$ of all operators A in Alg$\mathcal{F}$ satisfying

$$\|X - A\| \leq \sigma + \epsilon$$

is nonempty and is a compact set for the weak operator topology. Moreover, the sets $C_{\mathcal{F}}$ have the finite intersection property and so there exists an operator A in the intersection, and hence in Alg$\mathcal{E}$, with $\|X - A\| \leq \sigma + \epsilon$. This shows that $\text{dist}(X, \text{Alg}\mathcal{E}) \leq \sigma$. This reverse inequality is obvious in view of our opening remark, and so the proof is complete. ■

The distance formula is very useful in the general theory of nest algebras. In the next two theorems we shall give two rather different applications. The first involves best approximation in operator spaces.

A closed subspace of a Banach space is said to be *proximinal* if every element of the Banach space possesses at least one element of best approximation from the subspace.

Theorem 1.4.

(i) $T(N) \cap \mathcal{K}$ is a proximinal subspace of $\mathcal{K}$.

(ii) $\mathrm{dist}(K, T(N)) = \mathrm{dist}(K, T(N) \cap \mathcal{K})$ for K in $\mathcal{K}$.

(iii) $\mathrm{dist}(X, T(N) \cap \mathcal{K}) = \max\{\mathrm{dist}(X, \mathcal{K}), \mathrm{dist}(X, T(N))\}$ for X in $\mathcal{L}(\mathcal{H})$.

Proof: Let X belong to $\mathcal{L}(\mathcal{H})$ and let P_n denote the projection onto the span of the first n basis elements. Consider the finite number of subdiagonal block operator matrices in the following partition of X.

$$
A = \begin{bmatrix}
0 & 0 & \cdots & 0 & 0 & \cdots \\
x_{21} & 0 & \cdots & 0 & 0 & \cdots \\
x_{31} & x_{32} & \cdots & 0 & 0 & \cdots \\
\vdots & \vdots & \ddots & \vdots & \vdots & \\
x_{n1} & x_{n2} & \cdots & x_{n\,n-1} & 0 & \cdots \\
& & (I - P_n)X & & &
\end{bmatrix}.
$$

Assume first that X is a compact operator not in $T(N)$ and n is chosen so that

$$\|(I - P_n)X\| < \mathrm{dist}(X, T(N)).$$

Then the operator norms of the subdiagonal rectangles do not exceed $\mathrm{dist}(X, T(N))$. By Lemma 1.2 there is a (finite rank) replacement of the upper triangular part of X such that the resulting norm is no greater than $\mathrm{dist}(X, T(N))$. This completes the proof of (i) and a little more besides: a compact operator that is not in $T(N)$ has a best upper triangular finite rank approximant.

For a general operator X and $\epsilon > 0$, choose n so that

$$\|(I - P_n)X\| < \mathrm{dist}(X, \mathcal{K}) + \epsilon.$$

The subdiagonal rectangles have norm at most

$$\max\{\|(I - P_n)X\|, \|(I - P_k)XP_k\|; 1 \le k \le n - 1\}$$

which in turn is less than $\max\{\mathrm{dist}(X, \mathcal{K}), \mathrm{dist}(X, T(N))\} + \epsilon$. Again, by Lemma 1.2 there exists a finite rank upper triangular approximation F such that $\|X - F\|$ is no greater than this estimate, and so (iii) follows. Finally, (ii) is an immediate consequence of (iii). ∎

In the second application we obtain a fundamental theorem for Hankel operators due to Z. Nehari. To make the link with this function theoretic context we first construct an expectation mapping from $\mathcal{L}(\mathcal{H})$ onto L^∞.

Let $\mathcal{H} = L^2$, the Lebesgue space associated with normalised Lebesgue measure on the unit circle, with the standard orthonormal basis $e_n = z^n$, indexed by $\mathbb{Z}$. Each function ϕ in L^∞ gives rise to a multiplication operator M_ϕ on L^2, and the linear space $\mathcal{M} = \{M_\phi : \phi \in L^\infty\}$ is a weak operator topology closed commutative unital self adjoint algebra of operators. In fact, $\mathcal{M}$ is a maximal abelian von Neumann algebra that is equal to the commutant of the bilateral shift M_z. Since $\|\phi\|_\infty = \|M_\phi\|$ we regard $\mathcal{M}$ as a copy of L^∞.

Let Λ be a Banach limit on the sequence space $\ell^\infty(\mathbb{N})$, so that Λ is a positive linear functional with $\Lambda(1, 1, \ldots,) = 1$ and the following translation invariance property. If $(a_n) = (a_1, a_2, \ldots)$ is a sequence in $\ell^\infty(\mathbb{N})$ and (a'_n) is the translate $(0, a_1, a_2, \ldots)$, then $\Lambda(a_n) = \Lambda(a'_n)$.

For a given operator X in $\mathcal{L}(\mathcal{H})$ and vectors x, y in $\mathcal{H}$, define

$$[x, y] = \Lambda(U^{*n} X U^n x, y)$$

where $U = M_z$. Then $[,]$ is a bounded sesquilinear form and there is a unique operator $\pi(X)$ in $\mathcal{L}(\mathcal{H})$ such that for all x, y

$$(\pi(X)x, y) = [x, y].$$

Note that $\pi(X)$ commutes with the shift and so belongs to $\mathcal{M}$. It is now routine to verify that π is a positive linear map satisfying

$$\pi(I) = I, \pi(MX) = M\pi(X), \pi(XM) = \pi(X)M$$

for all M in $\mathcal{M}$ and X in $\mathcal{L}(\mathcal{H})$ (and is therefore an expectation onto $\mathcal{M}$).

Let H^∞ be the subspace of L^∞ consisting of functions ϕ whose negative Fourier coefficients vanish. Then, noting that it is only the kth diagonal of X that determines the kth (constant) diagonal of $\pi(X)$, we see that

$$\pi(T(\mathbb{Z})) = \{M_\phi : \overline{\phi} \in H^\infty\}$$

where $T(\mathbb{Z})$ is associated with the standard basis above.

Finally, if P is the orthogonal projection of L^2 onto the Hardy space H^2 (so that in the notation of section $0, P = I - P_{-1}$) then the *Hankel operator* H_ϕ associated with the function ϕ in L^∞ is defined by

$$H_\phi = (I - P)M_\phi P.$$

Theorem 1.5. (Z. Nehari) $\|H_\phi\| = \mathrm{dist}(\phi, H^\infty)$ for each function ϕ in L^∞.

Proof: (after S. Parrott) First observe that since π is contractive (exercise) and fixes multiplication operators, it follows that the distance from M_ϕ to $T(\mathbb{Z})$ is equal to the distance from M_ϕ to $\pi(T(\mathbb{Z}))$, and this in turn is equal to $\mathrm{dist}(\phi, \overline{H}^\infty)$, the distance from ϕ to $\overline{H}^\infty$ in L^∞. Also, if P_n is the orthogonal projection onto the subspace

$$\overline{\mathrm{span}}\{e_k : k \le n\},$$

then the operators $(I - P_n)M_\psi P_n$, corresponding to subdiagonal blocks of M_ψ, have identical norms. This is clear from the fact that the representing matrix of M_ψ is constant on diagonals. Since $H_\phi^* = (I - P_{-1})M_{\overline{\phi}}P_{-1}$ we have, using the distance formula,

$$\|H_\phi\| = \|H_\phi^*\|$$
$$= \|PM_{\overline{\phi}}(I - P)\|$$
$$= \|(I - P_{-1})M_{\overline{\phi}}P_{-1}\|$$
$$= \mathrm{dist}(M_{\overline{\phi}}, T(\mathbb{Z}))$$
$$= \mathrm{dist}(\overline{\phi}, \overline{H}^\infty)$$
$$= \mathrm{dist}(\phi, H^\infty). \quad \blacksquare$$

Remark. The usual proof of Nehari's Theorem makes use of the Riesz factorisation of an H^1 function f as a product of H^2 functions $f = f_1 f_2$, with

$$\|f\|_1 = \|f_1\|_1 \, \|f_2\|_2.$$

The Predual Approach:

Our last remark, together with the fact that H_0^1 is the predual space of L^∞/H^∞, suggest that there might be a predual analysis of $\mathcal{L}(\mathcal{H})/\mathrm{Alg}\mathcal{E}$ that leads to the distance formula. There is. This approach not only gives useful information concerning trace class operators in a nest algebra but provides a different perspective suitable for consideration in the context of more general reflexive operator algebras.

Once again the first lemma concerns 2×2 operator matrices and is the key. The constructions made in the proof seem to be rather fundamental and will be exploited greatly in Section 3.

Lemma 1.6. (E. C. Lance) *Let E be an orthogonal projection and let A be a trace class operator such that $(I-E)AE = 0$. Then there exists a decomposition $A = A_1 + A_2$ such that*

(i) $(I - E)A_1 = 0, A_2 E = 0$,
(ii) $\|A\|_1 = \|A_1\|_1 + \|A_2\|_1$.

Proof: We may suppose that $0 < E < I$. Let C be a positive operator such that, with respect to the decomposition $\mathcal{H} = E\mathcal{H} \oplus (I - E)\mathcal{H}$, we have the operator matrix representation

$$C = \begin{bmatrix} a & b \\ b^* & c \end{bmatrix}.$$

It is known that an operator matrix of this kind (with $c \geq 0$ and $a \geq 0$), in the special case when a is invertible, represents a positive operator if and only if $b^* a^{-1} b \leq c$. To see this fact observe that when a is invertible, so too is the operator

$$X = \begin{bmatrix} a^{1/2} & a^{-1/2}b \\ 0 & (c - b^* a^{-1} b)^{1/2} \end{bmatrix}.$$

Since

$$C = X^* \begin{bmatrix} I_1 & 0 \\ 0 & I_2 \end{bmatrix} X$$

the assertion follows. Applying this principle to the positive operators $n^{-1}I + C$, we see that

$$b^*(a + n^{-1} I_1)^{-1} b \leq c + n^{-1} I_2$$

and hence that the increasing sequence of positive operators $b^*(a + n^{-1} I_1)^{-1} b$ converges in the strong operator topology to an operator c_1 with $c_1 \leq c$.

Assume now that $A = UC$ is the polar decomposition of A where the partial isometry U is represented by

$$U = \begin{bmatrix} u_1 & u_2 \\ u_3 & u_4 \end{bmatrix}$$

and $u_3 a + u_4 b^* = 0$. Since C is positive we have $Pb = b$, where P is the range projection of a, and so

$$
\begin{aligned}
u_3 b + u_4 c_1 &= u_3 P b + u_4 c_1 \\
&= \lim_{n \to \infty} u_3 a (n^{-1} I_1 + a)^{-1} b + u_4 b^*(n^{-1} I_1 + a)^{-1} b \\
&= \lim_{n \to \infty} (u_3 a + u_4 b)(n^{-1} I_1 + a)^{-1} \\
&= 0.
\end{aligned}
$$

In particular, we define

$$A_1 = U \begin{bmatrix} a & b \\ b^* & c_1 \end{bmatrix}, \quad A_2 = U \begin{bmatrix} 0 & 0 \\ 0 & c - c_1 \end{bmatrix}$$

and see that (i) holds. Moreover, observe that $\|UD\|_1 = \mathrm{trace}(D)$ whenever D is positive and the range projection of D is dominated by the initial projection of U. From this we get (ii). ■

Remark. It is possible to obtain the final statement of Lemma 1.1 from Lemma 1.6 by standard duality arguments and the Hahn Banach theorem.

As we have already mentioned, the rank one operators in a nest algebra, and also in more general reflexive algebras, form a distinguished class. The last lemma can be used constructively to show that a finite rank operator R in a nest algebra can be written as a sum $R = R_1 + \ldots + R_n$ such that each R_j is a rank one operator in the nest algebra, $n = \mathrm{rank}\ R$, and

$$\|R\|_1 = \|R_1\|_1 + \ldots + \|R_n\|_1.$$

In the next lemma we see that an infinite series decomposition of this type can be made, at least approximately, in the case of a trace class operator.

Let $\mathcal{A}_1$ be the Banach space $(\mathrm{Alg}\mathcal{E} \cap C_1(\mathcal{H}), \| \ \|_1)$ associated with the nest algebra $\mathrm{Alg}\mathcal{E}$, and let

$$\mathcal{A}_1^+ = \{A \in \mathcal{A}_1 : QAQ = 0 \text{ for all atoms } Q \text{ of } \mathcal{E}\}.$$

Lemma 1.7. (S. C. Power)

(i) *The extreme points of the unit ball of $\mathcal{A}_1$ (resp. $\mathcal{A}_1^+$) are the rank one operators of unit norm in $\mathcal{A}_1$ (resp. $\mathcal{A}_1^+$)*

(ii) *For A in $\mathcal{A}_1$ (resp. $\mathcal{A}_1^+$) and $\epsilon > 0$ there exist rank one operators $R_1, R_1, \ldots$ in $\mathcal{A}$ (resp. $\mathcal{A}^+$) such that*

$$A = \sum R_k \text{ and } \sum \|R_k\|_1 \leq \|A\|_1 + \epsilon.$$

Proof: (i). Let A be an extreme point of the ball $\mathcal{A}_1$. We show that there is a projection E in $\mathcal{E}$ such that $A = E_+ A(I - E)$.

Suppose that $E \in \mathcal{E}, A \neq EA$ and $A \neq 0$. Let $A = A_1 + A_2$ be the decomposition of Lemma 1.7 associated with the invariant projection E. Then $A_1 E = AE$ and so

$A_1 \neq 0$. Also $(I - E)A = (I - E)A_2$ and so $A_2 \neq 0$. By Lemma 1.6 (ii) A is not an extreme point. This contradiction shows that we have the alternative: $A = EA$ or $AE \neq 0$ for E in $\mathcal{E}$. Now, let $E = \sup\{F : AF = 0\}$. Then $AE = 0$ and $A = GA$ for all $G > E$. Hence, $A = E_+A(I - E)$.

Let $A = \sum \lambda_k S_k$ be any Schmidt decomposition for A with $\lambda_1, \lambda_2, \ldots$ the singular number sequence of A and $S_1, S_2, \ldots$ rank one operators of unit norm.

Then

$$A = E_+A(I - E) = \sum \lambda_k E_+ S_k(I - E).$$

Since $\|A\|_1 = \sum \lambda_k$, it follows that $\lambda_k = \lambda_k \|E_+ S_k(I - E)\|_1$ for all k, and so $S_k = E_+ S_k(I - E)$ for all $\lambda_k \neq 0$. But this condition on S_k implies membership of $\mathcal{A}$. Since A is an extreme point, it follows that $\lambda_2 = \lambda_3 = \ldots 0$, and A has rank one.

In the case of an extreme point A of $\mathcal{A}_1^+$, we can again apply Lemma 1.6 to the operator $A = E_+A(I - E)$ and see that either $A = EA(I - E)$ or $A = E_+A(I - E_+)$. The condition $S = FS(I - F)$ with F in $\mathcal{E}$ guarantees membership of A^+, and so as before we see that A has rank one.

(ii). Let $\mathcal{S}$ denote the closed linear span in the operator norm of the rank one operator sR such that $R = ER(I - E)$ for some projection E in $\mathcal{E}$. Then an operator A in $\mathcal{C}_1(\mathcal{H})$ lies in the annihilator of $\mathcal{S}$ if, for every rank one operator X and projection E in $\mathcal{E}$,

$$0 = tr(AEX(I - E)) = tr((I - E)AEX).$$

That is, the annihilator of $\mathcal{S}$ coincides with $\mathcal{A}_1$ and $\mathcal{A}_1$ is a dual Banach space, canonically isomorphicto the dual space of $\mathcal{K}/\mathcal{S}$. Using the Krein Milman Theorem and additional arguments, it can be shown that the unit ball of $\mathcal{A}_1$ is the closed convex hull of its extreme points. Using (i) the proof of (ii), for $\mathcal{A}_1$, is completed by an elementary iterative argument. Similarly, $\mathcal{A}_1^+$ coincides with the annihilator of the operators $E_+X(I - E)$ with E in $\mathcal{E}$ and X of rank one, and the proof of this case is similarly completed. ■

The Second Proof of the Distance Formula (E. C. Lance)

The operator X in $\mathcal{L}(\mathcal{H})$ determines a coset $[X]$ in the quotient space $\mathcal{L}(\mathcal{H})/\mathrm{Alg}\mathcal{E}$. This quotient space is naturally the dual space of $\mathcal{A}_1^+$ (this being the preannihilator of

$\mathcal{A}$) and so, in view of Lemma 1.6 (ii),

$$\inf_{A \in \mathrm{Alg}\mathcal{E}} \|X - A\| = \|[X]\|_{\mathcal{L}(\mathcal{H})/\mathrm{Alg}\mathcal{E}}$$

$$= \sup\{|tr(XA^+)| : A^+ \in \mathcal{A}_1^+, \|A^+\|_1 \leq 1\}$$

$$= \sup\{|tr(XR)| : R = ER(I - E), E \in \mathcal{E}, \|R\|_1 \leq 1, R \text{ rank one}\}$$

$$= \sup\{|tr((I - E)XEY)| : E \in \mathcal{E}, Y \text{ rank one}, \|Y\|_1 \leq 1\}$$

$$= \sup_{E \in \mathcal{E}} \|(I - E)XE\|. \quad \blacksquare$$

We finish this section with a more concrete application of the decomposition of Lemma 1.7 to Volterra integral operators. The result may be interpreted as an integral operator version of Hardy's inequality for H^1 functions:

$$\sum_{n=0}^{\infty} \frac{|\widehat{h}(n)|}{n+1} \leq \pi \|h\|_1.$$

Theorem 1.8. (A. L. Shields, S. C. Power) *Let $h(x, y)$ be a complex valued square integrable function on $\mathbf{R}^2$ such that $h(x, y) = 0$ for $x > y$. If the integral operator $Inth$ determined by the kernel function $h(x, y)$ is of trace class then*

$$\iint_{y \geq x} \frac{|h(x, y)|}{y - x} dx dy \leq \pi \|Inth\|_1.$$

Proof: For almost every x we have

$$(Inth)f(x) = \int_x^{\infty} h(x, y)f(y)dy.$$

In particular if f vanishes on $(-t, \infty)$ then so too does $(Inth)f$. Thus $Inth \in T(\mathbf{R})$.

Suppose first that $Inth$ has rank one so that it assumes the form $u(x) \otimes v(y)$. Then, since the Hilbert transform on $L^2(\mathbf{R})$ with kernel $(x - y)^{-1}$ has operator norm π we have

$$\iint_{y \geq x} \frac{|h(x, y)|}{y - x} dx dy = \iint_{y \geq x} \frac{|u(x)v(y)|}{y - x} - dy dx$$

$$\leq \pi \|u\|_2 \|v\|_2$$

$$= \pi \|Inth\|_1.$$

In the general case let $\epsilon > 0$ and using Lemma 1.7 obtain the series

$$Inth = \sum_{k=1}^{\infty} Inth_k,$$

where $Inth_i \in T(\mathbf{R})$ is of the rank one form and where

$$\sum_{k=1}^{\infty} \|Inth_k\|_1 \leq \|Inth\|_1 + \epsilon.$$

Then $h(x,y) = \sum_{k=1}^{\infty} h_k(x,y)$ almost everywhere and

$$\iint_{y \geq x} \frac{|h(x,y)|}{y-x} dx dy \leq \sum_{k=1}^{\infty} \iint_{y \geq x} \frac{|h_k(x,y)|}{y-x} dx dy$$

$$\leq \pi \left(\sum_{k=1}^{\infty} \|Inth_k\|_1 \right)$$

$$\leq \pi \left(\|Inth\|_1 + \epsilon \right)$$

completing the proof. ∎

2. Density

We now examine some density properties of the finite rank operators of a nest algebra $\mathrm{Alg}\mathcal{E}$. A fundamental result is the following density theorem of J. A. Erdos which is reminiscent of the Kaplansky density theorem for self-adjoint operator algebras. It may be viewed as the nest algebraic parallel of the weak star density of the analytic polynomials in H^∞ and has similar uses.

Theorem 2.1. (J. A. Erdos) *The finite rank operators in the unit ball of a nest algebra are dense for the weak operator topology.*

The first proof of this is of a Banach space theoretic nature. It rests on an elementary Banach space principle, sometimes called Goldstine's Theorem, and a result of independent interest: the nest algebra $\mathrm{Alg}\mathcal{E}$ is the second dual of $\mathrm{Alg}\mathcal{E} \cap \mathcal{K}$ in a natural fashion. This duality is achieved with the help of Lemma 1.7(ii) and this represents the core of the proof.

First Proof (S. C. Power)

The collection of rank one operators in a nest algebra $\mathcal{A} = \mathrm{Alg}\mathcal{E}$, consists precisely of the non zero operators of the form $EX(I - E_-)$, where $X \in \mathcal{L}(\mathcal{H})$ has rank one, and $E \in \mathcal{E}$ (see the preliminaries). From the identity

$$tr(AEX(I - E)_-)) = tr((I - E_-)AEX)$$

we see that an operator A is in the annihilator of this collection, and hence of the closed linear span, $\mathcal{R}$ say, if and only if $(I - E_-AE = 0$ for all E in $\mathcal{E}$. Thus, using earlier notation, $\mathcal{A}_1^+$ is the annihilator of $\mathcal{R}$ in $\mathcal{C}_1(\mathcal{H})$. Denote the dual space of a Banach space X by X'. We have established now the (natural) identification

$$\mathcal{R}' = \mathcal{C}_1(\mathcal{H})/\mathcal{A}_1^+$$

for the dual space of $\mathcal{R}$ (with the operator norm).

We now compute the annihilator of $\mathcal{A}_1^+$ in $\mathcal{L}(\mathcal{H})$. By Lemma 1.7(ii) this annihilator, $(\mathcal{A}_1^+)^\perp$, agrees with the annihilator of the rank one operators of $\mathcal{A}_1^+$, and these have the form $EX(I - E)$ with $X \in \mathcal{L}(\mathcal{H})$ of rank one and $E \in \mathcal{E}$. Since

$$tr(AEX(I - E)) = tr((I - E)AEX)$$

it follows that $(\mathcal{A}_1^+)^\perp = \mathcal{A}$ and that we have the natural identification

$$(\mathcal{C}_1(\mathcal{H})/\mathcal{A}_1^+)' = \mathcal{A}.$$

Moreover, the weak * topology on $\mathcal{A}$ coincides with the ultraweak topology. Since the unit ball of a Banach space is weak * dense in the unit ball of the second dual (Goldstine's Theorem), we now see that the unit ball of $\mathcal{R}$ is ultraweakly dense in the unit ball of $\mathcal{A}$. The weak and the ultraweak operator topologies agree on the unit ball, and the theorem follows. ∎

We have observed in the last section that every finite rank operator of $\mathcal{A}$ belongs to the linear span of the rank one operators of $\mathcal{A}$ and we conclude from the theorem that $\mathcal{A} \cap \mathcal{K} = \mathcal{R}$. In particular, we have the successive dual spaces

$$\mathcal{A} \cap \mathcal{K}, \quad \mathcal{C}_1(\mathcal{H})/\mathcal{A}_1^+, \quad \mathcal{A}$$

in analogy with the successive dual sequence

$$H^\infty \cap C(T), \quad L^1/H_0^1, \quad L^\infty$$

associated with the Lebesgue and Hardy spaces of the unit circle, and where $C(T)$ is the subalgebra of continuous functions.

The next proof adopts a more direct spatial approach and provides quite different insights. The idea is to construct finite rank shifting operators in the unit ball which converge to the identity operator in the weak operator topology (WOT). If $R_\alpha \to I(WOT)$ then $XR_\alpha \to X(WOT)$, and so the theorem will follow.

Second Proof (K. R. Davidson)

Suppose first that $\mathcal{E}$ is continuous and consider a unit vector x in $\mathcal{H}$.

Given $n = 1, 2, \ldots$, use the continuity of the mapping $E \to \|Ex\|$ (exercise!) to select orthogonal intervals $E_{k,n}$ for $1 \le k \le 2^n$, such that the vectors $\chi_{k,n} = 2^{-n/2} E_{k,n} x$ have unit norm. Consider the finite rank shift operators

$$R_n = \sum_{k=2}^{2^n} \chi_{k-1,n} \otimes \chi_{k,n},$$

that belongs to Alg$\mathcal{E}$, and notice that for $m > n$,

$$\|R_m \chi_{k,n} - \chi_{k,n}\| = 2^{(n-m+1)/2}.$$

Consequently, $R_m y \to y$ for any vector y in the linear span of $\{\chi_{k,n} : 1 \le k \le 2^n\}$. The shift operators have unit norm and so $R_m y \to y$, as $m \to \infty$, for every vector y in the closed linear span of $\{E_{k,n} x : 1 \le k \le 2^n, n = 1, 2, \ldots\}$. We have shown that $R_m \to I$ pointwise on the subspace $\overline{\text{span}}\{\mathcal{E}x\}$. Indeed, we have $R_m \to P_x(WOT)$ with $P_x R_m P_x = R_m$, where P_x is the orthogonal projection onto this subspace.

Notice that since $\mathcal{E} = \mathcal{E}^*$, if z is a vector with $P_x z = 0$, then P_x and P_z are orthogonal projections. We may therefore construct a maximal orthogonal family of projections P_{x_ω} associated with unit vectors $x_\omega, \omega \in \Omega$, such that

$$\mathcal{H} = \sum_\omega \oplus P_{x_\omega} \mathcal{H}.$$

For each vector x_ω we have associated shift operators R_n^ω. Consider the operators

$$R_{\mathcal{F}}^n = \sum_{\omega \in \mathcal{F}} \oplus R_\omega^n,$$

indexed by the directed set $\widetilde{\Omega} \times \mathbb{N}$, where $\widetilde{\Omega}$ is the family of finite subsets $\mathcal{F}$ of Ω $((\omega, n) \le (\omega', n') \leftrightarrow \omega \subset \omega', n \le n')$. Then $\{R_{\mathcal{F}}^n\}$ is a net of finite rank operators in the unit ball of Alg$\mathcal{E}$ that converges to the identity in the weak operator topology.

Now in the case of a general nest $\mathcal{E}$ let P_a and $P_c (= I - P_a)$ be the orthogonal projections onto the subspaces

$$\mathcal{H}_a = \sum_{\text{atoms } Q} \oplus Q\mathcal{H}, \quad \mathcal{H}_c = \mathcal{H}_a^\perp.$$

respectively. The nest $\mathcal{E}_c = \{E|\mathcal{H}_c : E \in \mathcal{E}\}$ is a continuous nest in $\mathcal{H}_c$, and so, by our constructions, there is a net R_α in the unit ball of $\mathrm{Alg}\mathcal{E}_c$ such that $R_\alpha \to P_c(WOT)$. On the other hand, it is clear that there are finite rank orthogonal projections P_β say, in $\mathrm{Alg}\mathcal{E}_a$, similarly defined, with $P_\beta \to P_a(WOT)$. Hence the operators $P_\beta \oplus R_\alpha$ provide the desired net in the general case. ∎

The Density Theorem implies that there is a bounded approximate identity in a nest algebra for the weak operator topology that consists of finite rank operators. Indeed, the second proof actually displays finite rank shifting operators that are of unit norm and form a net (sequence if $\mathcal{H}$ is separable) that converges to the identity operator.

For purely atomic nests $(\mathcal{H} = \mathcal{H}_a)$ the theorem is elementary since an approximate identity of projections may be chosen in the diagonal algebra $\mathrm{Alg}\mathcal{E} \cap (\mathrm{Alg}\mathcal{E})^*$. In particular, $\|I - R_\alpha\| = 1$ holds in this case. In fact, through the following construction with convex averages we can improve the properties of a given approximate identity to get a useful substitute for this strong condition, namely $\lim_\alpha \|I - R_\alpha\| = 1$.

Proposition 2.2. (K. R. Davidson and S. C. Power) *Let R_k be a bounded sequence of compact operators that converges to the identity operator in the weak operator topology. Then there exist convex combinations E_n of $\{R_k\}$ such that*

$$\lim_{n\to\infty} \|E_n\| = \lim_{n\to\infty} \|I - E_n\| = 1$$

and E_n and E_n^* converge to the identity in the weak operator topology.

Proof: Let $Q_j, j \geq 1$ be a fixed increasing sequence of finite rank orthogonal projections with the identity as its supremum. Let $C = \sup \|R_k\|$, and let N be a given integer. Choose an integer $M \geq C^2 N^2$. Let $j_1 = N$, and alternatively choose j_i and $k_i, 1 \leq i \leq M$ such that

$$\|Q_{j_i} - R_{k_i}Q_{j_i}\| < 1/N$$

and

$$\|R_{k_i} - R_{k_i}Q_{j_{i+1}}\| < 1/N.$$

Now let

$$E_N = \frac{1}{M}\sum_{i=1}^{M} R_{k_i}$$

and

$$F_n = \frac{1}{M}\sum_{i=1}^{M} A_{j_i}.$$

Now, $Q_N \leq F_N \leq I$, so F_N tends to the identity in the weak operator topology and

$$\|F_N\| = \|I - F_N\| = 1.$$

Compute

$$\|E_N - F_N\| = \|\frac{1}{M} \sum_{i=1}^{M} R_{k_i} - Q_{j_i}\|$$

$$= \frac{1}{M} \|\sum_{i=1}^{M} (R_{k_i} - R_{k_i} Q_{j_{i+1}}) + R_{k_i} (Q_{j_{i+1}} - Q_{j_i}) + (R_{k_i} Q_{j_i} - Q_{j_i})\|$$

$$\leq \frac{2}{N} + \frac{1}{M} \|\sum_{i=1}^{M} R_{k_i} (Q_{j_{i+1}} - Q_{j_i})\|$$

$$\leq \frac{2}{N} + \frac{1}{M} C M^{1/2} \leq 3/N.$$

The last estimate holds since the operators $R_{k_i}(Q_{j_{i+1}} - Q_{j_i})$ have orthogonal domains, and norms bounded by C. It is now clear that $\{E_N\}$ has the desired properties. ∎

Theorem 2.3. *Let $\mathcal{A} = \mathrm{Alg}\mathcal{E}$ be a nest algebra on a separable Hilbert space $\mathcal{H}$. Then $\mathcal{A} \cap \mathcal{K}$ contains a bounded approximate identity R_k for $\mathcal{K}$ such that*

$$\lim_{k \to \infty} \|R_k\| = \lim_{k \to \infty} \|I - R_k\| = 1.$$

Moreover, for any operator X on $\mathcal{H}$ we have

$$\mathrm{dist}(X, \mathcal{A} + \mathcal{K}) = \lim_{k \to \infty} \mathrm{dist}(X(I - R_k), \mathcal{A})$$

$$= \lim_{k \to \infty} \sup_{E \in \mathcal{E}} \|(I - E)X(I - R_k)E\|.$$

Proof: By Theorem 2.2 and Proposition 2.2 there exists a sequence R_k of finite rank operator, with the required norm conditions, convergent to the identity in the weak operator topology. The first assertion follows.

The final identity is a consequence of the distance formula, and since $X(I - R_k)$ is a compact perturbation of X it remains to establish the inequality

$$\mathrm{dist}(X, \mathcal{A} + \mathcal{K}) \leq \lim_{k \to \infty} \mathrm{dist}(X(I - R_k), \mathcal{A}).$$

However, if $K \in \mathcal{K}$ and $A \in \mathcal{A}$, then

$$(X - A - K)(I - R_n) = X(I - R_n) - A(I - R_n) - K(I - R_n),$$

206

and since $K(I - R_n) \to 0$ in operator norm, and $A(I - R_n)$ belongs to $\mathcal{A}$, we have

$$\lim_{k \to \infty} \mathrm{dist}(X(I - R_k), \mathcal{A}) \leq \lim_{k \to \infty} \|X - A - K\| \, \|I - R_k\|$$
$$= \|X - A - K\|.$$

So the inequality holds and the proof is complete. ∎

The algebra $\mathcal{A} + \mathcal{K}$ associated with the nest algebra $\mathcal{A}$ is called the quasitriangular algebra associated with $\mathcal{A}$ (or $\mathcal{E}$), and the last theorem shows how the distance of an operator X to $\mathcal{A} + \mathcal{K}$ may be computed in terms of a good approximate identity and the nest $\mathcal{E}$. In the notes we state an improvement on this situation in which we identify the quasitriangular distance with an obvious lower bound given in terms of X and $\mathcal{E}$ alone.

We have not yet shown that $\mathcal{A} + \mathcal{K}$ is norm closed. This fact rests on the ultraweak density of $\mathcal{A} \cap \mathcal{K}$ in $\mathcal{A}$, which in turn follows from the density theorem. We shall now give a proof that underlines the parallel with function spaces, and we start with an elementary theorem used by W. Rudin to exhibit the closure of spaces of type $H^\infty + C$.

Theorem 2.4. (W. Rudin) *Suppose that Z and Y are closed subspaces of a Banach space X and let Φ be a collection of linear transformations of X such that*

(a) $\Lambda Z \subset Z$ *for all $\Lambda \in \Phi$.*
(b) $\Lambda X \subset Y$ *for all $\Lambda \in \Phi$.*
(c) $\sup\{\|\Lambda\| : \Lambda \in \Phi\} = M < +\infty.$
(d) *For every $y \in Y$ and $\epsilon > 0$ there exists a Λ in Φ such that $\|y - \Lambda y\| < \epsilon$.*

Then $Z + Y$ is closed.

Proof: Let $y \in Y, z \in Z, \epsilon > 0$ and choose Λ so that $\|\Lambda y - y\| < \epsilon$. Then

$$\|y - \Lambda z\| \leq \|y - \Lambda y\| + \|\Lambda y - \Lambda z\| < \epsilon + M\|y - z\|.$$

Since $\Lambda z \in Y \cap Z$ we see that $\mathrm{dist}(y, Y \cap Z) \leq M \, \mathrm{dist}(y, Z)$. It follows that the canonical embedding of $Y/Y \cap Z$ in Y/Z is bicontinuous. In particular, Y/Z is closed in X/Z, and hence $Y + Z$, the pre-image of Y/Z under the canonical quotient map, $X \to X/Z$, is closed.

First Corollary. (D. Sarason) *The space $H^\infty(T) + C(T)$ is closed.*

Proof: Let $X = L^\infty(T), Z = H^\infty(T)$ and $Y = C(T)$. Let $\Phi = \{\Lambda_n : n = 1, 2, \ldots\}$ where $\Lambda_n f$ is the Fejer polynomial formed by the arithmetic mean of the first n partial sums of the Fourier series of f. Properoties (a) and (b) are clear, and (d) is Fejer's Theorem. Since Λ_n is also defined by convolution with a Fejer kernel $k_n(e^{i\theta})$ and since $\|k_n\|_1 = 1$, we see that (c) holds, with $M = 1$. The maps are linear, the theorem applies, and $H^\infty(T) + C(T)$ is closed. ∎

Second Corollary. (T. Fall, W. Arveson, P. Muhly) *Let $\mathcal{A} = \text{Alg}\mathcal{E}$ be a nest algebra. Then the quasitriangular algebra $\mathcal{A} + \mathcal{K}$ is norm closed.*

Proof: Let $X = \mathcal{L}(\mathcal{H})$, $Z = \mathcal{A}$ and $Y = \mathcal{K}$. Let R_α be a net of finite rank operators in the unit ball of $\mathcal{A}$ which converge to the identity in the weak operator topology. Define $\Lambda_\alpha(T) = R_\alpha T$ for T in $\mathcal{L}(\mathcal{H})$. The collection of lienar maps $\Phi = \{\Lambda_\alpha\}$ satisfies the requirements of Theorem 2.4 and hence $\mathcal{A} + \mathcal{K}$ is norm closed. ∎

The original proof that $\mathcal{A} + \mathcal{K}$ is closed, for a general nest algebra $\mathcal{A}$, relied on duality methods and an important theorem of V. B. Lidskii. This theorem is a natural one and yet surprisingly difficult to prove. However, we shall see that it is a consequence of Lemma 1.7 and the invariant subspace theorem for compact operators.

The following notation will also be useful later. If $\mathcal{F}$ is a finite subset of $\mathcal{E}$, then for X in $\mathcal{L}(\mathcal{H})$ we write

$$E_{\mathcal{F}}(X) = \sum_{atoms\ of\ \mathcal{F}} QXQ,$$

where the sum is taken over all atoms Q in $\mathcal{F}$. In fact, the linear mapping is a expectation of $\mathcal{L}(\mathcal{H})$ onto the commutant of $\mathcal{F}$.

Theorem 2.5. (V. B. Lidskii) *The trace of a trace class operator is the sum of its eigenvalues, counted with their algebraic multiplicities.*

Proof: (S. C. Power) Let T be a trace class operator and let P_T denote the orthogonal projection onto the closed subspace generated by the principal vectors of T. Thus

$$\text{ran}P_T = \overline{\text{span}}\{x \in \mathcal{H} : (\lambda I - T)^n x = 0 \text{ for some } \lambda \neq 0, n > 0\}.$$

Using the compactness of T, and the Riesz-Schauder Theory, we can obtain an orthonormal basis $f_1, f_2, \ldots$ for $\text{ran}P_T$ by the Gram-Schmidt process applied to an exhaustive sequence of linearly independent principal vectors. We then have

$$tr(P_T T P_T) = \sum_{k=1}^{\infty}(Tf_k, f_k) = \sum_{k=1}^{\infty}\lambda_k$$

where $\lambda_1, \lambda_2, \ldots$ are the eigenvalues of T counted with their algebraic multiplicity. We now have

$$tr(T) = \sum_{k=1}^{\infty} \lambda_k + tr((I - P_T)T(I - P_T)).$$

The operator $T_1 = (I - P_T)T(I - P_T)$ can have no nonzero eigenvalues. It suffices then to establish the theorem in this quasinilpotent case. This is the crux of the proof.

By the invariant subspace theorem for compact operators, and a natural argument with Zorn's Lemma, to each compact operator T there exists a nest $\mathcal{E}$ of invariant subspace projections, such that for all E in $\mathcal{E}$, $E = E_-$ or $E - E_-$ is a one-dimensional projection. So T belongs to Alg$\mathcal{E}$, and for every projection E in $\mathcal{E}$ with $E_- < E$ we can associate the unique scalar $\alpha_E(T)$ in the spectrum of the one-dimensional operator $(E - E_-)T|(E - E_-)\mathcal{H}$. The scalars $\alpha_E(T)$ are called the diagonal coefficients of T and are eigenvalues of T. This follows from the fact that $T|E\mathcal{H}$ is a compact operator and $(\alpha_E(T) - T)|E\mathcal{H}$ has proper range. In view of our earlier argument the proof of the theorem will be completed if we show that when T is of trace class we have

$$\text{trace } T = \sum \alpha_E(T).$$

To see this, note first that if $R = e \otimes f$ is a rank one operator in Alg$\mathcal{E}$, so that there is a projection F in $\mathcal{E}$ with $Fe = e$ and $F_-f = 0$, then

$$\lim_{\mathcal{F}} E_{\mathcal{F}}(R) = (E - E_-)F(E - E_-),$$

where the limit is taken through the net $E_{\mathcal{F}}(R)$, directed by finite subsets of $\mathcal{E}$, with respect to any operator norm. In view of Lemma 1.7(ii), it follows that for every trace class operator A in Alg$\mathcal{E}$ the limit $\lim_{\mathcal{F}} E_{\mathcal{F}}(A)$ exists, *in the trace class norm*, and defines a contractive projection, $E_{\mathcal{E}}(\cdot)$ say, such that

$$E_{\mathcal{E}}(A) = \sum_{\text{atoms}} QAQ.$$

In particular, since $tr(A) = tr(E_{\mathcal{F}}(A))$ for every finite subset $\mathcal{F}$ of $\mathcal{E}$, we have

$$tr(T) = \lim_{\mathcal{F}} tr(E_{\mathcal{F}}(T))$$

$$= tr(\lim_{\mathcal{F}} E_{\mathcal{F}}(T))$$

$$= tr\left(\sum_{\text{atoms of } \mathcal{E}} QAQ\right)$$

$$= \sum_{\text{atoms of } \mathcal{E}} tr(QAQ)$$

$$= \sum_{\text{atoms of } \mathcal{E}} \alpha_E(T). \quad \blacksquare$$

Our next use of the Density Theorem is in the characterisation of the ideals of a nest algebra which are closed in the ultraweak topology. This topology agrees with the weak star topology inherited from the duality identified after the first proof of Theorem 2.1. A consequence of the proof below is that an ideal that is closed in any one of the weak operator topologies (ultrastrong, strong, weak, star strong, etc.) is closed in all of them.

From the characterisation we see that weakly closed ideals are singly generated. There is an echo of function theory here; the weak star closed ideals of H^∞ are also principal. In this sense a nest algebra can be thought of as a noncommutative principal ideal ring.

Let $\mathcal{E}$ be a projection nest on the Hilbert space $\mathcal{H}$, as usual, and suppose that $E \to \tilde{E}$ is an order homomorphism of $\mathcal{E}$ into itself. That is, $E \leq F$ implies that $\tilde{E} \leq \tilde{F}$. Then the set

$$\mathcal{U} = \{X \in \mathcal{L}(\mathcal{H}) : (I - \tilde{E})XE = 0 \text{ for all } E \in \mathcal{E}\}$$

is a two-sided $\mathcal{A}$-module and closed in the weak operator topology. Indeed, if $A \in \text{Alg}\mathcal{E}$ and $X \in \mathcal{U}$ then

$$(I - \tilde{E})XAE = (I - \tilde{E})XEAE = 0$$

and

$$(I - \tilde{E})AXE = (I - \tilde{E})A(I - \tilde{E})XE = 0.$$

If, in addition, $\tilde{E} \leq E$ for all E, then $\mathcal{U} \subset \text{Alg}\mathcal{E}$ and is a weakly closed (two-sided) ideal.

Theorem 2.6. (J. A. Erdos and S. C. Power) *Any ideal $\mathcal{I}$ of a nest algebra $\text{Alg}\mathcal{E}$ that is closed in any one of the weak, strong, ultraweak or ultrastrong topologies is of the form*

$$\mathcal{I} = \{X \in \mathcal{L}(\mathcal{H}) : (I - \tilde{E})XE = 0 \text{ for all } E \in \mathcal{E}\}$$

where $E \to \tilde{E}$ is a left order continuous homomorphism of $\mathcal{E}$ into $\mathcal{E}$ such that $\tilde{0} = 0$ and $\tilde{E} \leq E$ for each $E \in \mathcal{E}$.

Proof: Let $\mathcal{I}$ be any operator norm closed ideal of $\text{Alg}\mathcal{E}$. For $E \in \mathcal{E}$ define $\tilde{E}$ to be the orthogonal projection onto the closure of $\mathcal{I}E\mathcal{H}$. Since $\mathcal{I}$ is a left ideal, $\tilde{E}$ is invariant for $\text{Alg}\mathcal{E}$ and so belongs to $\mathcal{E}$. If $E = E_-$, then $\tilde{E} = \sup\{\tilde{F} : F < E\}$, and so the mapping $E \to \tilde{E}$ is a left continuous order homomorphism with $\tilde{0} = 0$.

210

Let $\mathcal{J}$ denote the weak operator topology closed ideal determined by the homomorphism $E \to \tilde{E}$. We need only show that the utlrastrong closure of $\mathcal{I}$ coincides with $\mathcal{J}$ to complete the proof. However, by Theorem 2.1 there is a net R_α of operators in the unit ball of Alg$\mathcal{E}$, convergent to the identity in the weak operator (and hence, in this case, the ultrastrong) topology, with each R_α a linear sum of rank one operators in Alg$\mathcal{E}$. (This latter aspect follows from Proof 2 directly or from the comments following Lemma 1.6.) Let $e \otimes f \in$ Alg$\mathcal{E}$ and let $G \in \mathcal{E}$ be such that $e \in G$ and $f \in I - G_-$.

It suffices now to show that the operator $(T(e \otimes f) = Te \otimes f$ belongs to $\mathcal{I}$ whenever $T \in \mathcal{I}$. Now, $T \in \tilde{I}$ implies $Te \in \tilde{G}$. So there is a vector g_ϵ and an operator $X_\epsilon \in \mathcal{I}$ such that

$$\|X_\epsilon(g_\epsilon \otimes f) - Te \otimes f\| = \|X_\epsilon g_\epsilon - Te\|_2 \|f\|_2 < \epsilon.$$

Since $\mathcal{I}$ is a norm closed right ideal for Alg$\mathcal{E}$, and $g_\epsilon \otimes f \in$ Alg $\mathcal{E}$, we are done. ∎

Corollary 2.7. *Let $\mathcal{H}$ be separable. Then the weak star closed ideals of a nest alegbra are principal.*

Proof: Let $E \to \tilde{E}$ be a left continuous order homomorphism of the nest. The idea is simply to construct an operator T in the ideal $\mathcal{J}$ corresponding to $E \to \tilde{E}$, such that for each E in $\mathcal{E}$

$$\tilde{E} = \inf\{F \in \mathcal{E}|(I - F)TE = 0\}.$$

Then the weakly closed ideal $\mathcal{J}_T$ generated by T will have the same determining homomorphism and so will coincide with $\mathcal{J}$.

Let $\mathcal{D}$ be a countable subset of $\mathcal{E}$ that is dense in the strong operator topology. Let $\tilde{Q}$ be the collection of quadruples $q = (E, E', F, F')$, in $\mathcal{D}$ such that $F \leq \tilde{E}, E < E'$ and $F > F'$. For such a q let

$$\mathcal{L}(\mathcal{H})_q = (F - F')\mathcal{L}(\mathcal{H})(E' - E)$$

so that $\mathcal{L}(\mathcal{H})_q \subset \mathcal{J}$. By Zorn's lemma there exists a maximal family $\mathcal{M} \subset \tilde{Q}$ whose associated spaces are disjoint:

$$\mathcal{L}(\mathcal{H})_q \cap \mathcal{L}(\mathcal{H})_q = \{0\} \text{ for all } q, q' \in \mathcal{M}.$$

For each $q \in \mathcal{M}$ let R_q be a rank one operator in $\mathcal{L}(\mathcal{H})_q$ with unit norm, such that

$$(F - F_1)R_q(E_1 - E) \neq 0 \text{ for all } F_1 < F, E_1 > E.$$

Let $R_1, R_2, \ldots$ be a enumeration of the operators $R_q, q \in \mathcal{M}$, and define

$$T = \sum_{k=1}^{\infty} 2^{-k} R_k.$$

From the disjointness property of $\mathcal{M}$ it is clear that any ideal containing T also contains $R_1, R_2, \ldots$. We claim that the weakly closed ideal $\mathcal{J}_T$ generated by T is equal to $\mathcal{J}$. For if this is not the case and $E \to \alpha(E)$ is the order homomorphism for $\mathcal{J}_T$, then there exists $E_o \in \mathcal{E}$ such that $\alpha(E_o) < \tilde{E}_o$. Thus, by left continuity, there exists a projection E_1 in $\mathcal{D}$ such that $\alpha(E_1) < \tilde{E}_1$. By left continuity again, there is a projection $E < E_1$ with $\alpha(E_1) < \tilde{E}$. But now the space

$$\mathcal{L}(\mathcal{H})_r = (\tilde{E} - \alpha(E_1))\mathcal{L}(\mathcal{H})(E_1 - E)$$

for the quadruple

$$r = (E, E_1, \tilde{E}, \alpha(E_1))$$

is disjoint from $\mathcal{L}(\mathcal{H})_q$ for $q \in \mathcal{M}$, contrary to the maximality of $\mathcal{M}$. This completes the proof. ∎

3. Factorisation

Let $\mathcal{E}$ be a complete nest of projections on a Hilbert space $\mathcal{H}$ with nest algebra $\mathcal{A} = \text{Alg}\mathcal{E}$.

Definition 3.1. (W. B. Arveson)

(i) An operator A in $\mathcal{A}$ is called *outer* if the range projection of A commutes with $\mathcal{E}$, and for every projection E in $\mathcal{E}$

$$(AE\mathcal{H})^- = (A\mathcal{H})^- \cap E\mathcal{H}.$$

(ii) An operator U in $\mathcal{A}$ is called *inner* if U is a partial isometry whose initial projection U^*U commutes with $\mathcal{E}$.

Although it is by no means apparent from these definitions, the inner and outer operators of a nest algebra play the role of the inner and outer functions in the Hardy algebra H^∞ associated with the unit circle, at least if the nest is well ordered. The

212

significance of these notions for general nest algebras, including the continuous nest algebra $T(\mathbf{R})$, is not yet clear.

In this section we adopt a unified approach to obtain operator analogues of the following phenomena in function theory:

1. The Szego (or outer) factorisation of a positive function f:

$$f = \overline{h}h \text{ with } h \text{ outer.}$$

2. The inner-outer factorisation of an H^∞ function g;

$$g = uh \text{ with } u \text{ inner and } h \text{ outer.}$$

3. The Riesz factorisation of H^1 functions:

$$h = h_1 h_2 \text{ with } \|h\|_1 = \|h_1\|_2 \|h_2\|_2.$$

We also briefly discuss weak factorisation and Hankel forms in the context of nest algebras and obtain Arveson's Interpolation Theorem. We see that the operator variants of 2 and 3 (namely Theorems 3.6 and 3.7) follow quickly from outer factorisation (Theorem 3.5).

Our approach seems to be the "right one" in the sense that the techniques provide a constructive approach to the outer function factorisation $f = h^*h$ of a positive matrix valued function on the unit circle, when this factorisation is known to exist (see the Notes). This matrical version of Szego factorisation is a crucial feature of the theory of prediction of multivariate stationary stochastic processes. Also, the construction leads to the canonical outer factor whose diagonal part is a positive operator.

It is instructive to pause and look at the natural argument for obtaining a factorisation $C = A^*A$ in the case when C is a given positive invertible operator on $\ell^2(\mathbf{N})$ and A is to be an outer operator in $T(\mathbf{N})$. Let N_k be the subspace spanned by the basis elements $e_1, \ldots, e_k$ and choose an orthonormal basis $f_1, f_2, \ldots$ so that the first k elements form a basis for $C^{1/2}N_k$. Clearly, such bases exist. Let V be the unitary operator such that $Vf_j = e_j, j = 1, 2, \ldots$, and let $A = VC^{1/2}$. Then $AN_k = N_k$ for all k, A is an outer operator, and $A^*A = C$.

The same argument works if C is merely positive definite, and a limiting argument can be used to obtain the outer factorisation of any positive operator. Nevertheless, there are difficulties in extending the argument to more general nests, and the proof is not particularly constructive.

Notice that if $A \in \mathcal{A}$ has the strict density property $AE\mathcal{H} = E\mathcal{H}$, for all E in $\mathcal{E}$, then A is outer. Also, if A is invertible in $\mathcal{L}(\mathcal{H})$ then A is an outer operator if and only if A is invertible in $\mathcal{A}$. On the other hand, every operator in the diagonal algebra $\mathcal{A} \cap \mathcal{A}^*$ is outer. The proof of Theorem 3.5 clarifies the precise nature of outerness in operator matrix terms, when the nest is well ordered.

The outer factorisation $C = A^*A$ of a positive *matrix*, which is unique if the diagonal of A is positive, is referred to as the Cholesky factorisation, particularly amongst numerical analysts.

We begin the analysis with the Arveson-Cholesky-Szego-type outer factorisation of 2×2 operator matrices.

Definition 3.2. *Let C be a positive operator and let E be a self-adjoint projection. Then C is said to be E-minimal if $E^\perp C E^\perp = C'$ where*

$$C' = \text{s-}\lim_{t \to 0} E^\perp C(tE + ECE)^{-1}CD^\perp$$

and where the inverse indicated is computed in $\mathcal{L}(E\mathcal{H})$.

The operator matrix arguments in the first paragraph of the proof of Lemma 1.6 show that the limit exists in the strong operator topology (s-lim) and is dominated by $E^\perp C E^\perp$. We use the notation R_X for the range projection of an operator X.

Lemma 3.3. *The following conditions are equivalent for an operator A in $\mathcal{L}(\mathcal{H})$ with invariant self-adjoint projection E, such that $R_A E = E R_A$.*

(i) $(AE\mathcal{H})^- = (A\mathcal{H})^- \cap E\mathcal{H}$.

(ii) $R_{AE} \geq R_{EA(I-E)}$.

(iii) A^*EA is E-minimal.

Moreover, a positive operator CE is E-minimal if and only if $C = A_1^*A_1$ where $A_1 = EA_1$ and A_1 satisfies condition (ii).

Proof: Since $(I - E)AE = 0$, the equivalence of (i) and (ii) is elementary.

Suppose now that (ii) holds and let

$$EA = \begin{bmatrix} a_1 & b_1 \\ 0 & 0 \end{bmatrix}$$

214

so that $R_{a_1} \geq R_{b_1}$. Then

$$(A^*EA)' = \text{s} - \lim_{t \to 0} b_1^* a_1 (tE + a_1^* a_1)^{-1} a_1^* b_1$$
$$= \text{s} - \lim_{t \to 0} b_1^* (tE + a_1 a_1^*)^{-1} a_1 a_1^* b_1$$
$$= b_1^* R_{a_1} b_1 = b_1^* b_1 = E^\perp A^* EAE^\perp,$$

and so (iii) holds. On the other hand, if (iii) holds then this computation shows that we must have $b_1^* b_1 = (A^*EA)' = b_1^* R_{a_1} b_1$ and so (ii) holds.

Now consider an E-minimal positive operator with operator matrix representation

$$C = \begin{bmatrix} a & b \\ b^* & c_1 \end{bmatrix}.$$

Let e_t denote the spectral projection for the operator a corresponding to the interval (t, ∞). Then, for $t > 0$,

$$\|b^* a^{-1/2} e_t\|^2 = \lim_{s \to 0} \|b^*(sE + a)^{-1/2} e_t (sE + a)^{-1/2} b\|$$
$$\leq \lim_{s \to -0} \|b^*(sE + a)^{-1} b\|$$
$$\leq \|c_1\|.$$

It follows that the operator $d_t = b^* a^{-1/2} e_t$ converges strongly to an operator d as $t \to o$. Since $c_1 \geq b^* a^{-1/2} e_t a^{-1/2} b$ it follows that $c_1 \geq dd^*$. On the other hand,

$$\begin{bmatrix} a & b \\ b^* & dd^* \end{bmatrix} = \begin{bmatrix} a^{1/2} & 0 \\ d & 0 \end{bmatrix} \begin{bmatrix} a^{1/2} & d^* \\ 0 & 0 \end{bmatrix} \geq 0,$$

and so, by minimality, $dd^* \geq c_1$. It is clear from the definition of d that the range projection of a (namely e_o) dominates the range projection of d^*. So C has the form required in the last part of the lemma. ∎

The previous lemma essentially characterises the outer operators associated with a projection nest triple and points to the relevance of minimal positive operators, in the sense of the last definition, in the structure of outer operoators. In fact, for the positive operator

$$C = \begin{bmatrix} a & b \\ b^* & c \end{bmatrix}$$

we obtain the canonical Cholesky-type factorisation $C = A^*A$ by taking, in the notation above,

$$A = \begin{bmatrix} a^{1/2} & d^* \\ 0 & (c - c_1)^{1/2} \end{bmatrix}.$$

Clearly, in view of the lemma, A is an outer operator with respect to the nest $\{0, E, I\}$.

A consequence of the computations made in the proof of Lemma 1.6 is the following important algebraic feature of the outer factor:

$$X \in \mathcal{L}(\mathcal{H}), XC = \begin{bmatrix} * & * \\ 0 & * \end{bmatrix} \Rightarrow XA^* = \begin{bmatrix} * & * \\ 0 & * \end{bmatrix}.$$

This can also be seen from the following more general lemma which echoes the essential property of an H^∞ outer function h, namely if $\phi \in L^\infty$ and $\phi h \in H^\infty$ then $\phi \in H^\infty$.

Lemma 3.4. (W. B. Arveson) *Let $A \in \mathcal{A}$ be an outer operator and let X be an arbitrary operator in $\mathcal{L}(\mathcal{H})$ such that $XA \in \mathcal{A}$ and $X = 0$ on $(A\mathcal{H})^\perp$. Then X belongs to $\mathcal{A}$.*

Proof: For a projection E in $\mathcal{E}$ with $E \neq 0, I$, consider the orthogonal decomposition $E\mathcal{H} = (AE\mathcal{H})^- \oplus (E\mathcal{H} \ominus (AE\mathcal{H})^-)$. By hypothesis the projection R_A commutes with E. Since $R_A \mathcal{H} \cap E\mathcal{H} = (AE\mathcal{H})^-$ it follows that $E\mathcal{H} \ominus (AE\mathcal{H})^-$ is orthogonal to $R_A \mathcal{H}$. Hence, $X = 0$ on $E\mathcal{H} \ominus (AE\mathcal{H})^-$, so that $XE\mathcal{H}$ is contained in $X(AE\mathcal{H})^- \subseteq (XAE\mathcal{H})^- \subset E\mathcal{H}$ as required. ∎

Theorem 3.5. *Let $\mathcal{E}$ be a well ordered nest of projections with nest algera $\mathcal{A}$ and let C be a positive operator. Then there exists a factorisation $C = A^*A$ with A an outer operator in $\mathcal{A}$. Moreover, the outer factor belongs to the von Neumann algebra generated by C and the nest.*

Proof: We shall indicate the E-minimal part of the positive operator D with the notation $[D]_E$.

The central idea is to obtain a decomposition

$$C = \sum_{\text{atoms } Q} C_Q$$

convergent in the strong operator topolooy, where C_Q is a positive operator associated with the atom $Q = E_+ - E$ such that

(3.1) $$C_Q E = O \text{ and } C_Q \text{ is } E_+ - \text{minimal.}$$

Such a decomposition, if it exists, must be unique since the E_+-minimality implies that C_Q is determined by the restrictin of C_Q to E_+. In fact, let $0 = E_0 < E_1 < E_2 < \dots$ be

216

the first segment of the well ordered nest, and let $Q_k = E_k - E_{k-1}, k = 1, 2, \ldots$. Then we must have

$$C_{Q_1} = [C]_{E_1},$$
$$C_{Q_k} = [C - (C_{Q_1} + \ldots + C_{Q_{k-1}})]_{E_k} \quad (k > 1).$$

In loose terms, we successively take the appropriate minimal part of what is left over. The construction achieves the desired series for C if $\mathcal{E}$ has order type N. For the general case of a well ordered nest, it is also clear how we must proceed. Suppose that for all atoms Q dominated by a fixed projection E' we have obtained positive operators C_Q satisfying (3.1), and such that

$$(3.2) \qquad\qquad \sum_{Q \leq E'} C_Q \leq C$$

$$(3.3) \qquad\qquad \sum_{Q \leq E'} C_Q E' = CD'.$$

Then we may define $C_{Q'}$, where $Q' = E'_+ - E'$, by setting

$$C_{Q'} = \left[C - \sum_{Q \leq Q'} C_Q \right]_{E'_+},$$

and thereby enlarge the class of constructed C_Q, maintaining property (3.1) and obtaining (3.2) and (3.3) with E'_+ replacing E'. By the principle of well ordered induction, the required series exists. (Note that in view of (3.2), all resulting series converge unconditionally in the strong operator topology.)

Using Lemma 3.3, we obtain a factorisation $C_Q = A_Q^* A_Q$ where A_Q has the form

$$A_Q \begin{bmatrix} 0 & 0 & 0 \\ 0 & a_Q & b_Q \\ 0 & 0 & 0 \end{bmatrix},$$

with respect to $E\mathcal{H} \oplus Q\mathcal{H} \oplus E_+^{\perp}\mathcal{H}$, and $R_{a_Q} \geq R_{b_Q}$. For any finite subset $\mathcal{F}$ of atoms we have

$$\left(\sum_{Q \in \mathcal{F}} A_Q \right)^* \left(\sum_{Q \in \mathcal{F}} A_Q \right) = \sum_{Q \in \mathcal{F}} A_Q^* A_Q \leq C,$$

from which it follows that the series $\sum_Q A_Q$ is unconditionally convergent in the strong operator topology to an operator A in $\mathrm{Alg}\mathcal{E}$ with $C = A^* A$.

From the conditions $R_{aQ} \geq R_{bQ}$ it follows, by a straightforward induction argument, that A is an outer operator.

Finally, from the proofs of Lemmas 3.4 and 1.6, we see that A belongs to the von Neumann algebra generated by C and $\mathcal{E}$. ∎

There is a converse to the last theorem. If every positive operator admits an outer factorisation, then the nest is well ordered. For example, if $e(x)$ is the constant function on $[0,1]$ with value 1, then $e \otimes e$ cannot be factored with respect to the nest algebra $T([0,1])$. This is because the outer factor A would necessarily have rank one, from which it follows that $EA^*AE = 0$ for some non-zero nest projectin E, a contradiction. More generally, let $\mathcal{E}$ be a nest with $E = E_+$ for some $\in \mathcal{E}$. Choose a vector e with $E^\perp e = e$ such that $Fe \neq 0$ for all $F > E$ and let $C = E + e \otimes e$. Similar reasoning, left to the reader, shows that C does not admit outer factorisation.

Theorem 3.6. *Let $\mathcal{E}$ be a complete projection nest such that $E \neq E_+$ for all non zero projections E, and let $T \in \mathrm{Alg}\mathcal{E}$. Then*

(i) *$T = UA$ where $U \in \mathrm{Alg}\mathcal{E}$ is an inner operator and $A \in \mathrm{Alg}\mathcal{E}$ is an outer operator.*

(ii) *If $T = UA = VB$ are two such factorisations, then there is a partial isometry $W \in (\mathrm{Alg}\mathcal{E}) \cap (\mathrm{alg}\mathcal{E})^*$ such that $W^*W = R_A, WW^* = R_B, B = WA$ and $V = UW^*$.*

(iii) *If $T = UA$, as in (i), then U and A belong to the von Neumann algebra generated by T and $\mathcal{E}$.*

Proof:

(i) Let $T = VC$ be the polar decomposition of T. If $0_+ \neq 0$, then $\mathcal{E}$ is well ordered and by Theorem 3.5, there is an outer factorisation $T^*T = C^2 = A^*A$ for the positive operator C^2. On the other hand, if $0_+ = 0$, then $[C^2]_E \leq T^*ET$ and hence $[C^2]_E \to 0$, in the strong operator topology as $E \to 0$. In view of this, the argument in the proof of Theorem 3.5 carries over, and once more $T^*T = C^2 = A^*A$ with A an outer operator. Now A has the polar decomposition $A = WC$ with $R_W = R_A$. Let $U = VW^*$ so that U is a partial isometry with initial projection $U^*U = WV^*VW^* = WR_CW^* = WW^* = R_A$, which commutes with $\mathcal{E}$ since A is outer. Also, by Lemma 3.4, since $UA = VW^*WC = T$, we have $U \in A$, so U is inner. $T = UA$ is the required factorisation.

218

(ii) Let $W_1 = V^*U$ so that $B = W_1A$ and $A = W_1^*B$. By Lemma 3.4 the partial isometry W_1 lies in the diagonal algebra. Thus, $W = R_BW_1R_A$ is the required partial isometry.

(iii) In view of (ii) and Theorem 3.5, it suffices to examine the inner factor of the canonical factorisation obtained in the proof of (i). Since V(resp.W) lies in the von Neumann algebra generated by T (resp. A), it now follows from Theorem 3.5 that $U = VW^*$ lies in the indicated von Neumann algebra. ∎

Let us introduce the following terminology to formulate the next theorem. A projection lattice $\mathcal{L}$ is said to *admit Riesz factorisation* if for each $T \in (\text{Alg}\mathcal{L}) \cap C_1$ there exist operators A_1, A_2 in $(\text{Alg}\mathcal{L}) \cap C_2$ such that

$$T = A_1A_2 \text{ and } \|T\|_1 = \|A_1\|_2\|A_2\|_2.$$

(of course many lattices qualify vacuously.)

Theorem 3.7. *Let $\mathcal{E}$ be a well ordered projection nest. Then $\mathcal{E}$ admist Riesz factorisation.*

Proof: Let $T = VC$ be the polar decomposition of T and let $C = A^*A$ be the outer factorisation given in Theorem 3.5. Note that $\|VA^*\|_2^2 = \|A\|_2^2 = \|T\|_1$. By Lemma 3.4 the operator $A_1 = VA^*$ belongs to $\text{Alg}\mathcal{E}$ and so setting $A_2 = A$ we obtain the desired factorisation of T. ∎

Since $(\text{Alg}\mathcal{E})^* = \text{Alg}\mathcal{E}'$ where $\mathcal{E}'$ is the nest of projections $E^\perp$ for $E \in \mathcal{E}$, we see that $\mathcal{E}$ admits Riesz factorisation if E' is well ordered. On the other hand, we have the following necessary condition.

Proposition 3.8. *If $\mathcal{E}$ admits Riesz factorisation, then $E_+ \neq E_-$ for all projections E in $\mathcal{E}$ with $0 < E < I$.*

Proof: Suppose that $E \in \mathcal{E}, 0 < E_- = E_+ < I$, and choose unit vectors e, f so that $Ee = e$, $E^\perp f = f$, $(E-F)e \neq 0$ for all $f < E$, and $(G-E)f \neq 0$ for all $G > E(F, G \in \mathcal{E})$. Suppose that $T = RS$ is a Riesz factorisation of the rank one operator $T = e \otimes f$. Then $R = ER$ and $S = SE^\perp$. Moreover, if

$$R = \begin{bmatrix} A & X \\ 0 & 0 \end{bmatrix}, \quad S = \begin{bmatrix} 0 & Y \\ 0 & B \end{bmatrix}$$

with respect to $E\mathcal{H} \oplus E^\perp\mathcal{H}$ then $AY = XB = e \otimes f$ and

$$1 = \|AY + YB\|_1$$
$$\leq \|AY\|_1 + \|XB\|_1$$
$$\leq \|A\|_2\|Y\|_2 + \|X\|_2\|B\|_2$$
$$\leq \left(\|A\|_2^2 + \|X\|_2^2\right)^{1/2} \left(\|Y\|_2^2 + \|B\|_2^2\right)^{1/2}$$
$$\leq \|R\|_2\|S\|_2 = 1.$$

Since the first inequality is an equality, and since the rank one operators of unit norm are the extreme points of the unit ball of $\mathcal{C}_1$ (reader's exercise), we conclude that $XB = \alpha(e \otimes f)$ where $\alpha = \|XB\|_1$. However, the above implies that $\alpha = \|X\|_2 \|B\|_2$, and so combining this with the inequality

$$\alpha = (XBf, e) \leq \|Bf\|_2\|X^*e\|_2$$

we see that $\|Bf\|_2 = \|B\|_2$ and hence that B is zero or a rank one operator. But if B has rank one then, in view of the special form of rank one operators in the nest algebra on $E^\perp\mathcal{H}$ for the nest $E^\perp\mathcal{E}$, we have $BF = 0$, for some $f > E$, and hence $\|Bf\|_2 < \|B\|_2$, a contradiction. Thus $B = 0$. Similar reasoning shows that $A = 0$, and hence that Riesz factorisation for T is not possibe. ∎

Problem: Characterise the projection lattices that admit Riesz factorisation.

For the Hardy space H^1 associated with the ball or sphere in several complex dimensions, it is known that Riesz factorisation fails, but that nevertheless a good substitute is available, namely weak factorisation ($f = \sum h_k g_k$) with norm control.

We now state a theorem that expresses a strong form of weak factorisation for nest algebras. The reader may construct a proof by first considering operators T of rank one and then using Lemma 1.7 for the general case.

Theorem 3.9. (S. C. Power) *Let $\mathcal{A}_1$ be the set of trace class operators of a nest algebra $\mathcal{A}$ and let $\epsilon > 0$. Then for each operator $T \in \mathcal{A}_1$ there exist rank one operators $R_1, R_2, \ldots$ and $S_1, S_2, \ldots$ in $\mathcal{A}$ such that*

(i) $T = \sum_k R_k S_k$,
(ii) $\sum_k \|R_k\|_2 \|S_k\|_2 < \|T\|_1 + \epsilon$.

As in function theory, this weak factorisation can be used to characterise the bounded Hankel forms on $\mathcal{A}_2$. More precisely, let $[,]$ be a complex bilinear form on

$\mathcal{A}_2$, the Hilbert Schmidt operators of a nest algebra $\mathcal{A}$, such that

$$[A_1 A_2, A_3] = [A_1, A_2 A_3]$$

for all 3-tuples A_1, A_2, A_3 in $\mathcal{A}_2$. Moreover, suppose that the norm

$$\|[,]\| = \sup\{|[A_1, A_2]| : \|A_k\|_2 \leq 1 \ k = 1, 2, \}$$

is finite. Then there exists an operator $X \in \mathcal{L}(\mathcal{H})$ such that $\|X\| = \|[,]\|$ and

$$[A_1, A_2] = \text{trace } (A_2 X A_1).$$

Factorisation theory in nest algebras is an extensive topic and we have not attempted to give a broad view. (In particular we have not discussed the connections with linear systems, we have not considered the methods of Gohberg and Krein in the factorisation theory of operators of type $I + K$, with K in an operator ideal, nor have we even touched upon the significant nonfactorisation phenomena that arise through a deeper understanding of the quasitriangular algebras $\text{Alg}\mathcal{E} + K$.) Perhaps it is appropriate, in view of our ill-disguised leitmotif, to close this section with what was in fact the first application of the distance formula; Arveson's corona theorem, or interpolation theorem.

In order to discuss Arveson's theorem in a wider context than the original (nests of order type the integers) let us introduce the following terminology. A projection nest $\mathcal{E}$ is said to have the *factorisation property* if every positive *invertible* operator C admits a factorisation $C = A^*A$, where A is invertible and A^{-1} belongs $\text{Alg}\mathcal{E}$. A projection nest $\mathcal{E}$ is said to have the *partial factorisation property* if every positive invertible operator of the form $C = A_1^* A_1 + \ldots A_n^* A_n$, where $A_n \in \text{Alg}\mathcal{E}$, admits the outer factorisation $C = A^*A$ as before. We have only shown so far that every well-ordered nest has the factorisation property, and (from the proof of Theorem 3.6) that every nest with $E \neq E_+$ for all $E \neq 0$ has the partial factorisation property. In fact, D. Larson has shown that a projection nest $\mathcal{E}$ has the factorisation property if and only if $\mathcal{E}$ is countable. Whilst the necessity of countability is a deep fact, the sufficiency of the condition is fairly elementary, and it is this direction that is relevant to Arveson's Theorem.

Theorem 3.10. (W. B. Arveson) *Let $\mathcal{E}$ be a projection nest with the partial factorisation property and let $A_1, \ldots A_n$ be operators in the nest algebra $\text{Alg}\mathcal{E}$ such that*

$$\sum_{k=1}^{N} \left\| E^{\perp} A_k x \right\|^2 \geq \epsilon^2 \left\| E^{\perp} x \right\|^2$$

for some $\epsilon > 0$ and every $E \in \mathcal{E}$ and $x \in \mathcal{H}$. Then there exist operators $B_1, \ldots, B_n$ in $Alg\mathcal{E}$ such that

$$\sum_{k=1}^{n} B_k A_k = I.$$

Moreover, if $\|A_k\| \leq 1$, then we can arrange that $\|B_k\| \leq 4n\epsilon^{-3}$ for all k.

Unfortunately, a characterisation of which nests have the partial factorisation property has not yet been given.

We leave it to the reader to verify that the norm condition on the n-tuple $A_1, \ldots A_n$ is indeed a necessary condition for the existence of the asserted internal factorisation. The simpler condition $\sum \|A_k x\| \geq \delta \|x\|$ for all x (which is perhaps a more literal analogue of the corona condition $|f_1(z)| + \ldots + |f_N(z)| \geq \delta$, for all $|z| < 1$) is *not* sufficient.

The key to the proof is the following factoriszation lemma for partial isometries. The proof requires the distance formula.

Lemma 3.11. *Let A be an inner operator in a nest algebra $Alg\mathcal{E}$ and suppose for some $\epsilon > 0$ that $\|E^{\perp} A x\| \geq \epsilon \|E^{\perp} x\|$ for all $E \in \mathcal{E}$ and for all x in the initial space of A. Then there exists an operator B in $Alg\mathcal{E}$ such that $BA = P$ and $\|B\| \leq 4\epsilon^{-2}$.*

Proof: If $E \in \mathcal{E}$ and $E^{\perp} x = x$ then $\|EAE^{\perp} x\|^2 = \|Ax\|^2 - \|E^{\perp} Ax\|^2 \leq (1 - \epsilon^2) \|x\|^2$. Hence by Theorem 1.3 dist$(A^*, Alg\mathcal{E}) \leq (1 - \epsilon^2)^{1/2}$ and there is an operator C in $Alg\mathcal{E}$ with $\|A^* - C\| \leq 1 - \epsilon^2$. Without loss we may assume that $C = PC$ where P is the initial projection of A. Now let $T = I - P + CA$. Since $\|I - T\| = \|(A^* - C)A\| \leq (1 - \epsilon^2)^{1/2} < 1$, it follows that T is invertible, $T^{-1} \in Alg\mathcal{E}$, and $\|T^{-1}\| \leq (1 - (1 - \epsilon^2)^{1/2})^{-1}$. Let $B = T^{-1}C$. Then $\|B\| \leq 4\epsilon^{-2}$ and

$$BA = (I - P + CA)^{-1}CA$$
$$= (I - P + CA)^{-1}(I - P + CA - (I - P))$$
$$= I - (I - P + CA)^{-1}(I - P)$$
$$= I - \sum_{k=0}^{\infty}(P - CA)^k(I - P)$$
$$= P.$$

Proof of Theorem 3.10. The hypotheses ensure that the operator $\sum A_k^* A_k$ is invertible and admits a factorisation $A^* A$ with A an invertible element in the algebra $Alg\mathcal{E}$. Let $A_k' = A_k A^{-1}$, for $k = 1, \ldots, n$. We check at the end of this proof that the n-tuple $A'_1, \ldots, A'_n$ satisfy the norm inequality of the theorem with ϵ replaced by $\epsilon n^{-1/2}$. Meanwhile, consider the operator

$$
U = \begin{bmatrix}
A'_1 & 0 & \cdots & 0 \\
A'_2 & 0 & \cdots & 0 \\
\vdots & \vdots & \ddots & \vdots \\
A'_n & 0 & \cdots & 0
\end{bmatrix}
$$

on the n-fold direct sum $K = \mathcal{H} \oplus \ldots \oplus \mathcal{H}$. Verify that

$$
U^*U = \begin{bmatrix}
I & 0 & \cdots & 0 \\
0 & 0 & \cdots & 0 \\
\vdots & \vdots & \ddots & \vdots \\
0 & 0 & \cdots & 0
\end{bmatrix}
$$

and that $\|(E - E^{(n)})Ux\| \geq \epsilon n^{-1/2}\|(I - E^{(n)})x\|$ for all x in K and for all projections $E^{(n)} = E \oplus E \oplus \ldots \oplus E$ with E in $\mathcal{E}$. Writing $\mathcal{E}^{(n)} = \{E^{(n)} : E \in \mathcal{E}\}$ we see that the operator U is a partial isometry in the nest algebra $\mathrm{Alg}\mathcal{E}^{(n)}$ and satisfies the hypotheses of Lemma 3.11. (Note that $\mathrm{Alg}\mathcal{E}^{(n)}$ is the algebra of operator matrices with entires from $\mathrm{Alg}\mathcal{E}$). Hence there exists an operator

$$
B = \begin{bmatrix}
B'_1 & B'_2 & \cdots & B'_n \\
* & * & & * \\
\vdots & \vdots & \ddots & \vdots \\
* & * & \cdots & *
\end{bmatrix}
$$

in $\mathrm{Alg}\mathcal{E}^{(n)}$ such that $\|B\| \leq 4n\epsilon^{-2}$ and $BU = U^*U$. Let $B_k = A^{-1}B'_k$. Then $\sum B_k A_k = I$, and $\|B_k\| \leq \|B'_k\| \leq \epsilon^{-1} \cdot 4n\epsilon^{-2} = 4n\epsilon^{-3}$, as required.

Now consider the operators $A'_k = A_k A^{-1}$, and without loss of generality assume that $\|A_k\| \leq 1$ for $k = 1, \ldots, n$ so that $\|A\| \leq \sqrt{n}$. Since A and A^{-1} belong to $\mathrm{Alg}\mathcal{E}$ we have $E^\perp A E^\perp A^{-1} E^\perp = E^\perp$, and hence

$$
\|E^\perp A^{-1}x\|^2 = \|E^\perp A^{-1}E^\perp x\|^2 \geq \frac{\|E^\perp x\|^2}{\|E^\perp A E^\perp\|^2} \geq \frac{\|E^\perp x\|^2}{n}.
$$

Thus, as claimed earlier, we have

$$
\sum_k \|E^\perp A'_k x\|^2 \geq \epsilon^2 \|E^\perp A^{-1}x\|^2 \geq \epsilon^2 \frac{\|E^\perp x\|^2}{n}
$$

for all $x \in \mathcal{H}$ and $E \in \mathcal{E}$. ∎

It is not known whether every nest has the partial factorisation property. Nevertheless, the necessary and sufficient condition of Theorem 3.10 is also valid for a general nest algebra. To see this, consider finite subnests $\mathcal{E}_1 \subseteq \mathcal{E}_2 \subseteq \ldots$ of $\mathcal{E}$ whose strongly closed union is equal to $\mathcal{E}$. By Theorem 3.10 there exist operators $B_{1,k}, \ldots, B_{n,k}$ in $\mathrm{Alg}\mathcal{E}_k$ such that $B_{1,k}A_1 + \ldots, B_{n,k}A_n = I$, and such that the norms of $B_{1,k}, \ldots, B_{n,k}$ are bounded independently of k. Take a subsequence k_r such that B_{i,k_r} converges σ-weakly to an operator B_i as $r \to \infty$, for $1 \leq i \leq 1$, and note that $B_1, \ldots, B_n$ lie in $\mathrm{Alg}\mathcal{E}$ and satisfy $B_1 A_1 + \ldots + A_n B_n = I$.

4. Beyond Total Order

We now show that there are commutative projection lattices whose reflexive algebras, known as CSL algebras, fail to have a distance formula in the sense indicated below. For this reason many CSL algebras must be excluded from an analysis along the lines developed in the previous sections. Just which CSL algebras possess a distance formula remains an interesting open problem.

Let $\mathcal{L}$ be a lattice of projections that determines the reflexive operator algebra $\mathrm{Alg}\mathcal{L}$. Then it is clear that

$$\mathrm{dist}(X, \mathrm{Alg}\mathcal{L}) \geq \sup_{E \in \mathcal{L}} \|(I - E)XE\|$$

for each operator X. Let us write $\beta(X)$ for this supremum and suppose, moreover, that for some constant k we have

$$\mathrm{dist}(X, \mathrm{Alg}\mathcal{L}) \leq k\beta(X)$$

for all operators X. Then we say that $\mathrm{Alg}\mathcal{L}$ has a distance formula and has distance constant k_1, where k_1 is the best possible such constant, namely

$$k_1 = \inf_X \frac{\mathrm{dist}(X, \mathrm{Alg}\mathcal{L})}{\beta(X)}.$$

We need a construction in which we can see that the operator distance for X is large compared with the norms of the sub operators $(I - E)XE$.

Consider first a space $\mathcal{S}$ of $n \times n$ matrices determined by a prespecification of the zero entries of members of $\mathcal{S}$. More precisely, fix $Z \subset \{1, \ldots, n\} \times \{1, \ldots, n\}$ and let

$$\mathcal{S} = \{X \in \mathcal{L}(\mathbb{C}^n) : X_{ij} = 0 \text{ for all } (i, j) \in Z\}.$$

Such a space is seen to be a bimodule for the set of diagonal matrices, and the set, A say, of $2n \times 2n$ matrices.

$$\begin{bmatrix} D_1 & S \\ 0 & D_2 \end{bmatrix}, S \in \mathcal{S} \quad D_i \text{ diagonal,}$$

is an algebra. In fact $\mathcal{A}$ is a reflexive algebra with commutative invariant projection lattice $\mathcal{L}$ consisting of all diagonal projections $P_1 \oplus P_2$ such that the range of P_1 is contained in $\{Sx : S \in \mathcal{S}, x \in P_2\}$, the range of SP_2. Moreover, for a given matrix of the form

$$T = \begin{bmatrix} 0 & X \\ 0 & 0 \end{bmatrix}$$

it can be seen that $\beta(X)$ is the maximum of the norms of the submatrices $Q_1 X Q_2$ where Q_1, Q_2 are diagonal projections supported by subsets $E, F \subset \{1, \ldots, n\}$ such that $E \times F \subset Z$. On the other hand, the distance $\text{dist}(T, \mathcal{A})$ is equal to $\text{dist}(X, \mathcal{S})$.

We now construct spaces $\mathcal{S}_n$ of $3^n \times 3^n$ matrices and operator X_n so that

$$\text{dist}(X_n, \mathcal{S}_n)/\beta(X_n) = (9/8)^{n/2}.$$

Let $X_o = [1]$ be a 1×1 matrix. For $n \geq 0$ let X_{n+1} be the $3^{n+1} \times 3^{n+1}$ matrix given by

$$X_{n+1} = \begin{bmatrix} 0 & X_n & X_n \\ X_n & 0 & X_n \\ X_n & X_n & 0 \end{bmatrix}.$$

Let $\mathcal{S}_n$ denote the set of all $3^n \times 3^n$ matrices S such that the zero entries of S include the nonzero entries of X_n. Let $\mathcal{D}_n$ denote the $3^n \times 3^n$ diagonal matrices. Thus, $\mathcal{S}_0 = \{[0]\}, \mathcal{S}_1 = \mathcal{D}_1$ and $\mathcal{S}_{n+1}$ consists of all matrices of the form

$$\begin{bmatrix} A_1 & S_{12} & S_{13} \\ S_{21} & A_2 & S_{23} \\ S_{31} & S_{32} & A_3 \end{bmatrix}$$

where A_1, A_2, A_3 are arbitrary $3^n \times 3^n$ matrices and the S_{ij} belongs to $\mathcal{S}_n$.

Theorem 4.1. (K. R. Davidson and S. C. Power) *With $\mathcal{S}_n$ and X_n as above, we have*

$$\text{dist}(X_n, \mathcal{S}_n) = \inf_{S \in \mathcal{S}_n} \|X_n - S\| = \left[\frac{3}{2}\right]^n.$$

The significance of this result lies in the fact that we can also compute $\beta(X_n)$; it is the maximum norm of the rectangular arrays of ones occuring in X_n. This quantity is $\max(k\ell)^{1/2}$ over all $k \times \ell$ arrays of ones. For X_1 this is seen to be $\sqrt{2}$ by inspection. As $X_{n+1} = X_1 \otimes X_n$, it readily follows that $\beta(X_{n+1}) = (\sqrt{2})^{n+1}$. Thus Theorem 4.1 provides the desired growing distance constant. Indeed,

$$\frac{\text{dist}(X_n, \mathcal{S}_n)}{\beta(X_n)} = \left[\frac{3}{2\sqrt{2}}\right]^n.$$

Proof: We first need a general operator matrical lemma to the effect that

$$\left\| \begin{bmatrix} A_1 & Y & Y \\ Y & A_2 & Y \\ Y & Y & A_3 \end{bmatrix} \right\| \geq \frac{3}{2}\|Y\|,$$

where A_1, A_2, A_3 and Y are operators on a fixed Hilbert space. Equality is achieved here only by taking $A_1 = A_2 = A_3 = 3/2Y$. The proof of these facts are elementary and so we leave this puzzle to the reader.

Suppose $S_n \in \mathcal{S}_n$ and $\text{dist}(X_n, \mathcal{S}_n) = \|X_n - S_n\|$. Think of X_n and S_n as

$$X_n = \begin{bmatrix} 0 & X_{n-1} & X_{n-1} \\ X_{n-1} & 0 & X_{n-1} \\ X_{n-1} & X_{n-1} & 0 \end{bmatrix} \quad S_n = \begin{bmatrix} A_1 & S_{12} & S_{13} \\ S_{21} & A_2 & S_{23} \\ S_{31} & S_{32} & A_3 \end{bmatrix}.$$

Given a permutation π of three elements, we can induce an action on 3×3 matrices by simultaneously permuting rows and columns. As π runs through all permutations, each diagonal position is taken to each diagonal position twice, and the off diagonal entries are cyclically permuted. By averaging we may assume that the best approximating operator S_n has the special form

$$S_n = \begin{bmatrix} A & S & S \\ S & A & S \\ S & S & A \end{bmatrix}.$$

Applying the inequality, we obtain

$$\|X_n - S_n\| \geq \frac{3}{2}\|X_{n-1} - S\| \geq \frac{3}{2}\text{dist}(X_{n-1}, \mathcal{S}_{n-1})$$

and, using induction, the theorem follows. ∎

There are now many ways in which we can make use of the key example above to exhibit reflexive algebras with commutative projection lattices, that fail to have distance formula. For example, if $\mathcal{E}_1, \mathcal{E}_2, \ldots$ are non-trivial projection nests, then an embedding argument can be used with the above example to prove that the infinite tensor product algebra $\otimes_{k=1}^{\infty}\text{Alg}\mathcal{E}_k$ fails to possess a distance formula. The next corollary is more elementary.

Corollary 4.2. (K. R. Davidson and S. C. Power) *There is a reflexive algebra with commutative subspace lattice which fails to possess a distance constant.*

Proof: We have seen that on the Hilbert space $\mathbf{C}^{3n} \oplus \mathbf{C}^{3n}$ there is a reflexive algebra $\mathcal{A}_n$ of operators of the form

$$\begin{bmatrix} D_1 & S_n \\ 0 & D_2 \end{bmatrix},$$

with D_1, D_2 diagonal operators and $S_n \in \mathcal{S}_n$ as before, such that the distance constant of $\mathcal{A}_n$ is equal to $(3/2\sqrt{2})^n$. The algebra

$$\mathcal{A} = \mathcal{A}_1 \oplus \mathcal{A}_2 \oplus \ldots$$

is a reflexive operator algebra with the desired properties.

Postscript

I believe that the following remark owes its origins to Man Duen Choi. For theorems we must examine 2×2 matrices, whilst for counterexamples we need only consider 3×4 matrices.

Notes

We have tried to present some of the more fundamental aspects of nest algebras whilst emphasising parallels with the complex function theory associated with H^p spaces. Needless to say, there are many related directions that have not even been indicated, and to some extent we shall correct this in these brief notes.

Nest algebras were first studied by Ringrose [48], [49] who, in particular, made use of the structure of finite rank operators in a nest algebra in order to characterise the Jacobson radical. A study of ideals related to the radical has been made by Erdos in [14], whilst in his survey [13] there is commentary concerning the commutation properties of a nest algebra relative to various ideals. The problem of characterising the radical in CSL algebras has been considered by Hopenwasser [23], but the most fundamental problems in this interesting area remain unresolved.

The original proof of the distance formula in Arveson [3] made use of an analysis of the invariant subspaces of the ultraweakly closed operator algebra $\mathbf{C} \otimes \mathrm{Alg}\mathcal{E}$ associated with a nest algebra. The first proof we have given (Power [39]) uses the crucial Lemma

1.1 due to Parrott [36] and others (see the footnotes in [36]). I learnt the beautifully simple proof of Lemma 1.1 from Ken Davidson who attributes it to Chandler Davis. Lance's proof of the distance formula, which is more direct than our treatment, can be found in [27]. Our approach, which characterises the $\| \ \|_1$ extreme points of the trace class ball of a nest algebra, and makes explicit the double duality, is in Power [44]. Theorem 1.4 is in Davidson and Power [10]. Theorem 1.5 seems to be a new way of considering Nehari's Theorem, but there is already a very close link between this theorem and Lemma 1.1 as revealed in Parrott [36]. The integral operator inequality of Theorem 1.8 was a speculation of Shields in [52]. A proof that remains valid in quite general contexts was given in Power [43].

The Erdos density theorem appeared first in [11] where use was made of a representation theory for nests. Despite the naturalness and usefulness of the result, it has only received alternative proofs very recently. The first proof given here is in Power [44], and the second is an unpublished argument of Ken Davidson. Proposition 2.2 and Theorem 2.3 are in Davidson and Power [10] where there is also obtained a formula for the quasitriangular distance, expressed purely in terms of the nest $\mathcal{E}$ and the operator X. It seems useful to state this result here.

Let $\mathcal{E}$ be a projection nest endowed with the strong operator topology, making $\mathcal{E}$ a compact Hausdorff space. Let $C_{S^*}(\mathcal{E}, \mathcal{L}(\mathcal{H}))$ denote the C^*-algebra of all $*$-strongly continuous functions from $\mathcal{E}$ to $\mathcal{L}(\mathcal{H})$. For example, if X is a fixed operator, then the function $\Phi_X(E) = E^\perp X E$ is strongly continuous. Let $C_n(\mathcal{E}, \mathcal{K})$ denote the norm closed two sided ideal of norm continuous functions from $\mathcal{E}$ to $\mathcal{K}$. Then the distance from the operator X to $\mathrm{Alg}\mathcal{E} + \mathcal{K}$ is equal to the distance from ϕ_X to the ideal $C_n(\mathcal{E}, \mathcal{K})$ in $C_{S^*}(\mathcal{E}, \mathcal{L}(\mathcal{H}))$. For nest of order type $\mathbb{N}$, this specialises to the known formula

$$\mathrm{dist}(X, T(\mathbb{N}) + \mathcal{K}) = \lim_{n \to \infty} \sup \|(I - P_N) X P_n\|.$$

The general result is obtained through a C^*-algebraic development of best approximation techniques of Axler, Berg, Jewell, and Shields. In [10] it is also shown that every quasitriangular algebra $\mathrm{Alg}\mathcal{E} + \mathcal{K}$ is proximinal. For discrete nests this was also obtained independently by Feeman [17].

Rudin's Theorem (Theorem 2.4) was originally used in [51] to generalise Sarason's Theorem to function space contexts associated with locally compact groups. The original proof of the Fall–Arveson–Muhly Theorem (our second corollary to Theorem 2.4) in [16] made use of Lidskii's Theorem. We have preferred to deduce this result from triangular density properties. Erdos [12] gave the first such triangularisation proof, using the density theorem. Our proof is analogous to this. For a conventional approach to Lidskii's Theorem see Gohberg and Krein [21] or Ringrose [50]. For Lidskii's Theorem in Banach space contexts, see Konig [26] and Leiterer and Pietsch [30].

The weakly closed ideals of a nest algebra were characterised in Erdos and Power [15], where other related facts may be found. Corollary 2.7 seems to be a new observation, but is perhaps part of a metatheorem to the effect that weak star closed ideals in asymmetric (nonself-adjoint) operator algebras should be principle. This is the case for nonself-adjoint crossed products. (See McAsey, Muhly, and Saito [32, Section 5], for example.) Ideals in certain nest subalgebras of C^*-algebras have been considered in Power [42], and again, principality is a common feature.

Theorems 3.5 and 3.6 were inspired by the work of Arveson [3] who considered nests of order type $\mathbf{Z}$ and used a natural, but nonconstructive, spatial approach. Our approach appears in Power [45]. Theorem 3.7 was originally obtained by Shields [52] for the standard multiplicity one nest of order type the natural numbers. The well-ordered case is in [45]. Proposition 3.8 and the Riesz factorisation problem are stated here for the first time. Hankel forms on triangular noncommutative L^2 spaces are considered in Power [44], [45].

We should say a few more words about the constructions that appear in the proofs of Lemmas 1.6 and 3.3, since these methods can be taken to be the core results of each of the three sections. Let us borrow the notation $[C]_E$ from the proof of Theorem 3.5 to indicate the E-minimal part of the positive operator C. It was shown in Power [40] that for any given projection nest $\mathcal{E}$ there is a unique positive operator valued measure $C(\Delta)$ defined on the Borel subsets of $\mathcal{E}$ such that $C([0, E)) = [C]_E$ where $[0, E)$ is the set of projections F in $\mathcal{E}$ with $0 \leq F < E$. For a purely atomic nest, and in particular for a well-ordered nest, one can be a little more explicit about the construction of $C(\Delta)$, as we were in the proof of Theorem 3.5. In the case of a trace class positive operator, use can be made of the Radon Nikodyn derivative of $C(\Delta)$ with respect to the scalar measure trace$(C(\Delta))$ to obtain an integral representation ([40] Theorem 5.1) for a trace class operator in a nest algebra that is (almost) an integral version of the decomposition of Lemma 1.7 (ii).

The omissions in factorisation theory indicated in the text leading up to Theorem 3.10 include the work of Gohberg and Krein [22, Chapter 4]. Here boundedness properties of the triangular projection are used to obtain series formulae for the Cholesky (and related) factorisation $I + K = A^*A$, where K is an operator in an ideal of smooth operators. Such a factorisation may *not* be possible if K is merely compact and the underlying nest is continuous. This important nonfactorisation result, the first of its kind, was obtained by Larson in the interesting paper [29]. However, such counterexamples remain quite mysterious.

We mentioned that our constructive approach to factorisation appears to be the *right one*. In fact, the explicit nature of the construction was exploited in Power [46] in the determination of the extremal factorisation $f = hh^* + g$, with h outer, of a positive matrix valued function on the circle. Extremal here refers to the fact that we seek an outer factor h so that the positive matrix function $g = f - hh^*$ is minimised (see Sz-Nagy and Foias [50], Chapter 5, for the usual approach via the Beurling–Lax–Halmos Theorem). This problem is significant in prediction theory since the matrix $h(0)h(0)^*$ coincides with the prediction-error matrix associated with the multivariate stationary stochastic process with spectral density function f.

Theorem 3.10 and its proof is a rerun of Arveson's work in [3]. The limiting argument following the proof is an observation of Ken Davidson.

The construction in Section 4 can be found in Davidson and Power [9] where it is also shown that the infinite tensor product of nontrivial nest algebras fails to possess a distance formula. The problem of determining which reflexive operator algebras possess a distance constant has been promoted by Arveson in his Lecture Notes [4]. More recently Larson has identified a general framework–dual product constructions–generalising and providing a context for the inflating distance constant argument of the final section. However, it remains unknown if there is a distance formula for $T(\mathsf{N}) \otimes T(\mathsf{N})$.

A reflexive algebra A that admits a distance formula is said to be hyperreflexive, and this concept can be formulated from a predual viewpoint in terms of the rank one structure of the preannihilator of A. (Loosely speaking, a version of Lemma 1.7 (ii) with constants must hold.) This viewpoint is discussed in Arveson [4], Larson [28], and Kraus and Larson [25]. Christensen [5] has shown that every type I von Neumann algebra is hyperreflexive, but the case of other types seems not to be known. It should be noted that we have not mentioned the important similarity theory of nests (Andersen [1], Davidson [7], Larson [29]) in which the distance formula and compact perturbations play an important role. A deeper understanding of hyperreflexivity is required to extend this rather complete theory to more general projection lattices.

In conclusion, we observe that there is now a growing body of literature on more general nonself-adjoint, or analytic (or asymmetric), operator algebras. The monographs of Sz-Nagy and Foias [55] and Radjavi and Rosenthal [47] provide useful sources for the student. Concerning more recent devlopments, analytic crossed products have been studied by McAsey and Muhly [34], McAsey, Muhly and Saito [31], [32], [33], and by Solel [53], [54]. Nest-subalgebras of von Neumann algebras have been investigated by Gilfeather and Larson ([19], [20] for example), and since the important seminal work of Arveson [2], many authors have confronted the CSL algebras. See, for example, the papers of Gilfeather, Hopenwasser, and Larson [18], Davidson [6], and Hopenwasser, Laurie, and Moore [24]. Davidson is preparing a research monograph [8] on nest algebras. A dilation and representation theory for nest algebras, and for their tensor products will appear in [37] and [38].

There are many reasons for the contemporary health of the subject; asymmetric operator theory is not merely an offshoot of the giant discourse on self-adjoint matters. It is hoped that these notes may serve as an introduction to nest algebras, which occupy a central (but by no means well-understood) position, and as a prelude to participation in what must lie beyond total order.

References

1. Andersen, N. T., Similarity of continuous nests, Bull. London Math. Soc. **15** (1983), 131-132.

2. Arveson, W. B., Operator algebras and invariant subspaces, Ann. of Math. **100** (1974), 433-532.

3. ———, Interpolation problems in nest algebras, J. Func. Anal. **3** (1975), 208-233.

4. ———, Ten lectures on operator algebras, CMBS Regional Conference Series No. 55, AMS, 1984.

5. Christensen, E., Perturbations of operator algebras II, Indiana Univ. Math. J. **26** (1977), 891-904.

6. Davidson, K. R., Commutative subspace lattices, Indiana Univ. Math. J. **27** (1978), 479-490.

7. ———, Similarity and compact perturbation of nest algebras, J. fur der Reine und Angew. Math. **348** (1984), 72-87.

8. ———, *Nest Algebras*, Pitman Research Notes in Mathematics, Longman, in preparation.

9. ———— and S. C. Power, Failure of the distance formula, J. London Math. Soc. **32** (1985), 157-165.

10. ———— and S. C. Power, Best approximation in C^*-algebras, J. fur der Reine und Angew. Math. **368** (1986), 43-62.

11. Erdos, J. A., Operators of finite rank in nest algebras, J. London Math. Soc. **43** (1968), 381-397.

12. ————, On the trace of a trace class operator, Bull. London Math. Soc. **6** (1947), 47-50.

13. ————, Non-self-adjoint operator algebras, Proc. Royal Irish Acad. **81** (1981), 127-145.

14. ————, On some ideals of nest algebras, Proc. London Math. Soc. **44** (1982), 143-160.

15. ———— and S. C. Power, Weakly closed ideals of nest algebras, J. Operator Th. **7** (1982), 219-235.

16. Fall, T., W. B. Arveson, and P. Muhly, Perturbations of nest algebras, J. Operator Th. **1** 137-150.

17. Feeman, T. A., Best approximation and quasi-triangular algebras, Trans. Amer. Math. Soc. **288** (1925), 179-187.

18. Gilfeather, F., A. Hopenwasser, and D. Larson, Reflexive algebras with finite width lattices; tensor products, cohomology, compact perturbations, J. Func. Anal. **55** (1984), 176-199.

19. ———— and D. Larson, Nest-subalgebras of von Neumann algebras, Advances in Math. **46** (1982), 171-199.

20. ———— and D. Larson, Nest subalgebras of von Neumann algebras: commutants modulo compacts and distance estimates, J. Operator Theory **7** (1982), 279-302.

21. Gohberg, I. C. and Krein, M. G., Introduction to the theory of linear non-self-adjoint operators (Izdat. 'Nauka', Moscow 1965); Transl. of Math. Monographs **24**, Amer. Math. Soc., Providence 1965.

22. ———— and Krein, M. G., Theory and applications of Volterra operators in Hilbert space (Izdat 'Nauka', Moscow 1967); Transl. of Math. Monographs **24**, Amer. Math Soc., Providence 1970.

23. Hopenwasser, A., The radical of reflexive operator algebra, Pac. J. Math. **65** (1976), 375-392.

24. _____, C. Laurie, and R. Moore, Reflexive algebras with completely distributive lattices, J. Operator Th. **11** (1984), 91-108.

25. Kraus, J. and D. R. Larson, Some applications of a technique for constructing reflexive operator algebras, J. Operator Th. **13** (1985), 227-236.

26. Konig, H., s-numbers, eigenvalues and the trace theorem in Banach spaces, Studia Math. **67** (1980), 157-171.

27. Lance, E. C., Cohomology and perturbations of nest algebras, Proc. London Math. Soc. **43** (1981), 334-356.

28. Larson, D. R., Annihilators of operator algebras, Topics in modern operator theory, Birkhäuser, Basel, vol. 6 (1982), 119-130.

29. _____, Nest algebras and similarity transformations, Ann. of Math. **121** (1985), 409-427.

30. Leiterer, H. and A. Pietsch, An elementary proof of Liskii's trace formula, Wiss. Z. **31** (1982), 587-594.

31. McAsey, M., P. S. Muhly, and K.-S. Saito, Non-self-adjoint crossed products (Invariant subspaces and maximality), Trans. Amer. Math. Soc. **248** (1979), 381-409.

32. _____, P. S. Muhly, and K.-S. Saito, Non-self-adjoint crossed products II, J. Math Soc. Japan, **33** (1981), 485-495.

33. _____, P. S. Muhly, and K.-S. Saito, Non-self-adjoint crossed products III: Infinite algebras, J. Operator Th. **12** (1984), 3-22.

34. _____ and P. S. Muhly, Representations of non-self-adjoint crossed products, Proc. London Math. Soc. **47** (1983), 128-144.

35. Nehari, Z., Bounded bilinear forms, Ann. Math., **65** (1957), 153-162.

36. Parrott, S., On a quotient norm and the Sz.-Nagy-Foias lifting theorem, J. Func. Anal. **30** (1978), 311-328.

37. Paulsen, V. I., S. C. Power, and J. D. Ward, Semi-discreteness and dilation theory for nest algebras, preprint, 1987.

38. _____ and S. C. Power, Lifting theorems for nest algebras, preprint, 1987.

39. Power, S. C., The distance to upper triangular operators, Math. Proc. Cambridge Phil. Soc. **88** (1980), 327-329.

40. _____, Nuclear operators in nest algebras, J. Operator Th. **10** (1983), 337-352.

41. _____, Another proof of Lidskii's theorem on the trace, Bull. London Math. Soc. **15** (1983), 146-148.

42. _____, On ideals of nest subalgebras of C^*-algebras, Proc. London Math. Soc. **50** (1985), 314-332.

43. _____, A Hardy-Littlewood-Fejer inequality for Volterra integral operators, Indiana Univ. Math. J. **33** (1984), 667-671.

44. _____, Commutators with the triangular projection and Hankel forms on nest algebras, J. London Math. Soc. **32** (1985), 272-282.

45. _____, Factorisation in analytic operator algebras, J. Func. Anal. **67** (1986), 413-432.

46. _____, Spectral characterisation of the Wold-Zasuhin decomposition and prediction-error operator, preprint 1985.

47. Radjavi, H. and P. Rosenthal, *Invariant Subspaces*, Springer Verlag, Berlin, 1973.

48. Ringrose, J. R., on some algebras of operators, Proc. London Math. Soc. **15** (1965), 61-83.

49. _____, On some algebras of operators II, Proc. London Math. Soc. **16** (1966), 385-402.

50. _____, Compact non-self-adjoint operators, Van Nostrand Mathematical Studies No. 35, Van Nostrand, London, 1971.

51. Rudin, W., Spaces of type $H^\infty + C$, Ann. Inst. Fourier Grenoble **25** (1975), 99-125.

52. Shields, A. L., An analogue of a Hardy-Littlewood-Fejer inequality for upper triangular trace class operators, Math. Z. **182** (1983), 473-484.

53. Solel, B., Analytic operator algebras (factorisation and an expectation), Trans. Amer. Math. Soc. **287** (1985), 799-817.

54. _____, Non-self-adjoint crossed products: Invariant subspaces, cocycles and subalgebras, Indiana Univ. Math J. **34** (1985), 277-298.

55. Sz.-Nagy, B. and C. Foias, *Harmonic Analysis of Operators on Hilbert Space*, North Holland, Amsterdam, 1970.

Stephen Power

Department of Mathematics

University of Lancaster

Lancaster LA1 4YL, United Kingdom

C*-algebras and a Single Operator

by

Derek W. Robinson

Introduction

In these lectures we survey some recent results in the developing theory of differential operators associated with dynamical systems. This theory was initially motivated by the analysis of unbounded derivations acting on operator algebras. Such derivations occur as the generators of continuous one-parameter groups of *-automorphisms and one of the main problems is to establish when a given derivation is indeed a generator. This problem can be viewed as one of integration of an infinitesimal equation of motion and it often occurs in this guise. The derivation property then corresponds to a conservation law, which is reflected by the isometric character of the generated group of *-automorphisms. More generally, dissipative systems are described by equations of motion involving operators with weaker algebraic properties. Thus one is led immediately to the analysis of a broader class of operators acting on the underlying operator algebra, or Banach space.

The next point to emphasize is that most systems involve additional structure over and above the Banach space, or operator algebra, and the evolution equation determined by the operator describing the infinitesimal motion. Typically, the symmetries and textures of the system are described by other groups of isometries and the properties of the evolution in relation to these groups are of fundamental importance. Thus one is led to the analysis of operators associated, by appropriate general principles, with dynamical systems consisting of a Banach space, or more specifically, an operator algebra, together with the action of a group describing the symmetries of the system. This is the topic of these lectures.

In the first lecture we describe in some detail the theory of operators associated with abelian C*-algebraic systems. The derivations are then recognizable as generalized first-order differential operators and dissipative operators correspond to second-order elliptic operators.

In the second lecture we turn to the analysis of the equations of motion determined by derivations and dissipations on the abelian systems. But this leads naturally to the examination of a broader class of integration problems on Banach space.

In the final lecture we examine the characterization by smoothness properties of semigroup generators on Banach dynamical systems.

1. Local Operators on Abelian C*-dynamical Systems

1.1 Introduction

In this lecture we describe some recent results characterizing differential operators acting on operator algebras. The framework of this theory is a C^*-dynamical system $(\mathcal{A}, \mathbf{R}^\nu, \tau)$ consisting of a C*-algebra $\mathcal{A}$ and an action τ of the group $\mathbf{R}^\nu$ as *-automorphisms of $\mathcal{A}$. We concentrate, however, on abelian algebras $\mathcal{A}$ both for simplicity and because the theory is most developed in this special case.

The main object of the theory is a linear operator H associated with the system $(\mathcal{A}, \mathbf{R}^\nu, \tau)$. There are various rules of association and we consider three common types.

1. **Domain properties.** Typically, H is defined on the smooth elements of $(\mathcal{A}, \mathbf{R}^\nu, \tau)$.

2. **Locality properties.** The general idea behind locality is that H is compatible with the infinitesimal structure of the system $(\mathcal{A}, \mathbf{R}^\nu, \tau)$.

3. **Invariance properties.** The simplest invariance hypothesis is that H commutes with the action τ but it is also natural to consider approximate commutation properties.

In order to motivate more precise definitions of these properties we first consider the basic example, the theory of differential operators. Let $\mathcal{A} = C_0(\mathbf{R}^\nu)$, the continuous functions on $\mathbf{R}^\nu$ which vanish at infinity, and suppose τ acts by translations, i.e.,

$$(\tau_y f)(x) = f(x - y)$$

for all $f \in C_0(\mathbf{R}^\nu)$ and all $x, y \in \mathbf{R}^\nu$. Thus the infinitesimal generators of τ are the operators $\partial_i = \partial/\partial x_i$ of partial differentiation and the smooth elements $\mathcal{A}_\infty = C_0^\infty(\mathbf{R}^\nu)$ are the infinitely often differentiable functions which vanish at infinity. Now consider a differential operator

$$(1.1) \qquad\qquad H = \sum_{\alpha; |\alpha| \le n} \ell_\alpha \partial^\alpha$$

where $\alpha = (\alpha_1, \ldots, \alpha_\nu)$, the α_i are positive integers, $|\alpha| = \alpha_1 + \ldots + \alpha_\nu$, and $\partial^\alpha = \Pi_i \partial_i^{\alpha_i}$.

236

If the coefficient functions ℓ_α of H are bounded and continuous then H is defined on the smooth elements $\mathcal{A}_\infty$ and

$$(1.2) \qquad\qquad H\mathcal{A}_\infty \subseteq \mathcal{A}.$$

Moreover, H is local in the sense that

$$(1.3) \qquad\qquad \mathrm{supp}(Hf) \subseteq \mathrm{supp}(f), \quad f \in \mathcal{A}_\infty.$$

What is less obvious is that a linear operator H with properties (1.2) and (1.3) is necessarily of the form (1.1) with the ℓ_α bounded and continuous. This is a consequence of Theorem 1.2.4 below.

Finally note that

$$\tau_x H \tau_{-x} - H = \sum_{\alpha;|\alpha|\leq n} ((\tau_x \ell_\alpha) - \ell_\alpha)\partial^\alpha$$

and hence bounds on the differences $\tau_x H \tau_{-x} - H$ correspond to bounds on $\tau_x \ell_\alpha - \ell_\alpha$, i.e., to smoothness properties of the coefficients ℓ_α. Thus properties of approximate invariance are reflected by smoothness.

We now describe general results along these lines, with $\mathcal{A}$ abelian.

1.2. Abelian Systems

Let $\mathcal{A}$ be a C*-algebra, τ an action of $\mathbf{R}^\nu$ as *-automorphisms of $\mathcal{A}$, i.e., a strongly continuous representation of $\mathbf{R}^\nu$ as *-automorphisms. Further, let $\delta_1, \ldots, \delta_\nu$ denote the infinitesimal generators of τ with respect to some fixed basis of $\mathbf{R}^\nu$. Then define

$$\mathcal{A}_n = \cap_{\alpha:|\alpha|=n} D(\delta^\alpha)$$

where again $\alpha = (\alpha_1, \ldots, \alpha_\nu), \alpha_i \geq 0, |\alpha| = \alpha_1 + \ldots + \alpha_\nu$, and $\delta^\alpha = \Pi \delta_i^{\alpha_i}$. Next set

$$\mathcal{A}_\infty = \cap_{n \geq 0} \mathcal{A}_n.$$

Thus $\mathcal{A}_n$ is the analogue of the n-times continuously differentiable functions, and $\mathcal{A}_\infty$ the infinitely differentiable functions. Note that $\mathcal{A}_n$ is a Banach space with respect to the norm

$$\|A\|_n = \max_{0 \leq |\alpha| \leq n} \|\delta^\alpha A\|$$

and $\mathcal{A}_\infty$ is a Fréchet space when equipped with the family of norms $\{\|\cdot\|_n; n \geq 0\}$. In fact, each of these spaces is a *-algebra.

Next let $\mathcal{A}$ be abelian, then it can be identified with the algebra $C_0(X)$ where X denotes the spectrum of $\mathcal{A}$. Now one can associate with each *-automorphism τ of $\mathcal{A}$ a homeomorphism T of X such that

(1.4) $$(\tau f)(\omega) = f(T\omega)$$

for all $\omega \in X$ and $f \in C_0(X)$. More generally one can associate a group $(\mathbf{R}^\nu, T)$ of homeomorphisms of X with the action $(\mathbf{R}^\nu, \tau)$ on $\mathcal{A}$. It is then not difficult to establish that strong continuity of $(\mathbf{R}^\nu, \tau)$ is equivalent to joint continuity of the map $(x, \omega) \mapsto T_x \omega$. Thus the abelian C^*-dynamical system $(\mathcal{A}, \mathbf{R}^\nu, \tau)$ determines the topological dynamical system $(X, \mathbf{R}^\nu, \tau)$, and conversely given a topological dynamical system $(X, \mathbf{R}^\nu, \tau)$ one can construct a C^*-dynamical system $(\mathcal{A}, \mathbf{R}^\nu, \tau)$ by setting $\mathcal{A} = C_0(X)$ and defining τ by the relation (1.4). For the subsequent analysis of $(\mathcal{A}, \mathbf{R}^\nu, \tau)$ it is useful to introduce a number of standard definitions associated with the topological system.

First define the *fixed point set* $X_0 \subseteq X$ by

$$X_0 = \{\omega \in X; T_x \omega = \omega \text{ for all } x \in \mathbf{R}^\nu\}.$$

Second define the *stabilizer subgroups* $\omega \in X \mapsto S_\omega \subseteq \mathbf{R}^\nu$ by

$$S_\omega = \{x \in \mathbf{R}^\nu; T_x \omega = \omega\}.$$

Thus, for example, $\omega \in X_0$ if, and only if, $S_\omega = \mathbf{R}^\nu$ but in general the stabilizer subgroups are strict subgroups of $\mathbf{R}^\nu$ which can vary with ω. If, however, $S_\omega = 0$ for all $\omega \in X$ then the action $(\mathbf{R}^\nu, \tau)$, or $(\mathbf{R}^\nu, T)$, is defined to be *free*. Now we return to consider operators on $\mathcal{A}$.

Let $\mathcal{A}$ be abelian, then the linear operator H on $\mathcal{A}$ is defined to be *local* if

$$\mathrm{supp}(Hf) \subseteq \mathrm{supp}(f), \quad f \in D(H).$$

In fact, this definition is independent of the action $(\mathbf{R}^\nu, \tau)$ but it has particularly strong consequences when $D(H) = \mathcal{A}_\infty$.

Proposition 1.2.1. *Let $(\mathcal{A}, \mathbf{R}^\nu, \tau)$ be an abelian C^*-dynamical system. If H is a local operator from $\mathcal{A}_\infty$ into $\mathcal{A}$ then there exists an integer $n \geq 0$, and a $C > 0$, such that*

$$\|Hf\| \leq C\|f\|_n, \quad f \in \mathcal{A}_\infty.$$

An equivalent way of restating this result is that each local operator from the Fréchet space $\mathcal{A}_\infty$ into the Banach space $\mathcal{A}$ is automatically continuous. Note that if H were closable, as an operator from $\mathcal{A}_\infty$ into $\mathcal{A}$, then continuity would follow from the closed graph theorem. In fact, continuity is equivalent to closability. But H is not assumed to be closable and this has to be deduced from locality and the domain assumption. We will not describe the proof, it is given by Observation 1 in Section 3 of [4] and Theorem 2.2 of [2], but remark that it is somewhat analogous to the proof, or at least one of the proofs, of the closed graph theorem.

Continuity of H has an immediate implication.

Proposition 1.2.2. *Let $(\mathcal{A}, \mathbf{R}^\nu, \tau)$ be an abelian C^*-dynamical system and H a local operator from $\mathcal{A}_\infty$ into $\mathcal{A}$. Then there exist an integer $n \geq 0$, a bounded continuous function ℓ_0 on X, and functions $\{\ell_\alpha; 0 < |\alpha| \leq n\}$ continuous on $X \backslash X_0$ and zero on X_0 such that*

$$Hf = \sum_{\alpha; |\alpha| \leq n} \ell_\alpha \delta^\alpha f, \quad f \in \mathcal{A}_\infty.$$

The existence of the ℓ_α follows from locality and the continuity established in Proposition 1.2.1. One uses these properties to argue that if $(\delta_\alpha f)(\omega) = 0$ for all α with $|\alpha| \leq n$ and some $\omega \in X \backslash X_0$, then $(Hf)(\omega) = 0$. Thus the joint kernel of $f \in \mathcal{A}_\infty \mapsto (\delta^\alpha f)(\omega), 0 \leq |\alpha| \leq n$, is contained in the kernel of $f \in \mathcal{A}_\infty \mapsto (Hf)(\omega)$ and the existence of the $\ell_\alpha(\omega)$, for $\omega \in X \backslash X_0$, follows by linear algebra. The continuity of the ℓ_α at each $\omega \in X \backslash X_0$ is deduced by constructing functions $f_\beta, 0 \leq |\beta| \leq n$ such that the simultaneous linear equations

$$Hf_\beta = \sum_{\alpha; |\alpha| \leq n} \ell_\alpha \delta^\alpha f_\beta$$

are soluble for the ℓ_α in a neighbourhood of ω, in terms of the continuous functions Hf_β and $\delta^\alpha f_\beta$.

Proposition 1.2.2 establishes that the local operators from $\mathcal{A}_\infty$ into $\mathcal{A}$ are polynomials in the generator of τ. Thus they are natural generalizations of differential operators. But the inherent weakness in the statement is the lack of boundedness properties of the ℓ_α. In particular, without some control on the growth of the ℓ_α, it is impossible to check whether an operator of the form

$$\sum_{\alpha; |\alpha| \leq n} \ell_\alpha \delta^\alpha$$

is in fact local. Nevertheless, it must be emphasized that locality does not imply boundedness of the ℓ_α. This is illustrated by the following example of a rotational flow.

Example 1.2.3. Let $X = \mathbf{R}^2$, choose radial co-ordinates (r, θ), and define the group $(\mathbf{R}, T)$ of homeomorphisms of X such that

$$T_t(r, \theta) = (r, \theta + 2\pi t/r), \quad r \neq 0,$$

and the origin is fixed. This corresponds to a rotational flow with angular velocity $1/r$. Thus the velocity increases as one approaches the centre of the flow. The example describes an idealized whirlpool. Now the generator δ of the corresponding action $(\mathbf{R}, \tau)$ on $\mathcal{A} = C_0(\mathbf{R}^2)$ has the form $(1/r)\partial_\theta$. One checks that $\mathcal{A}_\infty$ consists of those $f \in \mathcal{A}$ which are infinitely often differentiable with respect to θ and such that the derivatives $(\partial_\theta^n f)(r, \theta), n \geq 1$, vanish faster than any power of r as $r \to 0$. In particular, the operator $H = (1/r^m)\delta = (1/r^{m+1})\partial_\theta$ is a local operator from $\mathcal{A}_\infty$ into $\mathcal{A}$. But the coefficient $\ell_1(r) = 1/r^m$ is not bounded on $X \backslash X_0$; it diverges at the fixed point $r = 0$.

The existence of fixed points of $(\mathbf{R}^\nu, T)$, or at least some subgroup, is, however, necessary for unboundedness of the ℓ_α. In the absence of such fixed points, one can obtain a rather complete characterization of local operators from $\mathcal{A}_\infty$ into $\mathcal{A}$ as differential operators, and also establish the equivalence of various alternative forms of locality.

Theorem 1.2.4. Let $(\mathcal{A}, \mathbf{R}^\nu, \tau)$ be an abelian C^*-dynamical system and assume the action $(\mathbf{R}^\nu, \tau)$ is free. Further, let H be a linear operator from $\mathcal{A}_\infty$ into $\mathcal{A}$. The following four conditions are equivalent:

1. H is local,
2. If $f \in \mathcal{A}_\infty, \omega \in X$, and $(\tau_x f)(\omega) = 0$ for all x in a neighbourhood of the origin in $\mathbf{R}^\nu$, then $(Hf)(\omega) = 0$,
3. If $f \in \mathcal{A}_\infty, \omega \in X$, and $(\delta^\alpha f)(\omega) = 0$ for all α, then $(Hf)(\omega) = 0$,
4. There exist a positive integer n and a (unique) family of bounded continuous functions ℓ_α over X such that

$$(Hf)(\omega) = \sum_{\alpha; |\alpha| \leq n} \ell_\alpha(\omega)(\delta^\alpha f)(\omega), \quad f \in \mathcal{A}_\infty.$$

Moreover, each finite family of bounded continuous functions ℓ_α over X determines a linear operator from $\mathcal{A}_\infty$ into $\mathcal{A}$ which satisfies these conditions.

The complete proof of this theorem is given in [4]. Note that it applies in particular to the usual differential operators. Thus, if $\mathcal{A} = C_0(\mathbf{R}^\nu)$, and $(\mathbf{R}^\nu, \tau)$ acts by translations,

240

one concludes that a linear operator $H : C_0^\infty(\mathbf{R}^\nu) \mapsto C_0(\mathbf{R}^\nu)$ is local if, and only if, it is a differential operator of finite-order with bounded continuous coefficients.

If $(\mathbf{R}^\nu, \tau)$ does not act freely, the characterization of local operators is more complex but it can be completely described in the case $\nu = 1$. This description requires, however, some more details concerning the flow $(\mathbf{R}, T)$.

The *period* p of a point $\omega \in X$ under the flow $(\mathbf{R}, T)$ is defined by

$$p(\omega) = \inf\{t > 0; T_t\omega = \omega\}$$

and the *frequency* ν by $\nu(\omega) = 1/p(\omega)$. There are only three possibilities. Either $p(\omega) = 0$ which corresponds to $\omega \in X_0$, or $0 < p(\omega) < +\infty$ which implies that the orbit $\mathcal{O}_\omega = \{T_t\omega; t \in \mathbf{R}\}$ is periodic, or $p(\omega) = +\infty$ and the orbit is open.

The period p and the frequence ν are both constant on the orbits $\mathcal{O}_\omega$ because $p(T_t\omega) = p(\omega)$ for all $t \in \mathbf{R}$ by definition. The frequency is, in fact, a useful measure of growth, from orbit to orbit, of various functions which subsequently occur. We define a function ℓ over $X \backslash X_0$ to be *polynomially bounded* if there exist $c > 0$ and an integer $k \geq 0$ such that

$$|\ell(\omega)| \leq c(1 + \nu(\omega)^k), \omega \in X \backslash X_0.$$

A little care has to be exercised, however, because ν is not necessarily continuous. Nevertheless, it is upper semi-continuous and one can establish [3] that if $\omega_\alpha \to \omega, \nu(\omega_\alpha) \to \nu$, and $\nu < +\infty$, then $\nu(\omega)/\nu \in \{1, 2, 3, \ldots\}$.

Theorem 1.2.5. *Let* $(\mathcal{A}, \mathbf{R}, \tau)$ *be an abelian* C^*-*dynamical system and* H *a linear operator from* $\mathcal{A}_\infty$ *into* $\mathcal{A}$. *The following four conditions are equivalent:*

1. H *is local,*
2. *If* $f \in \mathcal{A}_\infty, \omega \in X$, *and* $(\tau_x f)(\omega) = 0$ *for all* x *in a neighbourhood of the origin in* $\mathbf{R}$ *then* $(Hf)(\omega) = 0$.
3. *If* $f \in \mathcal{A}_\infty, \omega \in X$, *and* $(\delta^m f)(\omega) = 0$ *for* $m = 1, 2, 3 \ldots$ *then* $(Hf)(\omega) = 0$.
4. *There exist a positive integer* n *and a (unique) family of functions* $\ell_0, \ell_1, \ldots, \ell_n$ *on* X *with* ℓ_0 *bounded and continuous and* $\ell_1, \ldots, \ell_n$, *equal to zero on* X_0 *and polynomially bounded and continuous on* $X \backslash X_0$ *such that*

$$H = \sum_{m=0}^{n} \ell^m \delta^m |_{\mathcal{A}_\infty}.$$

Moreover, each finite family of functions $\ell_0, \ell_1, \ldots, \ell_n$, with the boundedness and continuity properties specified in Condition 4 determines a local operator from $\mathcal{A}_\infty$ into $\mathcal{A}$.

Example 1.2.3 illustrated a one-dimensional flow and a local operator H with an unbounded coefficient. In this example one has $\nu(r, \theta) = 1/r$ and hence the frequency diverges as one approaches the origin, i.e. the fixed point of the flow. But $H = (1/r^m)\delta$ and so $\ell_1 = \nu^m$ and the coefficient diverges with the m-th power of the frequency. Theorem 1.2.5 establishes, however, that this pattern is the only possible source of divergence. Such behaviour only occurs at fixed points which are limit points of periodic orbits of ever increasing frequency. This simplicity is rather striking in the light of the complex topological structure which can occur for the orbits.

1.3 Dissipations and Derivations

In the discussion of evolution problems, stability properties are of paramount importance. In the theory of differential operators stability is usually reflected by some form of dissipativity of the generator of the operator. There are various disparate mathematical notions of dissipativity which reflect different structural features but the concept of locality provides a unification.

The first form of dissipation is a Banach space concept. If $H : D(H) \mapsto \mathcal{B}$ is a norm densely defined, linear operator on a Banach space $\mathcal{B}$ then H is defined to be *dissipative* whenever $A \in D(H), \omega \in \mathcal{B}^*$, and $\omega(A) = \|\omega\| \cdot \|A\|$, always implies

$$Re\, \omega(HA) \geq 0.$$

It is a standard result that this is equivalent to the condition

$$\|(I + \alpha H)A\| \geq \|A\|$$

for all $A \in D(H)$ and all $\alpha > 0$. The key example of a dissipative operator is the generator of a continuous semigroup of contractions. If S is such a semigroup and $\omega(A) = \|\omega\| \cdot \|A\|$, one has

$$Re\, \omega(S_t A) \leq \|\omega\| \cdot \|S_t A\|$$
$$\leq \|\omega\| \cdot \|A\| = \omega(A) = Re\, \omega(A).$$

Therefore,

$$\lim_{t \to 0+} Re((\omega(A) - \omega(S_t A))/t) \geq 0.$$

242

The second, closely related, concept is for operators on a real ordered Banach space $\mathcal{B}$. If $\mathcal{B}$ is ordered by a convex cone $\mathcal{B}_+$ (which is assumed to be normal and generating) and $A_+ \in \mathcal{B}_+$ denotes the positive part of $A \in \mathcal{B}$ then H is defined to be *dispersive* whenever $A \in D(H)$, $A_+ \neq 0$, $\omega \in \mathcal{B}^*$, and $\omega(A) = \|\omega\| \cdot \|A_+\|$, always implies

$$\omega(HA) \geq 0.$$

The generator of a continuous semigroup of contractions S which is positive, i.e. $S_t \mathcal{B}_+ \subseteq \mathcal{B}_+$ for all $t \geq 0$, is automatically dispersive.

The third form of dissipation is an algebraic concept. If $H : D(H) \mapsto \mathcal{A}$ is a norm densely defined linear operator from a *-subalgebra $D(H)$ into a C^*-algebra $\mathcal{A}$ then H is defined to be *real* if $(HA)^* = HA^*$, for all $A \in D(H)$, and to be a *dissipation* if

$$H(A^*A) \leq (HA)^*A + A^*(HA),$$

for all $A \in D(H)$. If H is the generator of a strongly positive continuous semigroup S, i.e. a continuous semigroup satisfying

$$S_t(A^*A) \geq (S_tA)^*(S_tA),$$

for all $a \in \mathcal{A}$, then H is automatically a dissipation.

The fourth form of dissipation is restricted to differential operators. It is the concept of ellipticity, which can also be expressed in various manners. Typically, if H is a second-order differential operator,

$$H = -\sum_{i,j=1}^{\nu} a_{ij}(x)\partial_i\partial_j,$$

then ellipticity corresponds to positive definiteness of the matrices $(a_{ij}(x))$, $x \in \mathbf{R}^\nu$.

The surprise is that for real local operators associated with an abelian C^*-dynamical system $(\mathcal{A}, \mathbf{R}^\nu, \tau)$ these notions coincide, at least if the action $(\mathbf{R}^\nu, \tau)$ is free, or if $\nu = 1$. Note that the positive cone $\mathcal{A}_+$ corresponds to the positive elements of $\mathcal{A}$, i.e. the positive functions in $C_0(X)$.

Theorem 1.3.1. *Let $(\mathcal{A}, \mathbf{R}^\nu, \tau)$ be an abelian C^*-dynamical system and assume the action $(\mathbf{R}^\nu, \tau)$ is free. Further, let H be a real local operator from $\mathcal{A}_\infty$ into $\mathcal{A}$. The following four conditions are equivalent:*

    1. H is dissipative.

2. H is dispersive.

3. H is a dissipation.

4. $H = \left(\ell_0 + \sum_{i=1}^{\nu} \ell_i \delta_i + \sum_{i,j=1}^{\nu} \ell_{ij} \partial_i \partial_j\right)|_{\mathcal{A}_\infty}$ where $(\ell_0, \ell_i, \ell_{ij})$ are bounded continuous functions over X, $\ell_0 \geq 0$, the ℓ_i are real, and $(-\ell_{ij}(\omega))$ is a real positive-definite matrix for each $\omega \in X$.

Since H is assumed to be a local operator from $\mathcal{A}_\infty$ into $\mathcal{A}$ it can be expressed as a polynomial in the δ_i with bounded continuous coefficients by Theorem 1.2.4. The major part of the proof consists of showing that the condition of dissipativity, or the property of dissipation, eliminates all coefficients above the second order. The proof that $3 \Rightarrow 4$ can be found in [4] and the proof that $1 \Rightarrow 4$ is given in [1]. (The proof outlined in [1] is for $\nu = 1$ but it is not difficult to extend it to the general case.) The proof that $4 \Rightarrow 3$ is a simple verification and the proofs of $3 \Rightarrow 2$ and $3 \Rightarrow 1$ were given in [3], at least for $\nu = 1$. These proofs also extend easily to the general case. (For example, if $f \in \mathcal{A}_\infty$ and $f(\omega) = \|f\|$ then since $\|\tau_x f\| = \|f\|$ the function $x \mapsto (\tau_x f)(\omega)$ must have a maximum at $x = 0$ and hence $(\delta_i f)(\omega) = 0$, and $\sum a_{ij}(\delta_i \delta_j f)(\omega) \geq 0$ for each positive-definite matrix (a_{ij}). Consequently, Condition 3 must imply Condition 1.) The proof that $2 \Rightarrow 1$ follows by general reasoning because $\|f\| = \|f_+\| \vee \|(-f)_+\|$.

If $\nu = 1$, then one can prove a version of Theorem 1.3.1 without the assumption that the action $(\mathbf{R}, \tau)$ is free. But then, H has the form

$$H = (\ell_0 + \ell_1 \delta + \ell_2 \delta^2)|_{\mathcal{A}_\infty}$$

with ℓ_0, ℓ_1, ℓ_2 satisfying the properties given in Condition 4 of Theorem 1.2.5 and in addition $\ell_0 \geq 0$, ℓ_1 is real, and $\ell_2 \leq 0$.

Finally, we consider derivations $H; \mathcal{A}_\infty \mapsto \mathcal{A}$. Thus

$$H(AB) = (HA)B + A(HB)$$

for all $A, B \in \mathcal{A}_\infty$. The key new feature is that such operators are automatically local.

Proposition 1.3.2. *Let $(\mathcal{A}, \mathbf{R}^\nu, \tau)$ be an abelian C^*-dynamical system and H a derivation from $\mathcal{A}_\infty$ into $\mathcal{A}$. Then*

$$\mathrm{supp}(Hf) \subseteq \mathrm{supp}(f), \quad f \in \mathcal{A}_\infty.$$

To establish this suppose $f \in \mathcal{A}_\infty$ and $f = 0$ in a neighbourhood U of ω. Then one can choose a $g \in \mathcal{A}_\infty$ with the property that $\mathrm{supp}(g) \subseteq U$ and $g(\omega) = 1$. Therefore,

244

$fg = 0$ and $H(fg) = 0$ by linearity. But then

$$(Hf)(\omega) = (Hf)(\omega)g(\omega)$$
$$= (Hfg)(\omega) - f(\omega)(Hg)(\omega) = 0$$

where we have used the derivation property, together with $H(fg) = 0$, and $f = 0$ on U. Thus, $\text{supp}(Hf) \subseteq \text{supp}(f)$.

Combining Theorem 1.3.1 and Proposition 1.3.2, one immediately obtains the last result.

Theorem 1.3.3. *Let* $(\mathcal{A}, \mathbf{R}^\nu, \tau)$ *be an abelian* C^*-*dynamical system and assume the action* $(\mathbf{R}^\nu, \tau)$ *is free. Let* H *be a real operator from* $\mathcal{A}_\infty$ *into* $\mathcal{A}$. *The following conditions are equivalent*

1. $\pm H$ *are dissipative,*
2. $\pm H$ *are dispersive,*
3. H *is a derivation,*
4. $H = \sum_{i=1}^\nu \ell_i \delta_i|_{\mathcal{A}_\infty}$, *where the* ℓ_i *are real bounded continuous functions over* X.

An analogous result is true if $\nu = 1$ without the assumption that $(\mathbf{R}, \tau)$ is free but then ℓ_i is only polynomially bounded in the frequency.

Notes and Remarks

The characterization of differential operators by locality properties was first suggested by Peetre [9]. The analysis of local operators associated with abelian C^*-dynamical systems was begun by Bratteli, Elliott, and Robinson [4].

Further results were then obtained by Batty and Robinson [1], [2]. Additional properties of local dissipations were also derived by Bratteli, Digernes, Goodman, and Robinson [3].

There are also extensions of some of the results described in this lecture to non-abelian algebras. Bratteli reviewed these results at the 1985 Iowa conference on Operator Algebras. The proceedings of this conference are to be published by the American Mathematical Society.

2. Integration in Abelian C^*-dynamical Systems

2.1. Introduction

One motivation for the general analysis of differential operators is the desire to understand evolution equations

(2.1)
$$\frac{dA_t}{dt} + HA_t = 0.$$

Typically, A_t is considered as an element of a Banach space, H is a linear operator on the space governing the infinitesimal evolution, t corresponds to the time, and one seeks solutions for $t \geq 0$ which agree with the given initial data $A_0 = A$. The solutions of interest must also satisfy a number of general physical criteria such as continuity of $t \mapsto A_t$ and the causality requirement $(A_s)_t = A_{s+t}$, i.e. the evolute of A_s at time t is equal to the evolute of A_0 at time $s + t$. Such solutions are necessarily of the form $A_t = \sigma_t A$ where σ is a continuous semigroup and hence solutions exist if, and only if, H is a semigroup generator. Other requirements such as positivity and contractivity of σ correspond to conservation laws and these are reflected by properties such as dispersivity and dissipativity of H. In particular, if the underlying space is a C^*-algebra, these conservation properties lead to the examination of derivations and dissipations. In this lecture we discuss the integration of equation (2.1) on an Abelian C^*-dynamical system with H a derivation, or a local dissipation, associated with the system. But as a preamble, we make some general remarks on the integration problem in the algebraic setting.

First, all operators H that we consider on the C^*-algebra $\mathcal{A}$ will be real operators with domain $D(H)$ a norm dense $*$-subalgebra of $\mathcal{A}$. Thus, if H generates a continuous semigroup $\sigma = \{\sigma_t\}_{t \geq 0}$, with $\sigma_0 = I$, then $\sigma_t(A)^* = \sigma_t(A^*)$ for all $A \in \mathcal{A}$ and $t \geq 0$ by reality. Moreover, if H is a derivation

$$
\begin{aligned}
\sigma_t(AB) - \sigma_t(A)\sigma_t(B) &= \int_0^t ds \frac{d}{ds} \{\sigma_{t-s}(\sigma_s(A)\sigma_s(B))\} \\
&= \int_0^t ds\, \sigma_{t-s}\{H(\sigma_s(A)\sigma_s(B)) - H(\sigma_s(A))\sigma_s(B) \\
&\qquad - \sigma_s(A)H(\sigma_s(B))\} \\
&= 0
\end{aligned}
$$

for all $A, B \in D(H)$. Hence, by a density argument,

(2.3)
$$\sigma_t(AB) = \sigma_t(A)\sigma_t(B),$$

246

for all $A, B \in \mathcal{A}$ and all $t \geq 0$. Thus, the σ_t are *-morphisms of $\mathcal{A}$. In particular, σ is positive, because

$$(2.4) \qquad \sigma_t(A^*A) = \sigma_t(A)^*\sigma_t(A) \geq 0,$$

and contractive, because (2.4) implies $\|\sigma_t\|^2 \leq \|\sigma_t\|$. But the σ_t are not necessarily *-automorphisms, because the kernel of σ_t can be nontrivial or the range of σ_t can be a strict subspace of $\mathcal{A}$. Left and right translations on $C_0(\mathbf{R}_+)$ illustrate these phenomena. In fact, σ is a semigroup of *-automorphisms if, and only if, it extends to a continuous one-parameter group, i.e. the evolution is reversible.

Similar, but weaker, observations are valid if H is a dissipation which generates a positive continuous semigroup σ. Then

$$\sigma_t(A^*A) - \sigma_t(A)^*\sigma_t(A) = \int_0^t ds\, \sigma_{t-s}\{H(\sigma_s(A)^*\sigma_s(A)) - H(\sigma_s(A)^*)\sigma_s(A)$$
$$- \sigma_s(A)^*H(\sigma_s(A))\}$$
$$\geq 0.$$

Thus positivity of σ implies the strong positivity condition

$$\sigma_t(A^*A) \geq \sigma_t(A)^*\sigma_t(A)$$

for all $a \in \mathcal{A}$, which in turn implies that σ is contractive.

2.2. Derivations and Automorphisms Groups

Let $(\mathcal{A}, \mathbf{R}, \tau)$ denote an abelian C^*-dynamical system and H a (real) derivation from $\mathcal{A}_\infty$ into $\mathcal{A}$. Our aim is to argue that smoothness conditions on the range of H ensure that H is norm-closable and its closure $\overline{H}$ generates a continuous group of *-automorphisms σ of $\mathcal{A}$. To this end it is convenient to introduce the Lipschitz elements of $(\mathcal{A}, \mathbf{R}, \tau)$ by

$$\mathcal{A}_{\text{Lip}} = \left\{ A \in \mathcal{A};\ \sup_{|x|>0} \|A - \tau_x(A)\|/|x| < +\infty \right\}.$$

It follows easily that $\mathcal{A}_{\text{Lip}}$ is a *-subalgebra of $\mathcal{A}$ and $\mathcal{A}_{\text{Lip}} \supseteq \mathcal{A}_1$. This last containment property implies that $\mathcal{A}_{\text{Lip}}$ is norm-dense in $\mathcal{A}$. But $\|A - \tau_x(A)\| \leq 2\|A\|$ and since

$$A - \tau_x(A) = \sum_{m=0}^{n-1} \tau_{mx/n}((I - \tau_{x/n})(A))$$

one also has

$$\|A - \tau_x(A)\| \leq n\|A - \tau_{x/n}(A)\|.$$

Hence

$$\mathcal{A}_{\text{Lip}} = \{A \in \mathcal{A}; \limsup_{|x|\to 0}\|A - \tau_x(A)\|/|x| < +\infty\}.$$

Thus the Lipschitz condition is a condition on $\tau_x(A)$ at $x = 0$.

Now we can state the main result.

Theorem 2.2.1. *Let $(\mathcal{A}, \mathbf{R}, \tau)$ be an abelian C^*-dynamical system and H a real derivation from $\mathcal{A}_\infty$ into $\mathcal{A}$. If $H\mathcal{A}_\infty \subseteq \mathcal{A}_{\text{Lip}}$ then H is norm-closable and its closure $\overline{H}$ generates a strongly continuous one-parameter group of $*$-automorphisms of $\mathcal{A}$.*

There are two different proofs of this result but both proofs contain two distinct arguments. First one constructs a group of $*$-automorphisms whose generator extends H. Second, one proves that $\mathcal{A}_\infty$ is a core of the generator, and hence the generator is the norm-closure of H.

The outline of the first proof is as follows.

Since $H : \mathcal{A}_\infty \to \mathcal{A}$ is a (real) derivation, $H = \ell_1\delta$ where ℓ_1 is a function over X which is zero on X_0 and is continuous and polynomially bounded on $X\backslash X_0$, by the results described in Section 1.4. Thus

(2.5)
$$|\ell_1(\omega)| \leq c(1 + \nu(\omega)^n)$$

for some $c > 0$ and integer $n \geq 0$. Now, since $H\mathcal{A}_\infty \subseteq \mathcal{A}_{\text{Lip}}$, it follows from the identity

$$[H, \tau_x]f = (Hf - \tau_x Hf) - H(f - \tau_x f)$$

that

$$\|[H, \tau_x]f\| = 0(|x|)$$

as $x \to 0$ for all $f \in \mathcal{A}_\infty$. Then, by the uniform boundedness principle

$$\|[H, \tau_x]f\| \leq c'|x| \, \|f\|_{n'},$$

for some $c' > 0$ and $n' \geq 0$, and all $f \in \mathcal{A}_\infty$. Thus,

$$|(\ell_1(\omega) - \ell_1(T_x\omega))(\delta f)(\omega)| \leq c'|x| \, \|f\|_{n'}$$

248

and it follows by the same argument used to obtain the polynomial growth of ℓ_1 that

(2.6)
$$|\ell_1(\omega) - \ell_1(T_x\omega)| \le c_1|x|(1 + \nu(\omega)^{n'-1}),$$

for all $x \in \mathcal{R}$ and $\omega \in X$. Therefore, if $\omega \in X\backslash X_0$, the function $x \in \mathbf{R} \mapsto \ell_1(T_x\omega)$ is a bounded Lipschitz function.

Second, if the closure of $\ell_1\delta$ generates an automorphism group σ, then each orbit of the homeomorphism group S on X asociated with σ should be contained in an orbit of T. Thus, if S exists, there should also exist functions $(x, \omega) \in \mathbf{R} \times X \mapsto \omega(x) \in \mathbf{R}$ such that

$$S_x\omega = T_{\omega(x)}\omega.$$

Then, formally
$$\ell_1(S_x\omega)(\delta f)(S_x\omega) = (H\sigma_x f)(\omega)$$
$$= \frac{d}{dx}(\sigma_x f)(\omega)$$
$$= \frac{d}{dx}f(T_{\omega(x)}\omega)$$
$$= \frac{d\omega(x)}{dx}(\delta f)(S_x\omega).$$

Consequently, ω satisfies the first-order differential equation

(2.7)
$$\frac{d\omega(x)}{dx} = \ell_\omega(\omega(x)),$$

where ℓ_ω is defined by $\ell_\omega(x) = \ell_1(T_x\omega)$, and one has the initial condition $\omega(0) = 0$. Now the Lipschitz property of ℓ_1 ensures that ℓ_ω is a Lipschitz function over $\mathbf{R}$, with Lipschitz constant $C_1(\omega)$ which is polynomially bounded in the frequency. Hence, the Picard-Lindelöf theorem of ordinary differential equations ensures that equation (2.7) has a unique solution $\omega(x)$ satisfying $\omega(0) = 0$, and this for each $\omega \in X\backslash X_0$. Then, setting $\omega(x) = 0$ if $\omega \in X_0$, one defines σ by

$$(\sigma_x f)(\omega) = f(T_{\omega(x)}\omega).$$

It follows readily that the σ form a group of *-automorphisms but the problem is to prove that σ is strongly continuous, or, equivalently, that the map $(x, \omega) \mapsto T_{\omega(x)}\omega$ is jointly continuous. A positive resolution of this problem, which has some analogue with stability problems in ordinary differential equations, can be given if the Lipschitz constants for ℓ_ω are uniformly bounded on sets of bounded frequency and if, in addition the ℓ_ω satisfy a certain boundedness property at low frequency points near X_0. But these conditions are satisfied, because $HA_\infty \subseteq A_{\text{Lip}}$. Hence, σ is a continuous group.

Finally, one must prove that A_∞ is a core for the generator of σ. This is technically the most difficult part of this proof and we will not describe it in detail. But one device which is important is to introduce the norm dense *-subalgebra

$$A_0 = \{f \in A; \text{there exists an } M > 0 \text{ such that}$$
$$f(T_x\omega) = f(\omega) \text{ whenever } \nu(\omega) \geq M \text{ and } x \in \mathbf{R}\},$$

i.e., the subalgebra of functions which are constant on orbits of high frequency. Then one in fact establishes that $A_\infty \cap A_0$ is a core for the generator of σ.

The endpoint of this first proof is in fact the starting point for the second. First we systematize the definition of the functions which cut off high frequencies.

If $N \geq 0$ define $A^{(N)}$ by

$$A^{(N)} = \{f \in A; f(T_x\omega) = f(\omega) \text{whenever } \nu(\omega) \geq N \text{ and } x \in \mathbf{R}\}$$

it follows that

1. each $A^{(N)}$ is a C^*-subalgebra of A,
2. $A^{(N_1)} \subseteq A^{(N_2)}$ whenever $N_1 \leq N_2$,
3. $A = \overline{\cup_{N \geq 0} A^{(N)}}$ where the bar denotes norm closure,
4. $\tau_t A^{(N)} \subseteq A^{(N)}$ for all $N \geq 0$ and $t \in \mathbf{R}$.

Thus the C^*-dynamical system $(A, \mathbf{R}, \tau)$ is the norm closure of the increasing family of C^*-dynamical systems $(A^{(N)}, \mathbf{R}, \tau), N \geq 0$.

Next, if $A_\infty^{(N)} = A^{(N)} \cap A_\infty$, then $A_\infty^{(N)}$ is a norm-dense *-subalgebra of $A^{(N)}$ and $HA_\infty^{(N)} \subseteq A^{(N)}$. Thus the idea behind the construction of the group σ, with generator $\overline{H}$, is to construct appropriate groups $\sigma^{(N)}$ on $A^{(N)}$ with the property $\sigma^{(N_1)} \subseteq \sigma^{(N_2)}$ if $N_1 \geq N_2$, then set $\sigma = \sigma^{(N)}$ on $A^{(N)}$ and extend σ to A by continuity. Thus the construction of σ is reduced to the construction of the $\sigma^{(N)}$ and we now concentrate on this problem.

Now if $f \in A^{(N)}$, then $(\delta f)(\omega) = 0$ whenever $\nu(\omega) \geq N$. Hence the bounds (2.5) and (2.6) on ℓ_1 lead to the estimates

(2.8)
$$\begin{cases} \|Hf\| & \leq C(N)\|\delta f\| \\ \|[H, \tau_x]f\| & \leq C_1(N)\|\delta f\| \ |x| \end{cases}$$

250

for $H = \ell_1 \delta$ restricted to $\mathcal{A}^{(N)}$, where $C(N) = C(1 + N^n)$ and $C_1(N) = C_1(1 + N^{n'-1})$. But these estimates are sufficient to deduce that $\overline{H}$ generates a group of *-automorphisms of $\mathcal{A}^{(N)}$. This observation will be the main object of the next lecture, but in the remainder of this lecture we explain a somewhat weaker result.

Suppose $H : \mathcal{A}_\infty \mapsto \mathcal{A}_2$, then ℓ_1 must satisfy more stringent conditions than in the case $H : \mathcal{A}_\infty \mapsto \mathcal{A}_{\mathrm{Lip}}$. In fact, this new range condition implies that $t \mapsto \ell_1(T_t \omega)$ is twice-differentiable and the derivatives, which we denote by $\delta \ell_1$ and $\delta^2 \ell_1$, are polynomially bounded in the frequency. Thus the derivation H must satisfy

(2.9)
$$\begin{cases} \|Hf\| & \leq C(N)\|\delta f\|, \\ \|[\delta, H]f\| & \leq C_1(N)\|\delta f\|, \\ \|[\delta, [\delta, H]]f\| & \leq C_2(N)\|\delta f\|, \end{cases}$$

for all $f \in \mathcal{A}^{(N)}$. Now we argue that these conditions imply that $\overline{H}$ generates a group of *-automorphisms of $\mathcal{A}^{(N)}$. Replacement of (2.9) by the weaker conditions (2.8) does not alter the main line of the argument but does require some additional regularization and approximation techniques.

The proof is a rather peculiar form of perturbation theory which occurs in various guises in analysis but does not appear to have been described systematically.

First, note that since δ generates a continuous group τ of *-automorphisms, $-\delta^2$ generates a continuous semigroup ρ of contractions, the heat semigroup associated with τ. The semigroup ρ can be constructed from τ with the aid of Gauss-Weierstrass integral

$$\rho_t = \frac{1}{(4\pi t)^{1/2}} \int_{-\infty}^{\infty} ds\, e^{-s^2/4t} \tau_s.$$

Second remark that δ is infinitesimally small in comparison with δ^2. Specifically,

(2.10)
$$\|\delta f\| \leq \epsilon \|\delta^2 f\| + \epsilon^{-1} \|f\|$$

for each $\epsilon > 0$ and all $f \in D(\delta^2)$. This follows from the Taylor series expansion

$$\tau_t f = f - t\delta f + \int_0^t ds(t - s)\tau_s \delta^2 f$$

which yields the bound

$$t\|\delta f\| \leq 2\|f\| + (t^2/2)\|\delta^2 f\|$$

for all $t > 0$.

Now it follows from the first condition in (2.9) that

$$\|Hf\| \leq \epsilon\|\delta^2 f\| + c(N)^2 \epsilon^{-1}\|f\|$$

for all $f \in D(\delta^2)$. Since $\pm H$ are dissipative, it then follows by perturbation theory for contraction semigroups that $\pm H - \epsilon\delta^2$ generate continuous contraction semigroups α^ϵ and β^ϵ.

Next we argue that α_t^ϵ, and $\beta_t^\epsilon, t \geq 0$ converge as $\epsilon \to 0$, and the limits α_t, and β_t form continuous semigroups α and β with generators $\pm H$. This is the peculiar part of the argument. The generator property is deduced by first adding the dominant term $-\epsilon\delta^2$ and then removing it. But once this step has been made, one can set $\sigma_t = \alpha_t, \sigma_{-t} = \beta_t$, for $t \geq 0$, and the σ_t form a continuous group with generator $\overline{H}$, which is automatically a group of *-automorphisms by the discussion in the Introduction.

Now the proofs of convergence of the α_t^ϵ, and the β_t^ϵ, are identical so we only consider the former. But since the α_t^ϵ form a bounded semigroup, it is only necessary to prove convergence for small t on a dense subset. This proof is based on the identity

$$\alpha_t^{\epsilon_1} f - \alpha_t^{\epsilon_2} f = \int_0^t ds \frac{d}{ds} \alpha_{t-s}^{\epsilon_2} \alpha_s^{\epsilon_1} f$$

$$= (\epsilon_1 - \epsilon_2) \int_0^t ds \, \alpha_{t-s}^{\epsilon_2} \delta^2 \sigma_s^{\epsilon_1} f,$$

where $f \in D(\delta^2)$. Hence,

$$\|\alpha_t^{\epsilon_1} f - \alpha_t^{\epsilon_1} f\| \leq |\epsilon_1 - \epsilon_2| \sup_{0 \leq s \leq t} \|\delta^2 \alpha_s^{\epsilon_1} f\|.$$

Now, if the supremum on the right hand side is uniformly bounded for all small $\epsilon_1 > 0$ and all small $t > 0$, for each f in the norm dense subset $D(\delta^2)$, then one concludes that the α_t^ϵ converge strongly as $\epsilon \to 0$. Thus convergence is reduced to a boundedness property. But

$$\delta^2 \alpha_s^\epsilon f - \alpha_s^\epsilon \delta^2 f = \int_0^s dr \frac{d}{dr} \alpha_{s-r}^\epsilon \delta^2 \alpha_r^\epsilon f$$

$$= \int_0^s dr \, \alpha_{s-r}^\epsilon [H, \delta^2] \alpha_r^\epsilon f.$$

Therefore, using the identity

$$[H, \delta^2] = -[\delta, [\delta, H]] - 2[\delta, H]\delta$$

and the second and third estimates in (2.9) one finds

$$\|\delta^2 \alpha_s^\epsilon f\| \le \|\delta^2 f\| + 2s\, C_1(N) \sup_{0 \le r \le s} \|\delta \alpha_r^\epsilon f\|$$
$$+ sC_2(N) \sup_{0 \le r \le s} \|\delta^2 \alpha_r^\epsilon f\|.$$

Finally, bounding $\|\delta \alpha_r^\epsilon f\|$ by the perturbation estimate (2.10) one obtains a bound of the form

$$\sup_{0 \le s \le t} \|\delta^2 \alpha_s^\epsilon f\| \le 2\|\delta^2 f\| + (a/b)\|f\|.$$

This establishes that the α_t^ϵ converge as $\epsilon \to 0$. But there is still potentially a problem because $t \mapsto \alpha_t$ could fail to be continuous. The foregoing estimates establish continuity, however, as a byproduct since

$$\|\alpha_t f - f\| \le \|\alpha_t^\epsilon f - f\| + \|\alpha_t f - \alpha_t^\epsilon f\|$$
$$\le \|\alpha_t^\epsilon f - f\| + \epsilon \sup_{0 \le s \le t} \|\delta^2 \alpha_s^\epsilon f\|$$

for $f \in D(\delta^2)$. To complete the argument one then needs to argue that the generator of α is in fact $\overline{H}$. But this is straightforward and we will not elaborate.

2.3. Local Dissipations

Let $H : \mathcal{A}_\infty \mapsto \mathcal{A}$ be a local dissipation. Then it follows from the discussion in Section 1.3 that

$$H = (\ell_0 + \ell_1 \delta + \ell_2 \delta^2)|_{\mathcal{A}_\infty}$$

where ℓ_0 is a bounded continuous positive function over X and ℓ_1, ℓ_2, are polynomially bounded continuous functions over $X \backslash X_0$, which vanish on X_0, and $-\ell_2 \ge 0$. In light of Theorem 2.2.1, it is natural to ask whether there is a range condition which ensures that $\overline{H}$ is the generator of a contraction semigroup. But the situation is more complicatd. The term ℓ_0 causes no difficulty since it is bounded. Moreover, each of the terms $\ell_1 \delta, \ell_2 \delta^2$, can be handled separately. Difficulties arise, however, with the sum of $\ell_1 \delta$ and $\ell_2 \delta^2$ and these cannot be controlled by a range condition.

The difficulty can be understood as follows.

Assume for simplicity that $H\mathcal{A}_\infty \subseteq \mathcal{A}_2$ and then $t \mapsto \ell_i(T_t \omega), i = 1, 2$, are both twice differentiable and all derivatives are polynomially bounded in the frequency. Now

$$0 \le -\ell_2(T_t \omega) = -\ell_2(\omega) + t(\delta \ell_2)(\omega) - \int_0^t ds(t-s)(\delta^2 \ell_2)(T_s \omega)$$

and consequently,

$$0 \le -\ell_2(\omega) + t(\delta\ell_2)(\omega) + (t^2/2)P(\nu(\omega))$$

where P is the polynomial bound for $|(\delta^2\ell_2)(\omega)|$. Since this inequality is valid for all $t \in \mathbf{R}$, one must have

(2.11) $$|(\delta\ell_2)(\omega)|^2 \le 2\ell_2(\omega)P(\nu(\omega)).$$

But this implies $t \mapsto \sqrt{-\ell_2(T_t\omega)}$ is differentiable. Hence the closures of both $\sqrt{-\ell_2}\delta$ and $\ell_1\delta$ are generators of continuous groups of *-automorphisms, by Theorem 2.2.1. Then, since

$$\ell_2\delta^2 = -(\sqrt{-\ell_2}\delta)^2 + (\delta\ell_2/2)\delta$$

it follows from (2.10) and (2.11) that the second term is a small perturbation of the first. Therefore, the closure of $\ell_2\delta^2$ generates a continuous semigroup of contractions. Thus $\ell_2\delta^2$ and $\ell_1\delta$ are both generators and the remaining question concerns their sum.

It follows from perturbation theory, applied to $\ell_1\delta$ and $\ell_2\delta^2$ in restriction to each of the subalgebras $\mathcal{A}^{(N)}$, that the closure of $\ell_2\delta^2 + \ell_1\delta$ is a generator if $|\ell_1|^2 \le -Q\ell_2$ where Q is a strictly positive function of the frequency. But difficulties arise if this condition fails, e.g. if there exists an $\omega \in X \backslash X_0$ such that $\ell_2(\omega) = 0$ but $\ell_1(\omega) \ne 0$. These problems do not arise if $-\ell_2 \ge \epsilon$ for some $\epsilon > 0$, because ℓ_1^2 is polynomially bounded in the frequency. But the condition $-\ell_2 \ge \epsilon$ is assured if $H + \epsilon\delta^2$ is a dissipation. This latter condition is a condition of stability which corresponds in some way to strict ellipticity.

By this form of reasoning, one obtains the following result.

Theorem 2.3.1. *Let $H : \mathcal{A}_\infty \mapsto \mathcal{A}_{1,\mathrm{Lip}}$, where*

$$\mathcal{A}_{1,\mathrm{Lip}} = \{f \in \mathcal{A}_1; \delta f \in \mathcal{A}_{\mathrm{Lip}}\},$$

be a real operator. If $H + \epsilon\delta^2$ is a dissipation for some $\epsilon > 0$, then the closure $\overline{H}$ of H generates a continuous semigroup of positive contractions.

Finally, note that if ℓ_2 is a constant function over X, then the above disscusion establishes that the closure $\overline{H}$ of the local dissipation H is a generator even under the weaker range condition $H\mathcal{A}_\infty \subseteq \mathcal{A}_{\mathrm{Lip}}$. If $\ell_2 = 0$, then this follows from Theorem 1.2.1, but if $-\ell_2 > 0$, then it is a consequence of the perturbation argument.

Notes

The first version of the proof of Theorem 2.1 was given in [3] but the full result was only obtained for derivations from $\mathcal{A}_\infty$ into $\mathcal{A}_1$. The extension to $\mathcal{A}_{\mathrm{Lip}}$ occurs in [11].

The second version of the proof was developed in [12].

3. Commutators and Generators

3.1 Introduction

In the previous lecture we described two approaches to the integration problem for derivations associated with an abelian C^*-dynamical system. The second approach consisted of restricting the problem to subsystems of bounded frequency and then applying general techniques based upon commutator estimates which reflect smoothness properties. In this last lecture we elaborate on these techniques for a general Banach dynamical system, or, more specifically, for a Hilbert dynamical system.

First let $\mathcal{B}$ denote a Banach space and $t \in \mathbf{R} \mapsto \tau_t \in \mathcal{L}(\mathcal{B})$ a strongly continuous one-parameter group of isometries of $\mathcal{B}$ with infinitesimal generator δ. Set $\mathcal{B}_p = D(\delta^p)$ and

$$\mathcal{B}_\infty = \cap_{p \geq 1} \mathcal{B}_p.$$

Then $\mathcal{B}_p$ is a Banach space with respect to the norm

$$\|a\|_p = \sup_{0 \leq q \leq p} \|\delta^q a\|$$

and $\mathcal{B}_\infty$ is a Fréchet space when equipped with the family of norms $\{\| \cdot \|_p ; p \geq 0\}$. It is also of interest to introduce the subspace $\mathcal{B}_{\mathrm{Lip}}$ of Lipschitz functions.

$$\mathcal{B}_{\mathrm{Lip}} = \left\{ a \in \mathcal{B}; \sup_{|t|>0} \|a - \tau_t a\|/|t| < +\infty \right\},$$

which is a Banach space with respect to the norm

$$\|a\|_{\mathrm{Lip}} = \|a\| \vee \sup_{|t|>0} \|a - \tau_t a\|/|t|.$$

Now we consider the analogue of the integration problem in this Banach space setting. The object of interest is an operator $H; \mathcal{B}_\infty \mapsto \mathcal{B}$ defined on the smooth elements $\mathcal{B}_\infty$ and the aim is to prove that dissipativity properties, combined with smoothness

255

conditions, and possibly supplemented by stability conditions, ensure that $\overline{H}$ is the generator of a continuous group, or semigroup.

The smoothness conditions can be expressed either as a range condition on H or by approximate commutation of H with τ. After examining the relation between these two types of expressions, we discuss generator theorems based upon dissipativity and approximate commutation properties. Finally, we give an example which shows that range conditions are not generally sufficient to ensure generator properties even for dissipative operators on Hilbert space.

3.2 Smoothness Properties

Theorem 2.1 of Lecture 2 established that the range condition $H\mathcal{B}_\infty \subseteq \mathcal{B}_{\mathrm{Lip}}$ is sufficient to ensure that the closure of a derivation associated with an abelian C^*-dynamical system generates a group of *-automorphisms. But this range condition is equivalent to an approximate commutation property.

Proposition 3.2.1. *Let* $H; \mathcal{B}_\infty \mapsto \mathcal{B}$ *be a dissipative operator. Then the following conditions are equivalent:*

1. *$H\mathcal{B}_\infty \subseteq \mathcal{B}_{\mathrm{Lip}}$,*
2. *$\|[\tau_t, H]a\| \leq c|t|\,\|a\|_q, a \in \mathcal{B}_\infty, t \in \mathbf{R}$, for some $c > 0$ and some integer $q \geq 0$.*

Proof: Since H is dissipative, it is norm closable, and hence closable as an operator from the Fréchet space $\mathcal{B}_\infty$ into the Banach space $\mathcal{B}$. Therefore,

$$(3.1) \qquad \qquad \|Ha\| \leq c_0\|a\|_p, \quad a \in \mathcal{B}_\infty,$$

for some $c > 0$ and some integer $p \geq 0$. But then

$$\|\|(I - \tau_t)Ha\| - \|[\tau_t, H]\|\| \leq \|H(I - \tau_t)a\|$$
$$\leq c_0\|(I - \tau_t)a\|_p$$
$$\leq c_0|t|\,\|a\|_{p+1},$$

and the equivalence of Condition 1 and 2 follows immediately.

The first point of interest in Proposition 3.2.1 is that the smoothing property of H, the property $H\mathcal{B}_\infty \subseteq \mathcal{B}_{\mathrm{Lip}}$, is equated to approximate τ-invariance of H. The operator H is *τ-invariant* if the order of action of H and τ is reversible, i.e. if $[\tau_t, H] = 0$ for all $t \in \mathbf{R}$. Thus the commutator characterization of smoothing given by Proposition 3.2.1 is an approximate τ-invariance. The second point of interest is that the invariance condition introduces an extra gradation of smoothness. The integer q occurring in the estimate

$$\|[\tau_t, H]a\| \leq c|t| \, \|a\|_q, \quad a \in \mathcal{B}_\infty,$$

is not fixed by the condition $H\mathcal{B}_\infty \subseteq \mathcal{B}_{\text{Lip}}$. All possible values of q can occur, but in some sense *the smaller the value of q, the smoother the action of H*. This rule of thumb is well-illustrated by the example of ordinary differential operators.

Example 3.2.2. Let $\mathcal{B} = C_0(\mathbf{R})$ and τ the action of $\mathbf{R}$ by translations. Thus, $\delta = d/dx$ and $\mathcal{B}_n = C_0^n(\mathbf{R})$. Moreover, $\mathcal{B}_{\text{Lip}}$ consists of the usual Lipschitz functions in $C_0(\mathbf{R})$. Now consider the ordinary differential operator

$$H = \sum_{m=0}^{n} \ell_m d^m / d^m x$$

where the coefficients ℓ_m are assumed to be bounded and continuous. Then $H\mathcal{B}_\infty \subseteq \mathcal{B}$ but $H\mathcal{B}_\infty \subseteq \mathcal{B}_{\text{Lip}}$ if the ℓ_m are all Lipschitz functions. But one readily verifies that H satisfies the commutator condition (3.2) if, and only if, the ℓ_m are Lipschitz functions and in addition ℓ_m is constant for $q < m \leq n$.

3.3. Semigroup Generators

Next we argue that approximate τ-invariance of a dissipative operator H associated with the Banach dynamical system $(\mathcal{B}, \mathbf{R}, \tau)$ is sufficient to ensure that $\overline{H}$ generates a contraction semigroup. The invariance property required is the condition (3.2), which reflects the smoothing action of H, but with the minimal value $q = 1$. Larger values of q are not generally sufficient to ensure the generator property.

Theorem 3.3.1. *Let $H; \mathcal{B}_\infty \mapsto \mathcal{B}$ be a dissipative operator satisfying*

(3.3) $$\|[\tau_t, H]a\| \leq c|t| \, \|a\|_1,$$

for all $a \in \mathcal{B}_\infty$, all $|t| \leq 1$, and some $c \geq 0$. Then the closure $\overline{H}$ of H generates a strongly continuous semigroup of contractions S. Moreover, if $H\mathcal{B}_\infty \subseteq \mathcal{B}_1$, then $S\mathcal{B}_1 \subseteq \mathcal{B}_1$ and $S|_{\mathcal{B}_1}$ is a $\|\cdot\|_1$-continuous semigroup satisfying

(3.4) $$\|S_t a\|_1 \leq e^{ct}\|a\|_1, \quad a \in \mathcal{B}_1, t \geq 0.$$

The proof of the first statement in this result is by singular perturbation theory and it is a long and complex argument which we cannot give in detail, but which we will outline.

257

First, the importance of the assumption $D(H) = \mathcal{B}_\infty$ is that it conceals the boundedness property (3.1). Explicitly

(3.1)
$$\|Ha\| \le c_0 \|a\|_p, \quad a \in \mathcal{B}_\infty$$

for some $c_0 > 0$ and some interger $p \ge 0$. Then since $\mathcal{B}_\infty$ is $\|\cdot\|_p$-dense in $\mathcal{B}_p$ the operator H has an extension to all of $\mathcal{B}_p$ satisfying the same bound. The assumption $D(H) = \mathcal{B}_\infty$ in the theorem could in fact be replaced by any other condition which ensures that (3.1) is valid for some $p \ge 0$ and all $a \in \mathcal{B}_p$.

Second, it follows from the Taylor series argument used in the discussion of the second proof of Theorem 2.2.1 that for each $\epsilon > 0$ there exists a k_ϵ such that

$$\|\delta a\| \le \epsilon \|\delta^2 a\| + k_\epsilon \|a\|, \quad a \in \mathcal{B}_2.$$

This has various applications. For example it can be used to deduce that the norm $\|\cdot\|_p$ is equivalent to the norm $a \in \mathcal{B}_p \mapsto \|\delta^p a\| + \|a\|$. Thus the bound (3.1) is equivalent to a bound

$$\|Ha\| \le c_0(\|\delta^p a\| + \|a\|), \quad a \in \mathcal{B}_p.$$

Then it can be further exploited to conclude that if $2n > p$ and $\epsilon > 0$ then there is a c_ϵ such that

(3.5)
$$\|Ha\| \le \epsilon \|\delta^{2n} a\| + c_\epsilon \|a\|, \quad a \in \mathcal{B}_{2n}.$$

This is the starting point of the perturbation argument.

The bound (3.5) states that H is infinitesimally small with respect to δ^{2n}. But since δ generates the group of isometries τ, it follows that $(-\delta^2)^n$ generates a holomorphic semigroup $S^{(n)}$ where

$$S_t^{(n)} = \int ds \mu_t^{(n)}(s) \tau_s$$

and the signed measures $\mu_t^{(n)}$ are given by

$$\mu_t^{(n)}(s) = \frac{1}{2\pi} \int dq\, e^{-tq^{2n}} e^{iqs}.$$

Therefore, by a standard result of conventional perturbation theory for holomorphic semigroups, the operators $H_\beta; \mathcal{B}_{2n} \mapsto \mathcal{B}$ defined for $\beta > 0$ by

$$H_\beta = \beta(-\delta^2)^n + H$$

are generators of holomorphic semigroups S^β.

Now the idea of singular perturbation theory is to establish the following three points:

1. the S_t^β converge strongly as $\beta \to 0$,
2. the strong limits S_t form a continuous semigroup S,
3. the generator of S is $\overline{H}$.

The proof of the first property is in two steps.

First, one must show that $\|S_t^\beta\|$ is uniformly bounded for all small positive β and t. Then, by the semigroup property, it suffices to prove the strong convergence for small t on a norm dense subspace of $\mathcal{B}$, e.g. on $\mathcal{B}_\infty$.

Sceond, one uses the identity

$$(S_t^{\beta_1} - S_t^{\beta_2})a = \int_0^t ds \frac{d}{ds}(S_{t-s}^{\beta_2} S_s^{\beta_1})a$$

$$= (\beta_2 - \beta_1) \int_0^t ds S_{t-s}^{\beta_2} (-\delta^2)^n S_s^{\beta_1} a$$

to deduce that

(3.6) $$\|(S_t^{\beta_1} - S_t^{\beta_2})a\| \leq |\beta_1 - \beta_2| t \sup_{0<r,s<t} \|S_r^{\beta_2}\| \cdot \|\delta^{2n} S_s^{\beta_1} a\|.$$

Thus convergence follows if $\|\delta^{2n} S_t^\beta a\|$ is uniformly bounded for all small β and t, for each $a \in \mathcal{B}_\infty$.

The scheme for establishing bounds on such quantities as $\|\delta^{2n} S_t^\beta a\|$ was already outlined in Section 2.2 in the special case $n = 1$. The general case is very similar once one has established the uniform bounds on $\|S_t^\beta\|$. Hence we next describe a method for obtaining these latter bounds.

First, the semigroup $t \mapsto S_{\beta t}^{(n)}$ generated by $\beta(-\delta^2)^n$ has the bound

$$\|S_{\beta t}^{(n)}\| \leq M_n = \frac{1}{2\pi} \int ds \left| \int dq \, e^{-\beta t q^{2n}} e^{iqs} \right|,$$

which can be seen to be independent of βt by a change of variables. Second, for $\gamma \geq \log M_n$ define the norm $\| \cdot \|_\beta$ by

(3.7) $$\|a\|_\beta = \sup_{t>0} \|S_{\beta t}^{(n)} a\| e^{-\gamma t}.$$

259

(The condition $\gamma \geq \log M_n$ is not critical but it does ensure that the supremum is attained for $t \in [0,1]$.) The choice of norm $\| \cdot \|_\beta$ is convenient for two reasons; the norms $\| \cdot \|$ and $\| \cdot \|_\beta$ are equivalent, and in particular

$$\|a\| \leq \|a\|_\beta \leq M_n\|a\|,$$

and the semigroup $t \mapsto S^{(n)}_{\beta t}$ satisfies the bounds

$$\|S^{(n)}_{\beta t} a\|_\beta \leq e^{\gamma t}\|a\|_\beta.$$

Therefore, $\beta(-\delta^2)^n + \gamma I$ is $\| \cdot \|_\beta$-dissipative, i.e. dissipative with respect to the new norm $\| \cdot \|_\beta$. Hence, if one can find $\omega_\beta \geq 0$ such that $H + \omega_\beta I$ is $\| \cdot \|_\beta$-dissipative, then $\beta(-\delta^2)^n + H + (\omega_\beta + \gamma)I$ will be $\| \cdot \|_\beta$-dissipative and the semigroup S^β generated $\beta(-\delta^2)^n + H$ will automatically satisfy the bounds

$$\|S^\beta_t a\|_\beta \leq e^{(\omega_\beta + \gamma)t}\|a\|_\beta.$$

Consequently, reverting to the original norm one finds

(3.8) $$\|S^\beta_t\| \leq M_n e^{(\omega_\beta + \gamma)t}.$$

Now to find ω_β, one can proceed as follows.

The condition of $\| \cdot \|_\beta$-dissipativity for $H + \omega I$ is equivalent to the statement that

$$\|(I + \alpha(H + \omega I))a\|_\beta \geq \|a\|_\beta$$

for all $a \in \mathcal{B}_\infty$ and $\alpha > 0$. But

$$
\begin{aligned}
\|(I + \alpha(H + \omega I))a\|_\beta &= \sup_{0<t<1} \|S^{(n)}_{\beta t}(I + \alpha(H + \omega I))a\|e^{-\gamma t} \\
&\geq \sup_{0<t<1} \|(I + \alpha(H + \omega I))S^{(n)}_{\beta t} a\|e^{-\gamma t} \\
&\quad - \alpha \sup_{0<t<1} \|[H, S^{(n)}_{\beta t}]a\|e^{-\gamma t} \\
&\geq (1 + \alpha\omega)\|a\|_\beta - \alpha \sup_{0<t<1} \|[H, S^{(n)}_{\beta t}]a\|
\end{aligned}
$$

where we have used the $\| \cdot \|$-dissipativity of H to obtain the last inequality and it is important that $S^{(n)}_{\beta t}\mathcal{B}_\infty \subseteq \mathcal{B}_{2n} \subseteq D(H)$. Hence, setting

(3.9) $$\omega_\beta = \sup\{\|[H, S^{(n)}_{\beta t}]a\|; a \in \mathcal{B}_\infty, \|a\| \leq 1, 0 < t < 1\},$$

260

one concludes that $H + \omega I$ is $\|\cdot\|_\beta$-dissipative whenever $\omega \geq \omega_\beta$. Moreover, $\beta \mapsto \omega_\beta$ is an increasing function. Hence, (3.8) gives the required uniform bound once one establishes that $\omega_1 < +\infty$. It is for this purpose that the approximate commutation property (3.3) is of paramount importance. This will be illustrated in the subsequent discussion of Hilbert space systems.

Unfortunately, things are not quite so easy as we have outlined them, and it is quite possible that the uniform boundedness of $\|S_t^\beta\|$ and the strong convergence of the S_t^β as $\beta \to 0$ are both false under the hypotheses of the theorem. Nevertheless, one can establish these properties for a slightly modified version of H. One replaces H by a regularization

$$H_f = \int dt \ f(t) \tau_t \ H \tau_{-t}$$

and then applies the singular perturbation theory arguments to H_f. Finally, one removes the regularization by an additional limiting argument. We will not elaborate further on the intricacies of the proof, but note that the second statement of the theorem, which we have not discussed, gives an interpretation of the measure (3.3) of invariance of H in terms of a growth property (3.4) of S.

3.3 Hilbert Space Theory

There is a Hilbert space version of Theorem 3.3.1 which can be proved directly without the use of singular perturbation theory. But it is of interest that the same basic estimates nevertheless enter. The Hilbert space theorem is partially weaker because it assumes that $\pm H$ are both dissipative and aims to establish that $\overline{H}$ generates a group of isometries, but it is partially stronger because the key commutator condition is expressed in terms of quadratic forms.

Let $\mathcal{H}$ denote a Hilbert space and $t \in \mathbf{R} \mapsto U_t \in \mathcal{L}(\mathcal{H})$ a strongly continuous one-parameter group of unitary operators on $\mathcal{H}$. Since each continuous one-parameter group of isometries on a Hilbert space is automatically unitary, the Hilbert dynamical system $(\mathcal{H}, \mathbf{R}, U)$ is the direct analogue of the Banach dynamical system $(\mathcal{B}, \mathbf{R}, \tau)$ considered previously.

The generator δ of U has the form $\delta = iK$ where K is self-adjoint and one can introduce the subspaces $\mathcal{H}_p = D(K^p)$ and $\mathcal{H}_\infty = \cap_{p \geq 1} \mathcal{H}_p$ as before. But it is also convenient to introduce fractional spaces. If $|K|$ denotes the positive self-adjoint modulus of K, then one defines $\mathcal{H}_{p/2} = D(|K|^{p/2})$ for all $p = 1, 2, \ldots$ and introduces the norms

$$\|a\|_{p/2} = \sup_{0 \leq q \leq p} \||K|^{q/2} a\|, \quad a \in \mathcal{H}_{p/2}.$$

Since $D(K^n) = D(|K|^n)$ for positive, integer, n, the new definitions agree with the old ones for integer powers.

Next consider an operator H on $\mathcal{H}$ such that $\pm H$ are both dissipative. It follows readily from the definition of dissipativity that this is equivalent to skew-symmetry of H. Thus the statement that $\overline{H}$ generates a group of isometries can be rephrased as a statement that H is essentially skew-self-adjoint. Hence, if one multiplies H by i, the statement concerns essential self-adjointness of a symmetric operator. We will express the Hilbert space theorem in these terms.

Let $H; \mathcal{H}_\infty \mapsto \mathcal{H}$ and define the quadratic form h associated with H by

$$h(a) = (a, Ha), \quad a \in \mathcal{H}_\infty.$$

The smoothness conditions on the range of H considered in Section 2 can then be generalized by consideration of smoothness properties of the functions $t \in \mathbf{R} \mapsto h(\tau_t a)$. But such properties can again be expressed by approximate commutation relations. For example, it is straightforward to prove the following analogue of Proposition 3.2.1.

Proposition 3.2.1. *Let $H; \mathcal{H}_\infty \mapsto \mathcal{H}$ be a norm-closable operator. Then the following conditions are equivalent;*

1. $\sup_{|t|>0} |h(\tau_t a) - h(a)|/|t| < +\infty$, $a \in \mathcal{H}_\infty$,
2. $|(a, [\tau_t, H]a)| \leq c|t| \, \|a\|_{q/2}^2$, $|t| < 1, a \in \mathcal{H}_\infty$,

for some $c > 0$ and some integer $q \geq 0$,

3. $|(Ka, Ha) - (Ha, Ka)| \leq c\|a\|_{q/2}^2$, $a \in \mathcal{H}_\infty$,

for some $c > 0$ and some integer $q \geq 0$.

Again, it is the commutation conditions with $q = 1$ which are the most useful for the discussion of generator properties, i.e. self-adjointness properties. Note that the norm $a \mapsto \|a\|_{1/2}$ is in fact equivalent to the norm $a \mapsto (a, (I + |K|)a)^{1/2}$ and hence the bound $\|a\|_{1/2}^2$ corresponds to a bound by the quadratic form associated with $I + |K|$.

After these preliminary explanations the Hilbert space theorem can be expressed as follows.

Theorem 3.4.2. *Let K be a self-adjoint operator on the Hilbert space $\mathcal{H}$ and set $\mathcal{H}_\infty = \cap_{p \geq 1} D(K^p)$. Further, let H be a symmetric operator on $\mathcal{H}$ with $D(H) = \mathcal{H}_\infty$. If*

(4.1) $$|(Ha, Ka) - (Ka, Ha)| \leq c(a, (I + |K|)a)$$

for some $c \geq 0$ and all $a \in \mathcal{H}_\infty$, then H is essentially self-adjoint.

Proof: It suffices to establish that the range of $I + i\alpha H$ is norm-dense in $\mathcal{H}$ for all small real α. Now assume

$$(b, (I + i\alpha H)a) = 0$$

for all $a \in \mathcal{H}_\infty$. Then, setting $S_t = \exp\{-tK^2\}$ and noting that $S_t \mathcal{H} \subseteq \mathcal{H}_\infty$, for $t > 0$, by spectral theory, one has

(4.2) $$Re(b, (I + i\alpha H)S_t b) = 0.$$

Now set

$$\omega_1 = \sup\{|(a, HS_t a) - (HS_t a, a)|; \|a\| \leq 1, 0 < t < 1\}.$$

Then, (4.2) gives

$$\|b\|^2 = \lim_{t \to 0}(b, S_t b) \leq |\alpha|\omega_1\|b\|^2/2.$$

Thus, if $\omega_1 < +\infty$, one must have $b = 0$ for $|\alpha| < 1/2\omega_1$, i.e. H is essentially self-adjoint.

Therefore, the proof is reduced to showing that $\omega_1 < \infty$. Note that this is the quadratic form analogue of the estimate needed in the Banach space discussion, see (3.9). Now we deduce that $\omega_1 < +\infty$ by proving that the symmetric operators $C_t; \mathcal{H}_\infty \mapsto \mathcal{H}$ defined by

$$C_t = i[H, S_t], \quad t > 0,$$

have bounded closure $\overline{C}_t$ and, moreover, $\|\overline{C}_t\|$ is uniformly bounded on $[0, 1]$. It then follows that

$$\omega_1 = \sup_{0 < t < 1} \|\overline{C}_t\| < +\infty$$

and the proof is complete.

In order to deduce the boundedness of $\overline{C}_t$, take $a, b \in \mathcal{H}_\infty$ and note that

$$\begin{aligned}
(a, C_t b) &= \int_0^t ds \frac{d}{ds}(a_s, Hb_s) \\
&= \int_0^t ds\{(K^2 a_s, Hb_s) - (Ha_s, K^2 b_s)\} \\
&= \int_0^t ds\{((KKa_s, Hb_s) - (HKa_s, Kb_s)) \\
&\quad + ((Ka_s, HKb_s) - (Ha_s, KKb_s))\}
\end{aligned}$$

263

where $a_s = S_{t-s}a$ and $b_s = S_s b$. Now, it follows from (4.1), by polarization, that

$$|(KKa_s, Hb_s) - (HKa_s, Kb_s)| \leq 4c\|(I + |K|)^{1/2}Ka_s\| \, \|(I + |K|)^{1/2}b_s\|$$

$$\leq 4cc_1c_2(t-s)^{-3/4}s^{-1/4}\|a\| \, \|b\|$$

for $0 < s < t \leq 1$ where

$$c_1 = \sup_{\lambda > 0}(1 + \lambda)^{1/2}\lambda e^{-\lambda^2}, \quad c_2 = \sup_{\lambda > 0}(1 + \lambda)^{1/2}e^{-\lambda^2}.$$

The second pair of terms in the integrand have a similar bound, but with s and $t - s$ interchanged. Thus

$$|(a, C_t b)| \leq 8cc_1c_2\|a\| \, \|b\| \int_0^t ds(t-s)^{-3/4}s^{-1/4}$$

$$= 8cc_1c_2\|a\| \, \|b\| \int_0^1 ds(1-s)^{-3/4}s^{-1/4}$$

where the last equality follows by a change of variables. Therefore, one concludes that

$$\omega_1 \leq 8cc_1c_2 \int_0^1 ds(1-s)^{-3/4}s^{-1/4}.$$

Theorem 3.4.2 is of interest because it has had a variety of applications in quantum field theory and the theory of Schrödinger operators. Details of these applications can be found in the references, but we cite one simple example as an illustration.

Example 3.4.3. Let $\mathcal{H} = L^2(\mathbf{R}^\nu)$. Set $p_j = i\partial/\partial x_j, q_j$ multiplication by x_j, and $K = p^2 + q^2$. Let V be a real continuous function on $\mathbf{R}^\nu$ such that $|\nabla V(x)| \leq c_1|x| + c_2$ for some $c_1, c_2 \geq 0$. It then follows that $|V(x)| \leq a_1|x|^2 + a_2$ for some $a_1, a_2 \geq 0$. Now it is well-known that K is self-adjoint on $D(p^2) \cap D(q^2)$ and hence one has an estimate

$$\|p^2 f\| + \|q^2 f\| \leq k(\|(p^2 + q^2)f\| + \|f\|)$$

for all $f \in D(K)$. Therefore, $H = p^2 + V(q)$ can be defined on all of $D(K)$. In particular H is a symmetric operator from $\mathcal{H}_\infty$ into $\mathcal{H}$. But verification of the commutator condition (4.1) is a routine calculation and hence Theorem 3.4.2 establishes that H is essentially self-adjoint on $\mathcal{H}_\infty$.

Finally, we give an example of a symmetric operator which maps $\mathcal{H}_\infty$ into $\mathcal{H}_\infty$, but which is not essentially self-adjoint.

Example 3.4.4. Let $\mathcal{H} = L^2(\mathbf{R}), p = i\partial/\partial x, q$ multiplication by x, and $K = p^2 + q^2$. Then $\mathcal{H}_\infty$ can be identified as the Schwartz space $\mathcal{S}(\mathbf{R})$. Now, qpq is a symmetric operator which maps $\mathcal{H}_\infty$ into $\mathcal{H}_\infty$, but it is not essentially self-adjoint. It has deficiency indices (1,1).

Notes

The first commutator-generator theorem was proved by Glimm and Jaffe [6]. The Glimm-Jaffe Theorem was a Hilbert space result which they applied to self-adjointness problems in quantum field theory. Subsequent Hilbert space generalizations were given by Nelson, McBryan, and Faris and Lavine [8], [7], [5]. Nelson and McBryan gave further applications to quantum field theory. Faris and Lavine, on the other hand, use these methods to study Schrödinger operators. A survey of these developments can be found in Section X.5 of the book by Reed and Simon [10].

All the generator theorems contained in these references differ from Theorem 3.4.2 insofar that they assume that K is positive, and in this respect Theorem 3.4.2 appears to be new.

The Banach space theory of commutator-generator theorems began in the paper [11] cited in Section 2. But the general theory was really developed in a joint paper with Batty [1]. The theory was further developed in [11], and this latter reference contains a full proof of Theorem 3.3.1.

References

1. Batty, C. J. K. and D. W. Robinson, Commutators and generators. Math. Scand. (to appear).

2. ———, The characterization of differential operators by locality: Abstract Derivations Erg. Th. Dyn. Syst. **5** (1985), 171-183.

3. Bratteli, O., T. Digernes, F. Goodman, and D. W. Robinson, Integration in Abelian C^*-dynamical systems, Publ. RIMS (Kyoto) **21** (1985), 1001-1030.

4. Bratteli, O., Elliott, G. A., and D. W. Robinson, The characterization of differential operators by locality; classical flows. Compositio Mathematica **58** (1980), 279-319.

5. Faris, W. and R. Lavine, Commutators and self-adjointness of Hamiltonian operators, Commun. Math. Phys. **35** (1973), 39-48.

6. Glimm, J. and A. Jaffe, The $(\lambda\Phi^4)_2$ quantum field theory without cutoffs, IV Perturbations of the Hamiltonian, J. Math. Phys **13** (1972), 1568-1584.

7. McBryan, O., Local generators for the Lorentz group in the $P(\Phi)_2$ model, Nuovo Cimente **18A** (1973), 564-662.

8. Nelson, E., Time-ordered products of sharp-time quadratic forms, J. Funct. Anal. **11** (1972), 211-219.

9. Peetre, J., Une caractérisation abstraite des opérateurs différentiels, Math. Scand. **7** (1959), 211-218; Rectification a l'article "Une caractérisation , . .", Math. Scand **8** (1960), 116-120.

10. Reed, M. and B. Simon, *Methods of Modern Mathematical Physics II*, Academic Press, 1975.

11. Robinson, D. W., Smooth cores of Lipschitz flows. Publ. RIMS **22** (1986), 659-669.

12. ———, Smooth derivations on Abelian C^*-dynamical systems. J. Austral. Math. Soc. **42** (1986), 247-264.

Derek W. Robinson

Department of Mathematics

Institute of Advanced Studies

Australian National University

Canberra, Australia